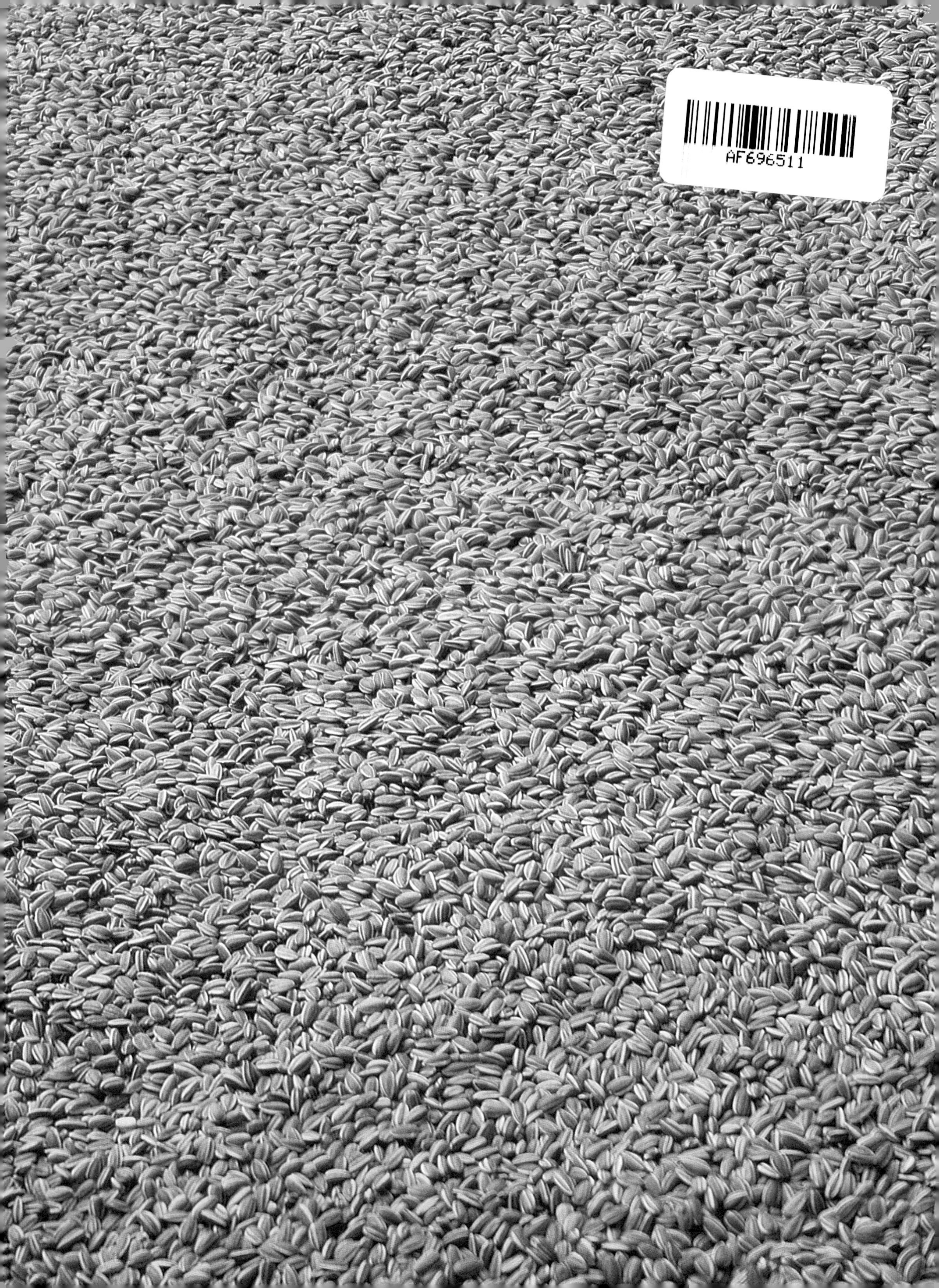

AF696511

Ceramics in America

CERAMICS IN AMERICA 2011

Edited by Robert Hunter

Published by the CHIPSTONE FOUNDATION

Distributed by University Press of New England

Hanover and London

CHIPSTONE FOUNDATION BOARD OF DIRECTORS
Edward S. Cooke Jr.
Charles Hummel
Peter M. Kenny
W. David Knox II *President*
Ted D. Kellner
John S. McGregor
Jonathan Prown *Executive Director*
Alison Stone
Stanley Stone III

EDITOR
Robert Hunter

PHOTOGRAPHER
Gavin Ashworth

NEW DISCOVERIES EDITOR
Merry Abbitt Outlaw

BOOK AND EXHIBITION REVIEWS EDITOR
Amy C. Earls

EDITORIAL ADVISORY BOARD
David Barker, *Former Keeper of Archaeology, City Museum and Art Gallery, Stoke-on-Trent*
Luke Beckerdite, *Editor,* American Furniture
Ellen Denker, *Museum and Decorative Arts Consultant*
David Gaimster, *Director, The Hunterian, University of Glasgow*
Graham Hood, *Chief Curator Emeritus, Colonial Williamsburg Foundation*
Ann Smart Martin, *Stanley and Polly Stone Chair, American Decorative Arts, University of Wisconsin-Madison*
George L. Miller, *Independent Ceramic and Glass Researcher*
Susan Myers, *Curator (retired), Ceramic and Glass Collection, Smithsonian Institution*

Cover Illustration: Ai Weiwei at the opening of "Sunflower Seeds," Unilever Series, Tate Modern, London, October 2010–May 2011. (© Ai Weiwei; photo, Tate Photography, 2010.)

Endpapers: Detail of Ai Weiwei, *Sunflower Seeds*, installation view, Turbine Hall, Tate Modern, London, 2011. Porcelain. (Courtesy, Tate.)

Design: Wynne Patterson, VT
Copyediting: Mary Gladue, CT
Proofreading: Fronia W. Simpson, VT
Type: Aardvark Type, CT
Printing: Meridian Printing, RI

Published by the Chipstone Foundation
Distributed by University Press of New England
1 Court Street
Lebanon, New Hampshire 03766

© 2011 by the Chipstone Foundation
All rights reserved
Printed in United States of America
ISSN 1533-7154
ISBN 978-0-9767344-8-2

Contents

Editorial Statement

Ceramics in America is an interdisciplinary journal intended for collectors, historical archaeologists, curators, decorative arts students, social historians, and studio potters. We seek articles on the broad role of historical ceramics in the American context, including essays on ceramic history, archaeological research, technology, social history, studio pottery, and collecting. Further information about this process can be found at the Chipstone Foundation's website, www.chipstone.org. Authors are encouraged to contact the editor before submitting materials.

Authors of submissions regarding new ceramic discoveries should contact New Discoveries editor Merry Abbitt Outlaw directly, at xkv8rs@aol.com.

The Chipstone Foundation offers significant honoraria for full-length manuscripts accepted for publication and reimburses authors for photography that is approved in writing by the editor.

If an article is accepted for publication, the author will be asked to submit photocopies of the title page and copyright page for all cited published sources. Authors are encouraged to provide digital files for all images, along with permission to reprint. Low-resolution digital images or photographs of poor quality will not be accepted for publication.

Questions and comments from readers and authors are welcome.

Robert Hunter
Editor, Ceramics in America
P.O. Box 114
Yorktown, Va. 23690
CeramicJournal@aol.com

Robert Hunter

Introduction

▼THE STORY of world ceramics encompasses thousands of years and vast geographic distances. Ceramics often served as a medium of exchange and were key players in the interaction between the cultures of the East and West. The clay vessels at the intersection of these were powerful transmitters of mythological and symbolic meaning. Learning how to decipher the beliefs, customs, and traditions of a society that are embodied in fired clay both ancient and modern is an ongoing challenge for the student of ceramic history.

The language and symbols of any given culture differ throughout time. Ceramics, too, have crossed social and geographic boundaries with relative frequency. Most interchanges occurred on a local or regional level, but by the eighth century A.D., Arab and Persian caravans and trading ships were making their way to China and bringing home luxury goods, including the distinctive tricolor *sancai* ware and, later, blue-and-white porcelain.

Global exploration by Western European countries in the sixteenth and seventeenth centuries opened up extensive trade networks that brought the much-coveted Chinese porcelains to a world hungry for the exotic. Kings and queens set the standards for taste that filtered down to commoners, who, ultimately, became mass consumers. For those that could not afford or obtain "true porcelain," earthenware industries sprang up in Portugal, Spain, the Netherlands, Germany, and England, making rude copies by appropriating the blue-and-white palette and Asian ornamentations.

In the opening article of this volume, Sarah Fayen Scarlett examines the most popular decorative motif—a robed Chinese man among grasses and rocks—used in the last quarter of the seventeenth century by English, Dutch, and German potters. The ubiquity of this pattern cannot be overstated; hundreds of surviving examples and archaeological fragments have been found on virtually every Anglo-American site from this period. Despite its widespread popularity, no period term exists for the motif, and collectors and antique dealers have been content to use a variety of descriptive names, such as Chinaman on a Rock. Using a combination of historiographic and literary techniques, Fayen Scarlett offers a more specific name for the pattern, the Chinese Scholar, which sets the new terminology standard for the field. She further examines the mysterious symbolic content of the Chinese Scholar and the implication of its widespread adoption in England.

The Chinese Scholar pattern is just one example of how the Western world absorbed Chinese art into its own decorative repertoire under the mantle of exoticism and novelty, with little awareness or concern for the

Chinese culture and its people. The impact of Chinese ceramics has been immense, yet the potters themselves have remained an anonymous aggregate of peasant artisans. In the late twentieth century, artist Ai Weiwei stormed onto the world stage as an avatar of the long-suppressed artistic voice in China. Garth Clark, the leading writer and thinker on contemporary ceramics, pulls back the curtain on this revolutionary artist and political dissident in his well-considered critique, "Mind Mud: Ai Weiwei's Conceptual Ceramics." The article reviews Weiwei's recent ceramic-based installations, including the widely publicized "Sunflower Seeds" held at the Tate Modern in London—one hundred million sunflower seeds produced and hand-painted by Jingdezhen potters. Clark examines the historical power of ceramics, and whether one views Weiwei as a guru or a gadfly, currently he is the preeminent spokesperson for the plight of the Chinese people, and his life force is a mighty presence in the contemporary discourse of ceramic art.

For archaeologists digging through the forgotten detritus of early American home sites, physical reminders of the Chinese contribution are often just the turn of a spade away. In "Digging Up Salem's Golden Age: Ceramic Use Among the Merchant Class," George Schwartz reminds us of the seaport's central place in the so-called China trade. Schwartz presents a thorough analysis of a huge assemblage of ceramics and glass excavated from the home of Captain Stephen Phillips (1764–1838), one of the many participants in the development of the American trade with China. The Phillips House assemblage is perhaps the best-preserved collection of archaeologically recovered ceramics from a New England context. It offers a remarkable snapshot of ceramic tastes of the mercantile class of the time: more than one-third of the vessels recovered—among English earthenware imports and locally produced coarsewares—were Chinese porcelain.

With the advent of mass-produced transfer printing in the early nineteenth century, the Staffordshire ceramic industry was fully engaged in the race to win potential customers with the latest and most fashionable dishware. Printed sources were often adapted from the popular travel books being printed in England and America. Scenic views from Europe, India, Africa, and even the Arctic found their way onto ceramic dinnerware and tea sets. Period consumers and later collectors have been mesmerized by the assortment of patterns and colors produced by the Staffordshire potters. Dick Henrywood, a leading collector and scholar of these printed patterns, provides new revelations in his article "The States Border Series by Ralph and James Clews." The States Border series is one of the most recognizable "American" patterns, with rich iconography that includes the printed names of the first fifteen American states. Henrywood has discovered that the central views used in this series, clearly meant to elicit patriotic sentiment, were copied from William Marshall's *Select Views in Great Britain* (1825–1828), a popular travel book. Whether American customers knew or cared that these dishes actually contained images of fine British "Great Houses," it is a somewhat ironic revelation for today's collectors of historical Staffordshire.

For collectors of American ceramics, the nineteenth-century products of the pottery of the Thompson family in Morgantown, West Virginia, have always had an element of mystique. This distinctive stoneware, remarkable for its diverse and innovative technological and decorative elements, is highly sought. In the 2004 volume of *Ceramics in America*, authors Don Horvath and Dick Duez discussed the earthenware produced in Morgantown between 1796 and 1854. For this issue, the authors pick up the story of the Thompsons, covering the family's stoneware production from 1854 to 1890. Particularly important to collectors are the "people pots" attributed to the family patriarch, John Thompson. These pots, whose images of men, women, and animals have an appealing yet primitive aesthetic reminiscent of prehistoric cave paintings, reflect a unique regional manifestation of American stoneware decoration.

The next article, by frequent contributors Al Luckenbach and John Kille, reminds us that much of America's ceramic history remains hidden in the ground. This is especially true for objects made in the seventeenth century—virtually nothing of our ceramic heritage from that time has survived aboveground. In their excavation of a 1660s cellar at a house site in Providence, Maryland, the authors encountered a number of British and Dutch ceramic forms that, though once common, are now considered decorative arts treasures. Most notable was the recovery of an English delft mid-drip candlestick, perhaps the first documented evidence of this form used in an American context. The continued careful documentation of these specific ceramic forms is crucial to the reconstruction of the material world of these early American inhabitants, both rich and poor. The authors recognize, moreover, that archaeologists and decorative arts curators benefit from shared information of this type and that both fields are best served by collaboration.

The final article, by Ivor Noël Hume, is yet another wonderful example of how diverse sources can be mined to present new insights from seemingly mundane objects. When approached by a curious collector with an odd, cylindrical brown stoneware jar, Noël Hume took on the challenge of deciphering its origins and function. Drawing on historic sources, auction records, and other extant examples, his detective's tale takes the reader on a journey from the mud flats of Guyana to the early-nineteenth-century industrial landscape of lower Manhattan. In the process, the author reintroduces us to the economic importance of the Virginia oyster as a nineteenth-century foodstuff and the ingenious vessels produced by early New York stoneware potters to ship these perishable delectables far from their source. In addition to reconstituting the biographies of the watermen who used these jars to hold their catches, Noël Hume determines that Manhattan potter and African-American Thomas Commeraw was the maker of these once common but now exceedingly rare objects.

As in previous issues, short submissions on ceramic topics are included in the New Discoveries section edited by Merry Outlaw. New Discoveries offers collectors and researchers an opportunity to publish brief research notes on a latest revelation or a most recent find. In the future, however, many of these contributions will be published on the Chipstone website, www.chipstone.org. Readers are also encouraged to take advantage of today's social networking resources by posting ceramic news and discoveries on, for example, *Ceramics in America*'s Facebook page.

For the foreseeable future, printed books will remain the lifeblood of scholarship, and, toward that end, book reviews are crucial to evaluate the advancement of the field. Amy Earls has assembled seven reviews of recent notable publications that covered a diverse range of topics relating to ceramics. Readers are always encouraged to suggest recent publications for review.

Ceramics in America

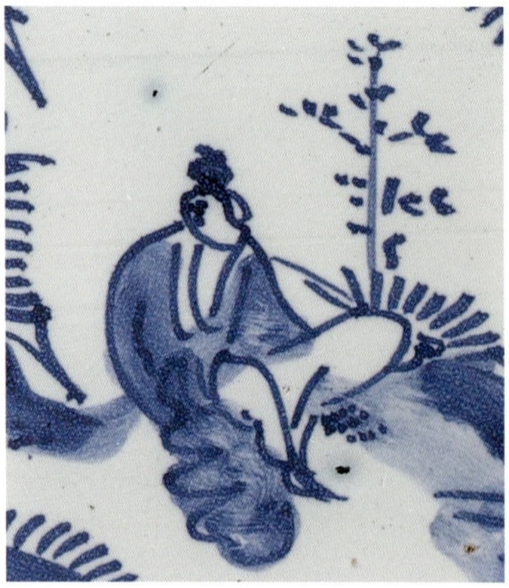

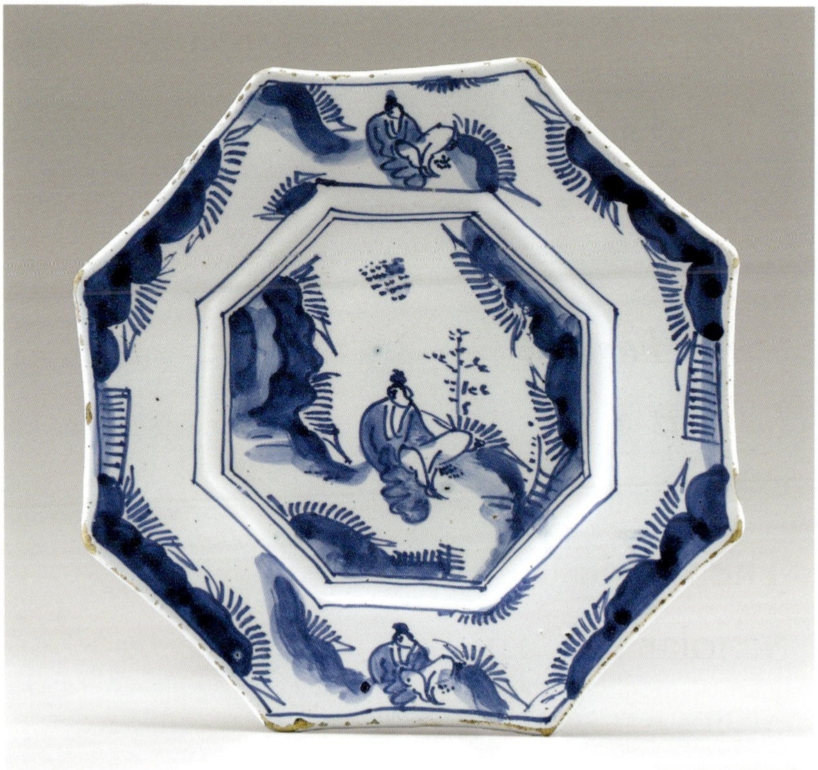

Figure 1 Plate, England, 1675–1695. Tin-glazed earthenware. W. 8". (Courtesy, Frederick and Anne H. Vogel; unless otherwise noted, all photos by Gavin Ashworth.) Figures 1–8 demonstrate the consistency of the English Chinese Scholar pattern and suggest the types of plates and vessels on which it survives. The man wears a robe and unusual headgear—perhaps a feather, a floppy hat, a turban, or a single strand of hair. Sketchily painted in blue or purple on a white tin-glazed ground, he is surrounded by spiky bushes, budding branches, and craggy rocks. The horizon descends from one side or the other made up of jagged cliffs, clouds, or other indeterminate swaths of color.

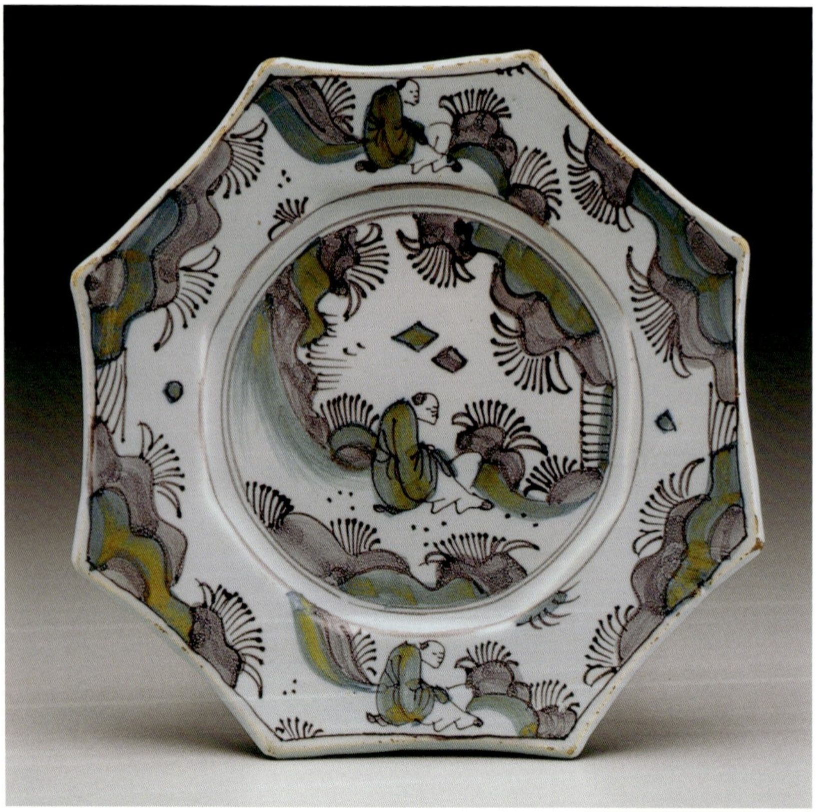

Figure 2 Plate, England, 1675–1695. Tin-glazed earthenware. W. 7¼". (Chipstone Foundation.)

Sarah Fayen Scarlett

The Chinese Scholar Pattern: Style, Merchant Identity, and the English Imagination

▼ THE IMAGE OF a robed figure seated in a scrubby landscape appears on hundreds of surviving English tin-glazed earthenware plates and vessels made between circa 1675 and 1695 (figs. 1–8). This exotic-looking pattern was created by potters in England to mimic the expensive porcelain imported from China and Japan and the similar Dutch copies flowing in from the Continent. No period term for the pattern is known, so historians traditionally have called it "Chinaman-in-grasses" and have recently updated the name to "Chinese-figure-in-grasses" or "squatting Chinese figure."[1] A new, more precise name—"Chinese Scholar"—offers a better understanding of this pattern's ultimate source in the narratives found on late Ming dynasty porcelain.[2] Spread by an extensive system of global trade, the image of a robed man in a wooded landscape originated in China and was then copied by potters in Japan, the Netherlands, and, ultimately, England. Through this process of visual translation, it was transformed into a hybrid pattern whose mixture of exotic novelty and subdued familiarity gained great popularity among seventeenth-century English consumers just becoming conscious of Asian aesthetics.

The Chinese Scholar pattern's widespread popularity throughout English settlements on both sides of the Atlantic and its compositional consistency over a twenty-year period make it a particularly potent source for exploring the English imagination during a time of great cultural change. The last quarter of the seventeenth century saw not only dramatic turnover in the monarchy but also enormous population growth in London and the establishment of promising trading posts in India and the Americas. With high prospects in trade and empire, the British people began to see themselves as viable players on the global stage. This shifting worldview altered trade, class relations, and aesthetic tastes throughout England and its colonies. Ceramic wares with the Chinese Scholar pattern were among the most popular decorative items owned during this time and can offer important insights into the era when read in the context of three major cultural shifts.

First, the Chinese Scholar pattern's free-flowing composition marks an early moment in the shift in English tastes from the order of the baroque toward the asymmetry and exotica of the rococo. Potters copied imagery from imported Asian porcelain as part of the pan-European fascination with Asian design that exploded in the late seventeenth century. Examined in relation to the writings of Sir William Temple, the first Englishman to praise Chinese asymmetry, the Chinese Scholar pattern emerges as an important

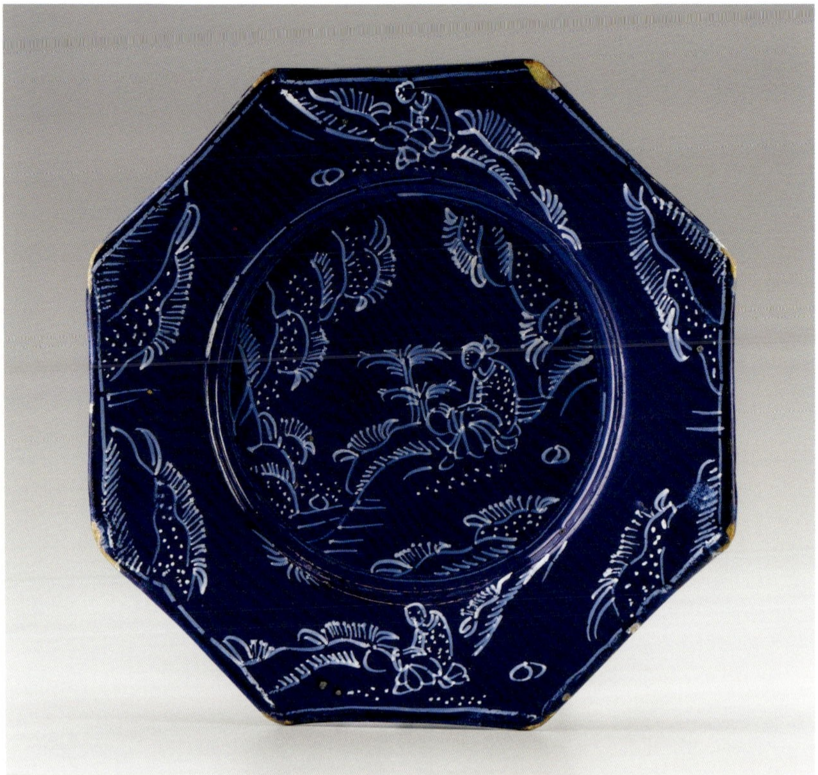

Figure 3 Plate, England, 1675–1695. Tin-glazed earthenware. W. 8". (Courtesy, Frederick and Anne H. Vogel.)

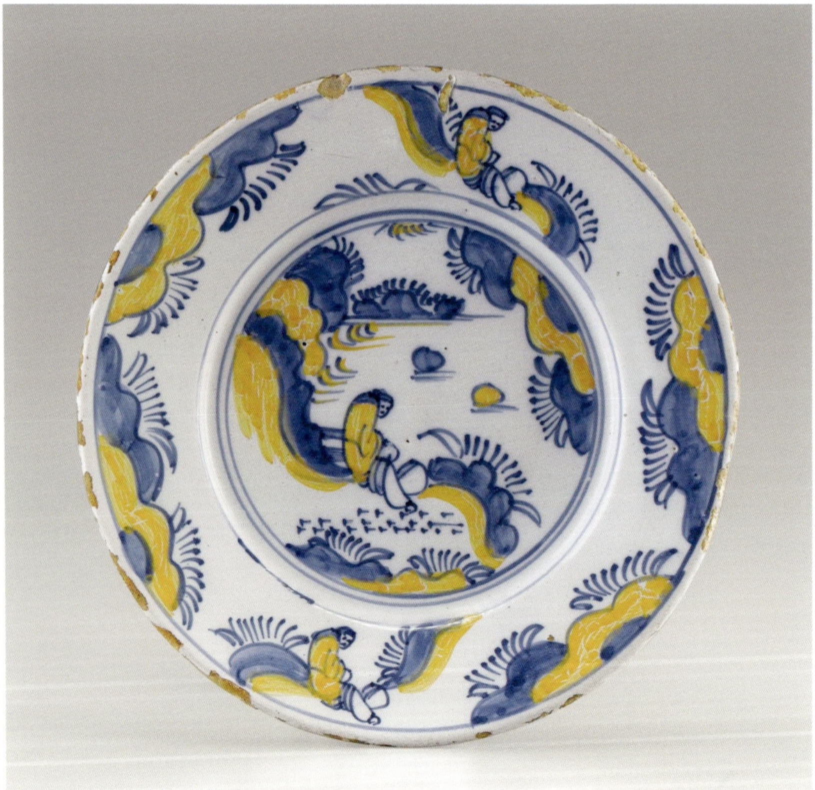

Figure 4 Plate, England, 1675–1695. Tin-glazed earthenware. D. 8½". (Courtesy, John H. Bryan.)

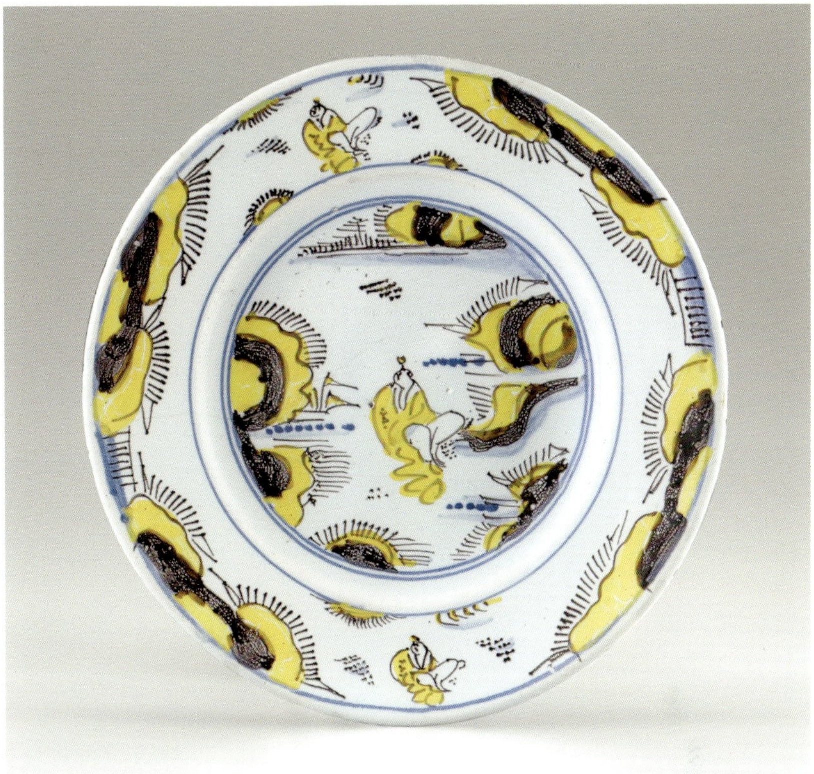

Figure 5 Plate, England, 1675–1695. Tin-glazed earthenware. D. 8½". (Courtesy, John H. Bryan.)

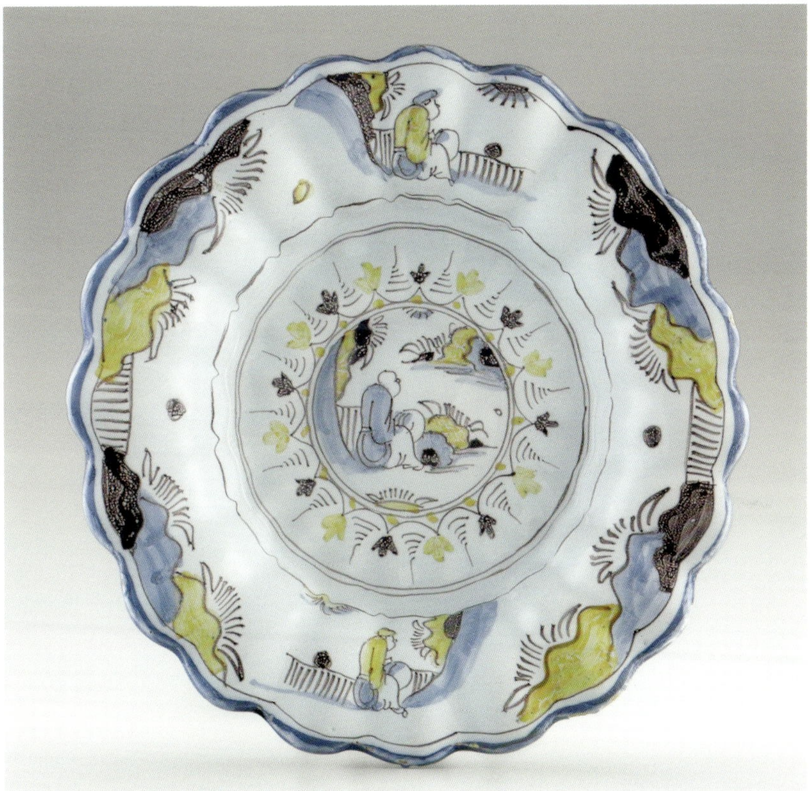

Figure 6 Dish, England, 1675–1695. Tin-glazed earthenware. D. 8½". (Private collection.)

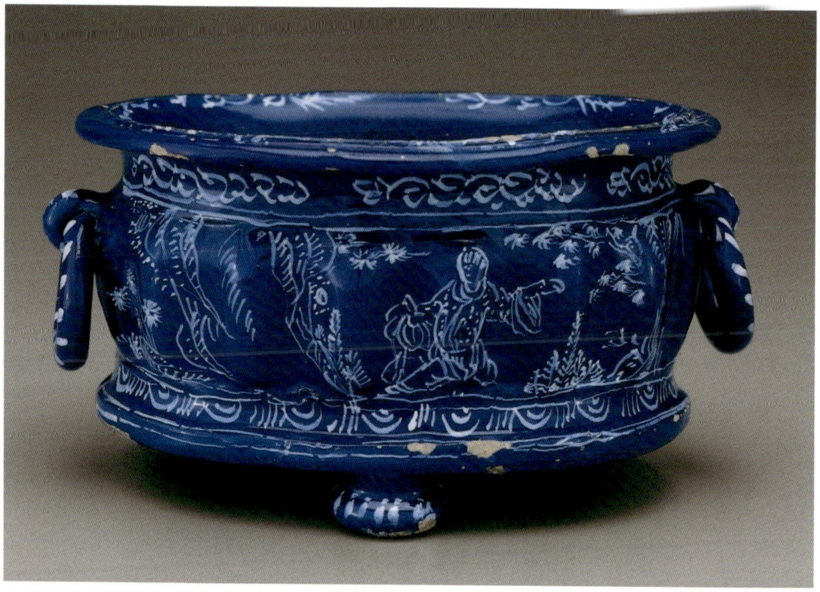

Figure 7 Cistern, England, 1675–1695. *Bleu persan* tin-glazed earthenware. L. 10". (Chipstone Foundation.)

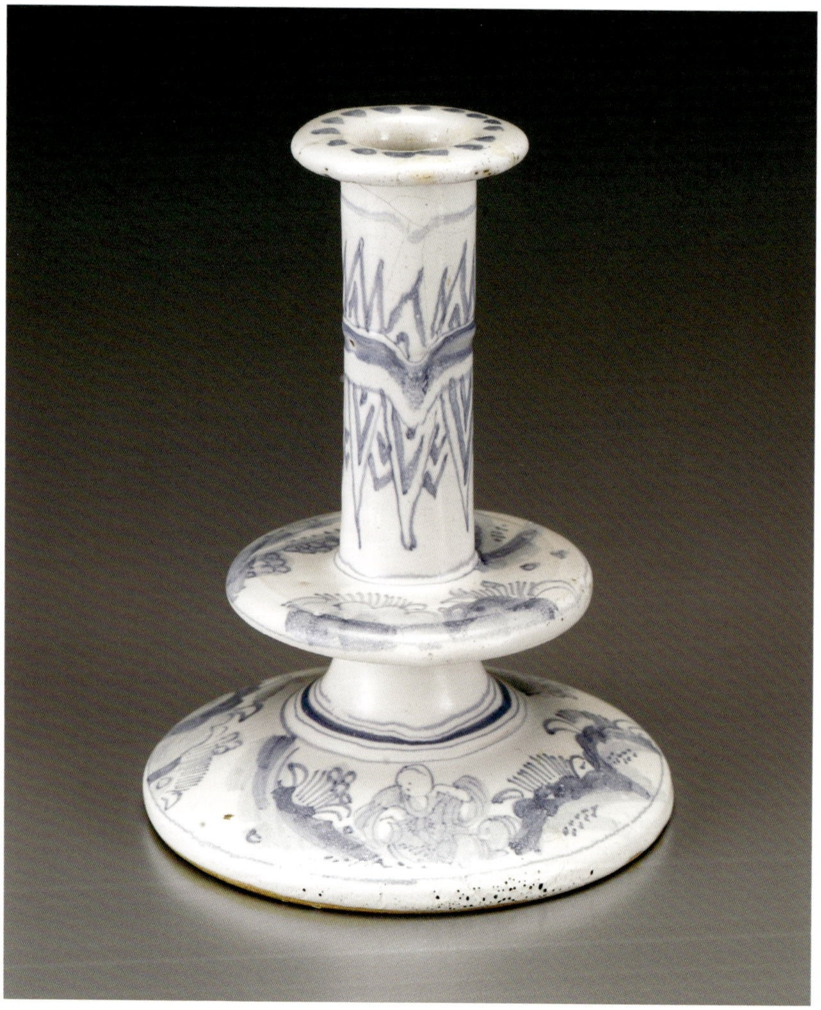

Figure 8 Candlestick, England, 1675–1695. Tin-glazed earthenware. H. 6". (Chipstone Foundation.)

early step toward creating the hybrid style known today as chinoiserie.[3] Ignorant of what the Chinese scholars and their accessories symbolized in their original contexts, English potters borrowed the images piecemeal, simplified them almost beyond recognition, and added them to conventionally Western forms such as drinking vessels, plates, candlesticks, and flower vases. Thus codified into a generic symbol of Eastern exoticism, this simplified image of a Chinese scholar spread throughout English settlements.[4]

Second, the Chinese Scholar pattern can be seen as part of a common visual language shared among English merchants and their customers all around the Atlantic Rim. Many of the merchants and the professionals who served the burgeoning port cities in North America and the Caribbean may have associated the Chinese Scholar wares—their blue-and-white decoration as well as the pattern's content—with the English mercantile endeavor that was stretching not only east but also west. When considered in its American context, the Chinese Scholar pattern highlights both the extent of international communication around the Atlantic Rim and the shifting national allegiances of a quickly changing marketplace.

Third, English opinions about Asia were changing rapidly with an influx of information and images. Travel accounts and trade goods exposed people to unfamiliar religions, philosophies, and aesthetics, and the English people began to reimagine themselves within an ever-expanding world. The Chinese Scholar pattern thus offers a glimpse into popular notions about China and suggests a simultaneous fascination with and condescension toward the far-off country. The Chinese Scholar pattern's assimilation of Asian aesthetics into European conventions preserved the newness of the image but also created prejudices that presaged the cross-cultural interactions of the nineteenth and twentieth centuries.

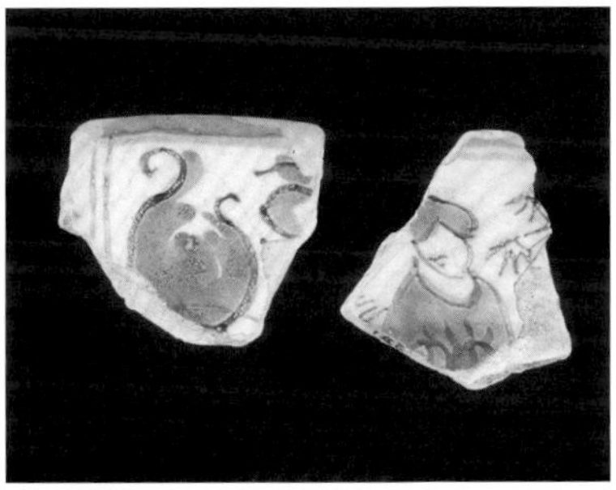

Figure 9 Dish fragments, probably England, 1675–1695. Tin-glazed earthenware. (Courtesy, Colonial Williamsburg.) Excavated at the William Finnie House Site, Williamsburg, Virginia.

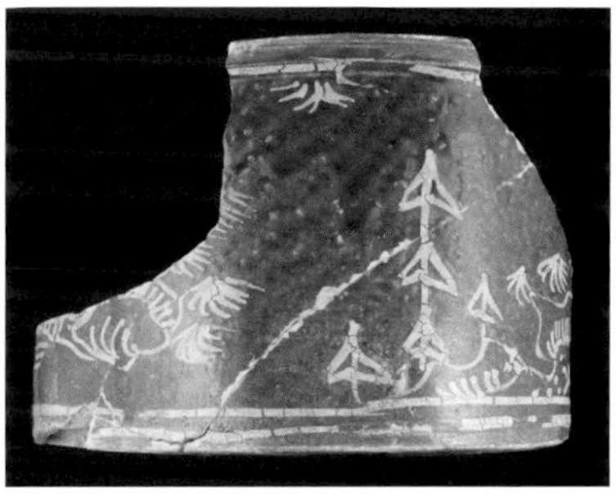

Figure 10 Meat pot fragment, England, 1680–1700. Tin-glazed earthenware. (Courtesy, Colonial Williamsburg.) Excavated at the domestic site adjacent to the Chiswell-Bucktrout House Site, Williamsburg, Virginia.

Archaeological evidence of the Chinese Scholar pattern in American colonial contexts poses difficult questions related to production, distribution, and origin. Sherds with the Chinese Scholar pattern and related imagery have been excavated along the coast of Maine; at Fort Orange, New York; Williamsburg, Hampton, and Jamestown, Virginia; St. Mary's City, Maryland; Port Royal, Jamaica; and other American merchant cities (figs. 9–14). While many of the sherds represent the products of English potteries in London and Bristol, others were made in the Netherlands, France, and German-speaking regions of northern Europe. Differentiating

Figure 11 Dish fragment, the Netherlands, 1670–1700. Tin-glazed earthenware. (Courtesy, New York State Office of Parks, Recreation and Historic Preservation; photo, Joseph Levy.) Excavated at Fort Orange, New York.

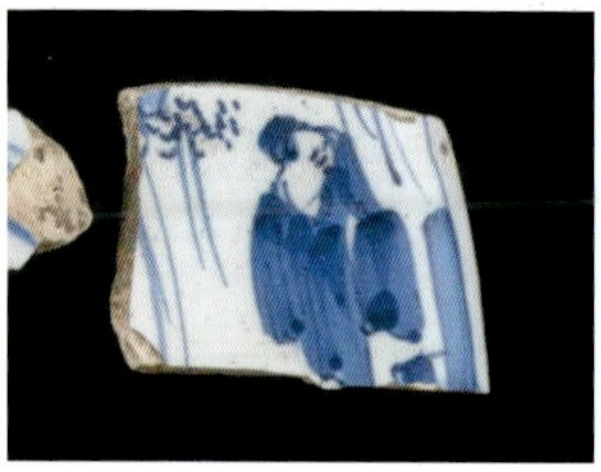

Figure 12 Dish fragment, the Netherlands, 1600–1650. Tin-glazed earthenware (majolica). (Courtesy, New York State Office of Parks, Recreation and Historic Preservation; photo, Joseph Levy.) Excavated at Fort Orange, New York. This dish is tin-glazed on the front but features a lead glaze on the back. The edges of this large fragment have been carefully ground down, suggesting that early owners discarded the rim, perhaps after part of it broke, and salvaged the central circular reserve as wall decoration.

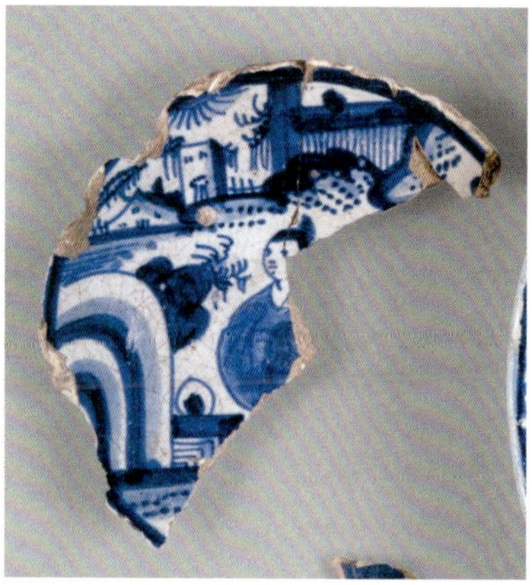

Figure 13 Dish fragments, England, 1675–1690. Tin-glazed earthenware. (Courtesy, Old Berwick Historical Society.) Excavated at the Lt. Humphrey Chadbourne Jr. Site, South Berwick, Maine, by Emerson W. Baker. The extensive Chadbourne estate was destroyed in 1690 by French and Native forces in the Salmon Falls Raid, an early encounter in King William's War.

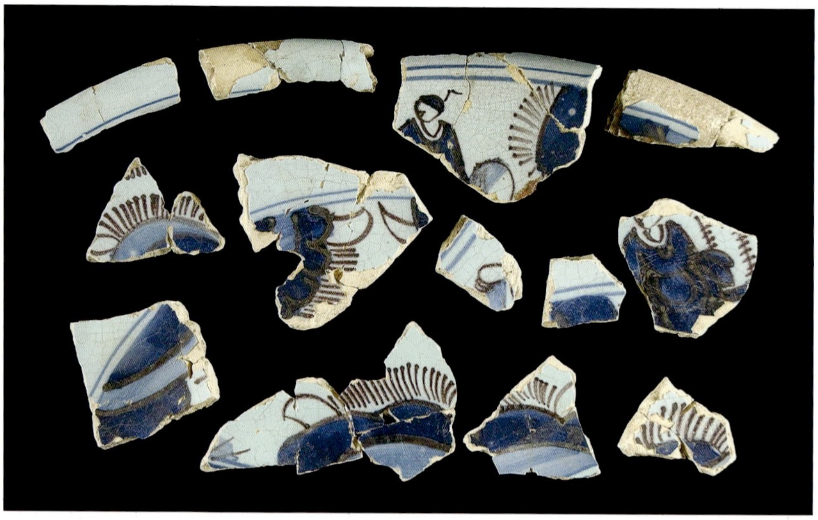

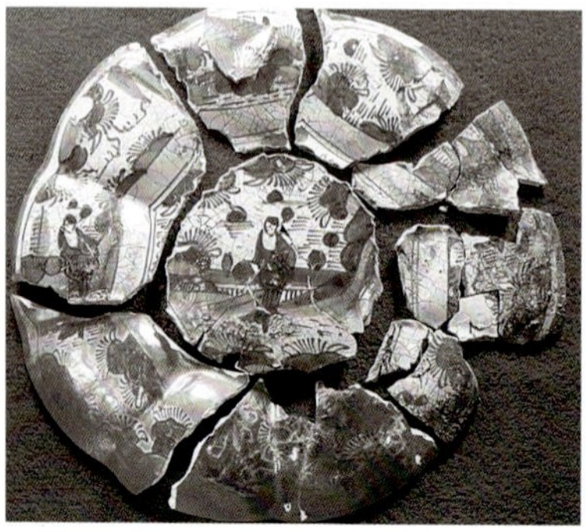

Figure 14 Lobed dish, England or the Netherlands, 1675–1690. Tin-glazed earthenware. (Courtesy, Saco Museum.) Excavated at the Richard Hitchcock Homestead at Biddeford Pool, Maine, by Emerson W. Baker. Discolored by fire during the Salmon Falls Raid of 1690, this dish was owned by a family who, though much less wealthy than the Chadbournes, also owned several pieces of tin-glazed earthenware, among other imported goods.

between English and Continental versions is extremely difficult, and future research might create typologies to identify the origins of similar pieces. This essay, however, traces the appearance of the specific Chinese Scholar pattern and suggests the initial insights that can be gained by thinking about why this pattern—whether made in London, Bristol, or the Continent—was of such interest to English consumers in the late seventeenth century. This particular configuration of a lone Chinese figure in a scrubby Walandscape played an important role in the perception of China in the English popular imagination.[5]

The Original Chinese Scholars

The Chinese Scholar pattern ultimately derives from Chinese porcelain of the early seventeenth century, but the pattern passed through several cultural filters before appearing, much altered, in England. During the first half of the seventeenth century, porcelain arriving in England mostly came through the Dutch East India Company (or VOC). The pattern most commonly imported at first featured blue-on-white hand-painted patterns with rectangular panels alternating in size decorating the rim, and foliage, birds, and other natural motifs in the center. Developed specifically for export, this pattern was called "Wan Li" after the sitting Ming emperor (1572–1620) (fig. 15). Many Europeans also called it *kraak* ware, after the Portuguese carrack ships that had first transported porcelain into Europe.[6]

Starting in the 1630s, the exported patterns began to change due to shifts in China's pottery industry. For centuries, the potteries in Jingdezhen primarily had fulfilled porcelain commissions for the Ming imperial court. Facing threats of military attack from the northern Manchus, the emperor redirected funds earmarked for porcelain production toward defense. Forced to find new buyers outside the court, the Jingdezhen potters began producing for China's educated elite and prosperous merchants. Their new porcelain featured landscapes with tall craggy mountains, streams, clouds,

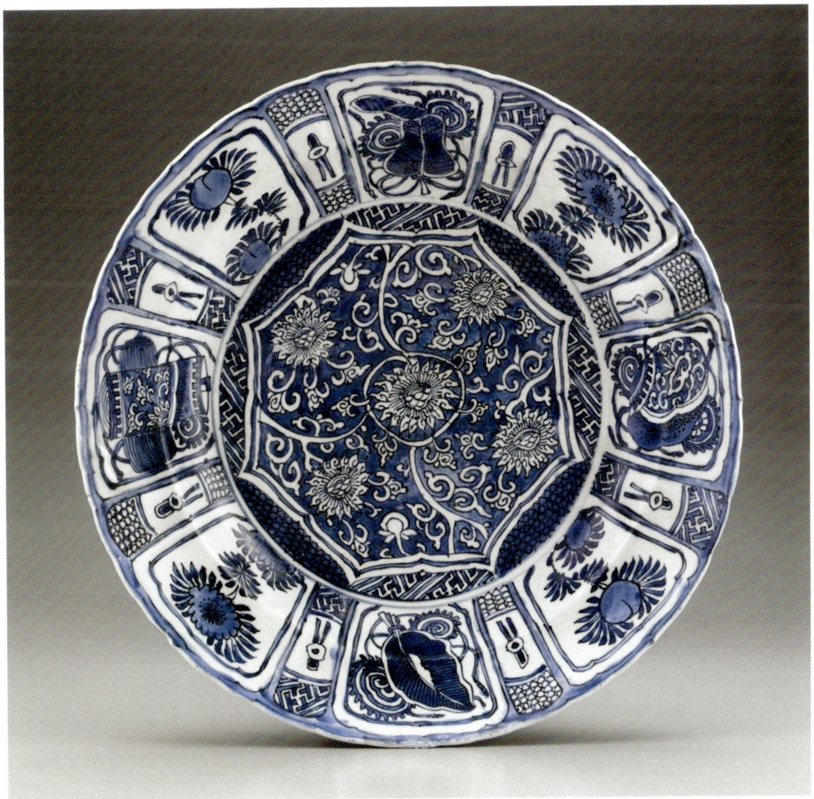

Figure 15 Dish, China, 1573–1619. Hard-paste porcelain, decorated in the so-called Wan Li or *kraak* pattern. D. 15". (Chipstone Foundation.)

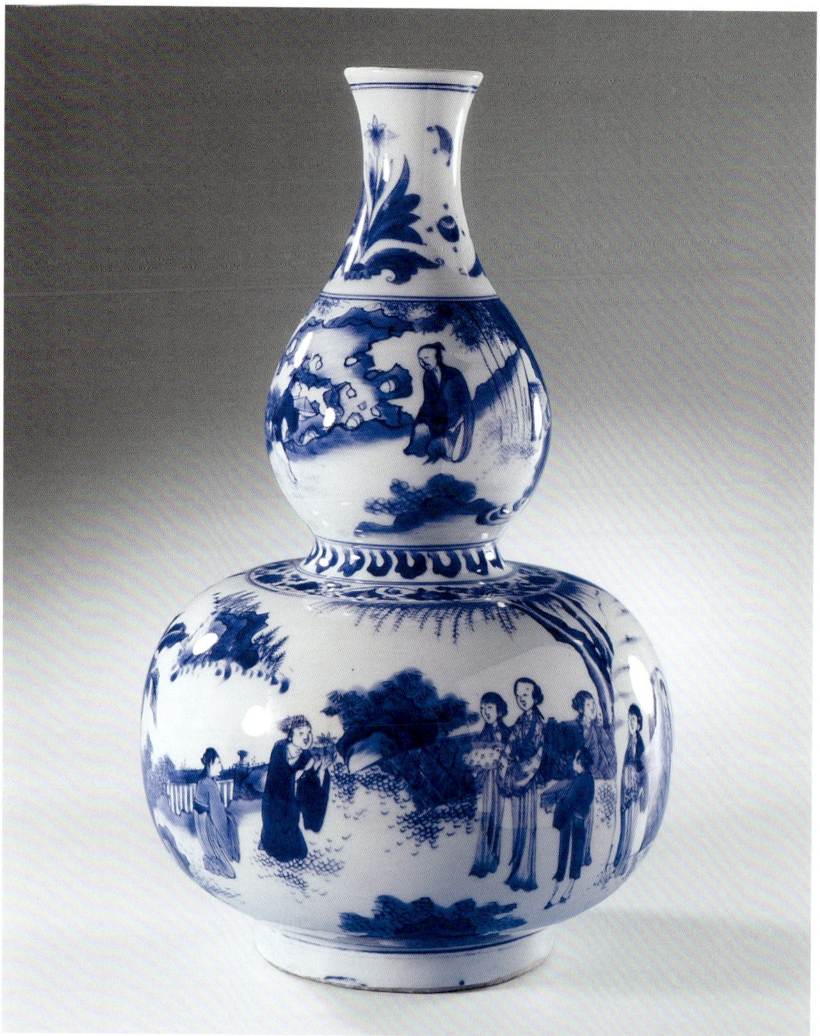

Figure 16 Vase, China, 1635–1645. Hard-paste porcelain. H. 12¾". (Courtesy, Rijksmuseum, Amsterdam; Gift of J. G. Fontein, 1924.)

trees, and human figures from folk tales, novels, and legends (fig. 16). The potters largely copied these motifs from popular paintings and woodblock prints. Known today as Ming Transitional ware, this is the porcelain that was imported in great quantity into Europe and sparked the craze among the wealthy for collecting blue-and-white "china."[7]

The change in taste for porcelain decoration was part of a larger cultural transformation in Chinese society that shifted power outside the imperial court system and opened up ancient aesthetic traditions to new audiences. Warring factions within the Ming imperial court were disrupting traditional power structures, creating a period of unprecedented socioeconomic fluidity. Due to inflation and expanded international trade, ambitious merchants and the educated elite were amassing large fortunes and gaining social power in the vacuum left by the weakened Ming court. To solidify their newfound status, many of these men began imitating the aristocratic lifestyle, often by collecting art that featured iconography formerly known only in higher cultural circles. Many developed special interest in the paintings of the Southern Song dynasty (1127–1279), widely regarded today as the zenith of landscape realism. To fulfill demand for these ancient paintings, Ming painters and woodblock printers resurrected the old-style landscapes, often passing off their own work as antique. During the same period, painters also began copying more mundane scenes from folk tales, novels, and plays that featured recognizable characters in animated scenes. The porcelain painters responded to the changing tastes of their new clientele and began decorating their wares with similar landscapes and narrative scenes (figs. 17, 18).[8]

Figure 17 Dish, China, 1645–1650. Hard-paste porcelain. D. 6⅝". (Private collection.)

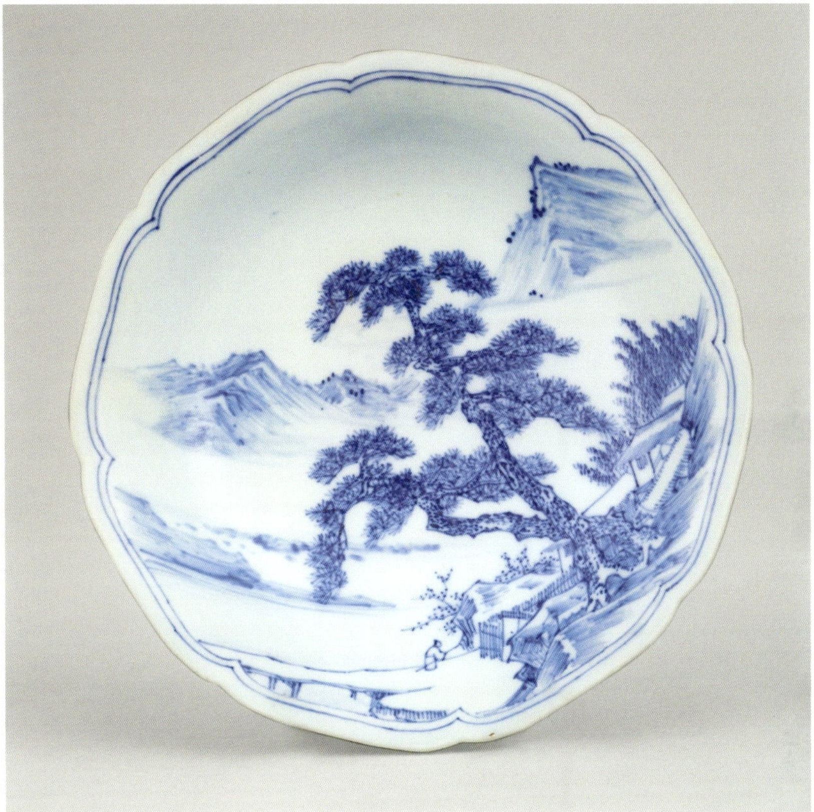

Figure 18 Vase, China, 1645–1660. Hard-paste porcelain. H. 15⅜". (Private collection.)

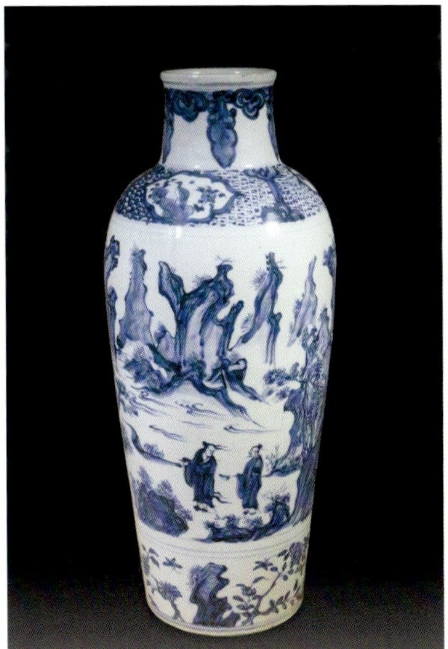

Another type of imagery that became popular in late Ming painting—on scrolls as well as on porcelain—portrayed the lifestyle of the Chinese scholar. For centuries the prestigious imperial education system had culminated in academic examinations that allowed those who passed to serve the government in positions that ensured a certain degree of comfort and power. The robed hermit-scholar alone in his pagoda-roofed study surrounded by accoutrements of learning had been a common trope in painting, poetry, and familiar lore for centuries. In most cases the scholars were anonymous, identified by recognizable motifs—scrolls, brushes, brush holders, and inkwells. Pine, prunus, and bamboo branches, collectively known as the "Three Friends of Winter," also connoted the scholar's lifestyle because they symbolized the revered traits of flexible strength, perseverance, and uprightness. In the 1630s, however, the civil-servant scholar-gentry educated in this system began to lose power as the late Ming emperors, threatened by Manchu uprisings in the north, were increasingly influenced by palace favorites and insider politicking. Many people lamented the loss of this long-standing tradition of academic rigor and worried about the future of the Ming dynasty. In this atmosphere of political uncertainty, iconography related to the scholar's study took on additional meaning. In paintings as well as on porcelain, hermit-sages became not only a reference to the stability and tradition of bygone eras but also a political statement against the current imperial court (fig. 19).[9]

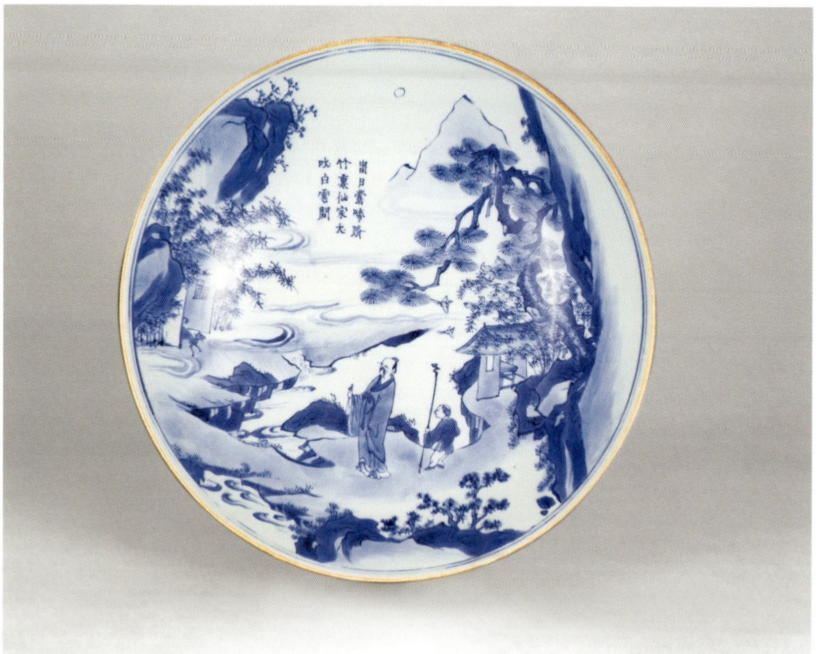

Figure 19 Dish, China, 1650–1660. Hard-paste porcelain. D. 13⅞". (Private collection.)

Among the many scholar motifs that appeared on Ming Transitional porcelain, one in particular may be the ultimate source of the Chinese Scholar pattern in England. Anonymous Ming porcelain painters often portrayed a single scholar seated on a cliff or outcropping, looking over a misty landscape with craggy rocks. Today the pattern is generally called "Master of the Rocks" (fig. 20). This popular image may have derived in part from the eighth-century poet Wang Wei, whose most famous verse reads: "I walk to the place where the water ends / and sit and watch the time when clouds rise." Images of Wang Wei and other sages seated on the edge of a rock looking into an abstract emptiness appeared for centuries in Chinese art and were especially popular among the Song painters (fig. 21). As late Ming painters resurrected Song motifs, they generated renewed interest in Wang Wei's ancient image, which may have helped inspire the Master of the Rocks pattern. During his lifetime, Wang Wei himself was a hermit farmer and remained closely associated with eremitic themes after his death. In the late Ming era, images of a lone seated scholar would have connoted not only the famous poet and the quality of Song painting but also the scholarly lifestyle that carried political resonance among the merchant elite. Some historians have also suggested that the image of a secluded natural retreat may have been especially appealing in a period of significant social upheaval.[10]

Figure 20 Dish, China, ca. 1670. Hard-paste porcelain, decorated in the Master of the Rocks style. D. 10⅞". (Private collection.)

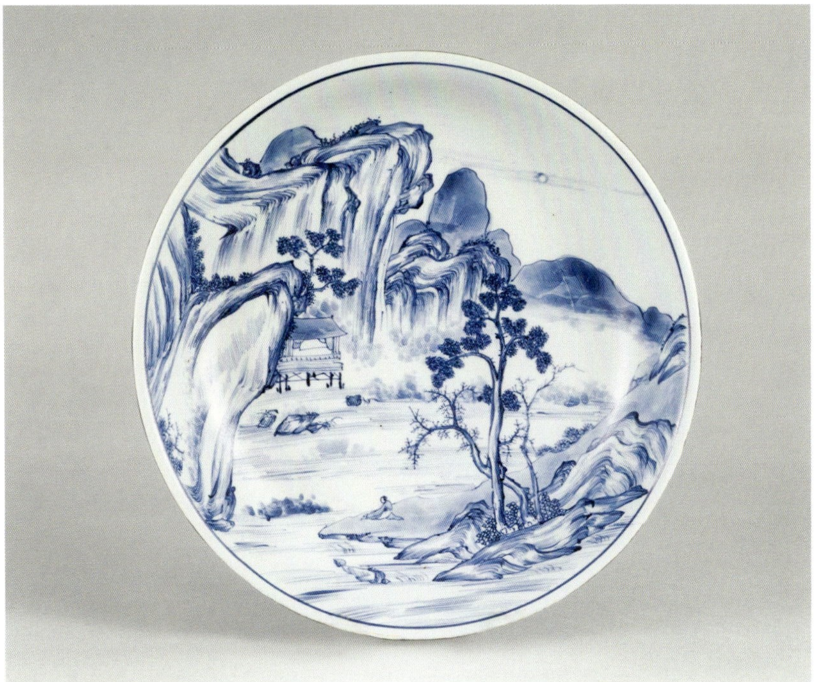

Figure 21 Ma Yuan (Chinese, active ca. 1190–1225), *Viewing Plum Blossoms by Moonlight*, Southern Song dynasty (1127–1279), fan mounted as an album leaf. Ink and color on silk. 9⅞" x 10½". (Courtesy, The Metropolitan Museum of Art, Gift of John M. Crawford Jr., in honor of Alfreda Murck, 1986 [1986.493.2].)

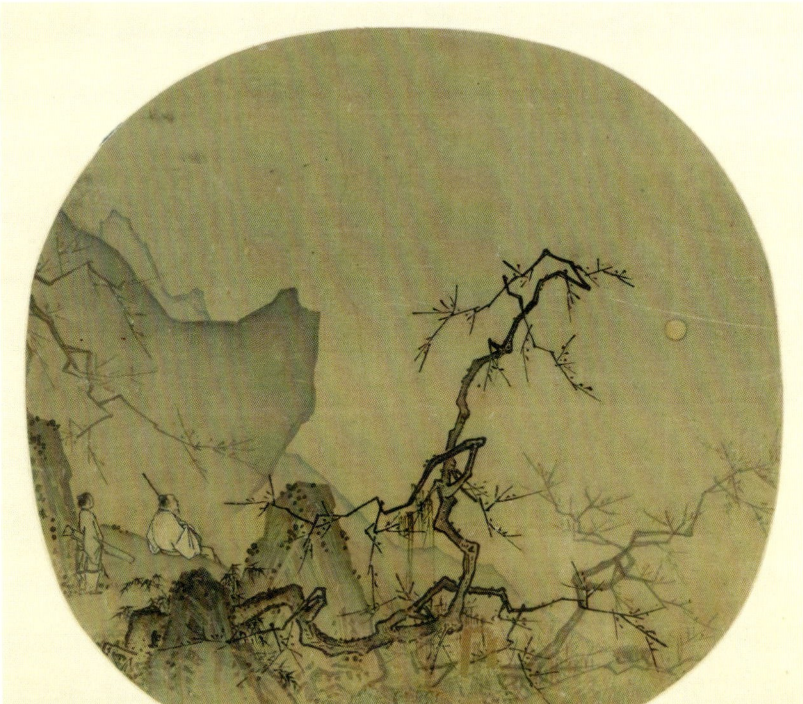

It is not entirely understood how Ming Transitional porcelain with the Master of the Rocks pattern and other scholar motifs came to be shipped to Europe by the VOC. Before the 1630s, Chinese merchants shipped only patterns developed specifically for export rather than patterns made for their own local market. It seems that as established trade networks broke down in the face of political turmoil, Chinese traders seized the opportunity to provide their wares—no matter their intended audience—to eager European shippers. Chinese traders already involved with inland networks of commerce, by which they supplied the domestic market, simply may have added the Dutch to their list of clients. The first shipment containing Ming Transitional wares left China for Amsterdam in 1634 and was described by a Dutch official in Taiwan as "a good lot of different porcelains of the old assortment and some new paintings with Chinese figures." Ming Transitional porcelain proved highly profitable for the Dutch, who traded millions of pieces through their factories in Southeast Asia, India, and several regions in the Near East.[11]

Shortly after the first arrival of Ming Transitional wares in Europe, the supply was cut off. The northern Manchus finally toppled the Ming dynasty in 1644 and decades of civil war ensued. The imperial kilns at Jingdezhen and many other production centers were closed, and external trade dramatically decreased. Demand in Europe for blue-and-white had risen to such a frenzy, however, that the Dutch asked potters in Japan to imitate the "old Chinese porcelain" to which their customers had become accustomed. Prodigious scholarship over the last twenty years has revealed that most of the blue-and-white arriving in Europe in the middle decades of the seventeenth century was indeed Japanese rather than Chinese. At first the Dutch imported only small numbers of Japanese porcelain, but as the storehouses of Chinese wares became exhausted, the VOC encouraged the potters in Arita to expand production. Beginning in 1657 the VOC sent prototypes—Chinese Ming Transitional porcelain as well as their own delftware copies—to Japan to be re-created. In one shipment the Dutch also sent turned wooden vase forms for the Japanese potters to reproduce.[12]

Figure 22 Jar, Arita, Japan, 1660–1680. Hard-paste porcelain. H. 6⅝". (Courtesy, Ashmolean Museum of Art and Archaeology, Oxford, Reitlinger Gift.)

Japanese copies of Ming Transitional wares included robed figures and rocky landscapes similar to those on the Chinese originals but in a freer painting style and with several telltale adaptations (fig. 22). Vertical bushes

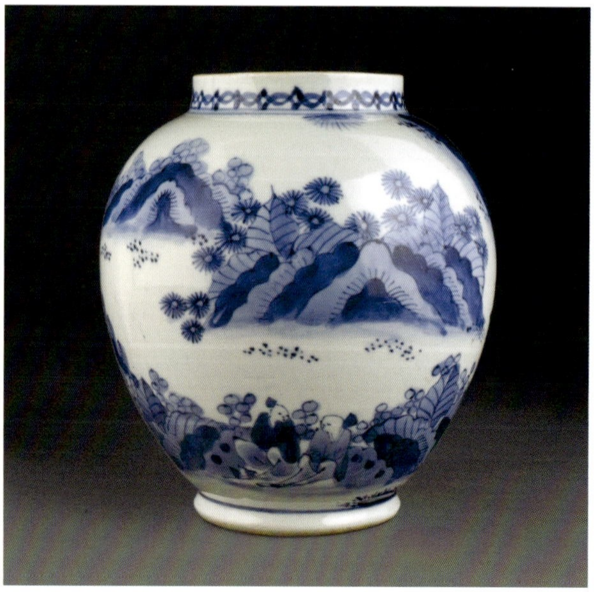

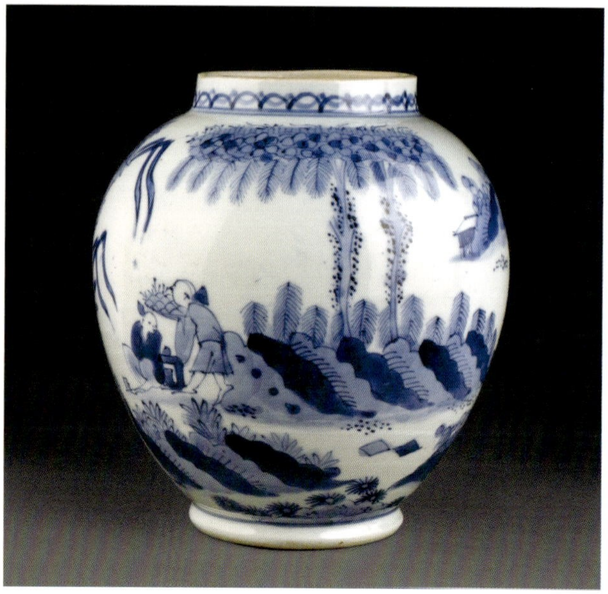

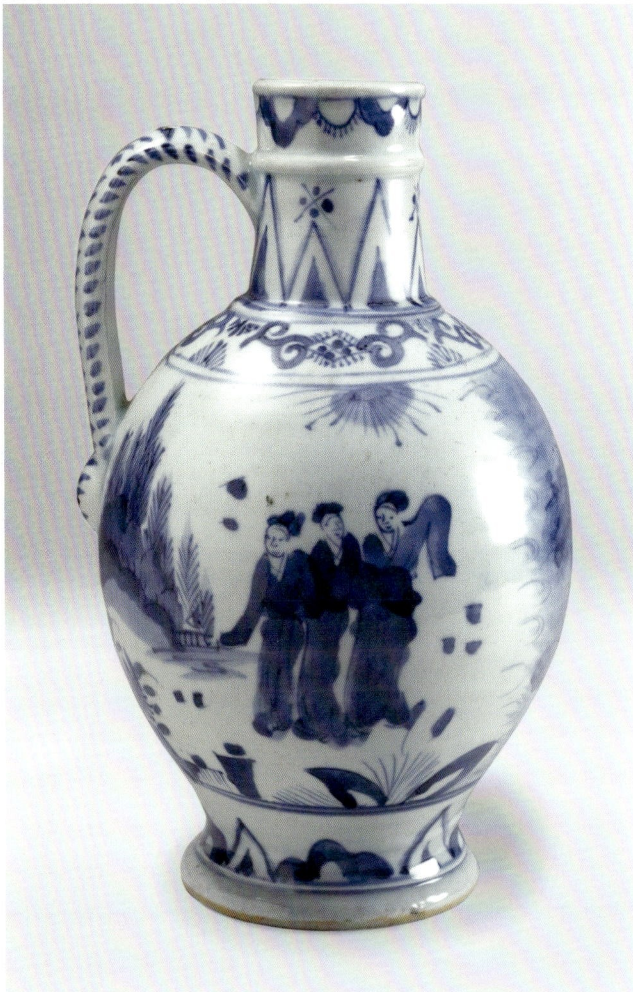

Figure 23 Ewer, Japan, 1660–1680. Hard-paste porcelain. H. 10". (Courtesy, Ashmolean Museum of Art and Archaeology, Oxford, Reitlinger Gift.)

painted with spiky V-shaped strokes, sometimes described by modern historians as "bottle-brush" trees, seem to have been a Japanese interpretation of Chinese foliage. Japanese-style *karakusa* scrolls appear alongside European-looking floral borders. The forms of the Japanese vessels generally adhere to Dutch tastes for Western jars, ewers, bowls, and plates (fig. 23). Some examples survive, however, with distinctively Chinese forms, such as the tall and thin *rolwagen* vase and the double-gourd-shaped vase. For about twenty-five years, the trade between the VOC and Arita potters created a distinctive aesthetic, filtering Chinese imagery through the tastes of Dutch merchants and the artistic impulses of Japanese potters.[13]

While the Dutch merchants were distributing Asian porcelain throughout northern Europe, their local potters responded quickly to the rising popularity of blue-and-white ceramics by creating copies of their own. Potters in Haarlem, Delft, and Rotterdam had been copying Italian maiolica for decades by applying an opaque white tin glaze and colorful oxides to their local earthenwares. In the second quarter of the seventeenth century, they began to shift their decorative schemes from baroque imagery rendered in blue, yellow, orange, and green to Asian-inspired scenes painted primarily in cobalt-blue underglaze. These approximations of imported

Figure 24 Dish, the Netherlands, ca. 1660. Tin-glazed earthenware. D. 11¾". (Courtesy, Philadelphia Museum of Art, Bequest of Emmeline Reed Bedell for the Bradbury Bedell Memorial Collection.)

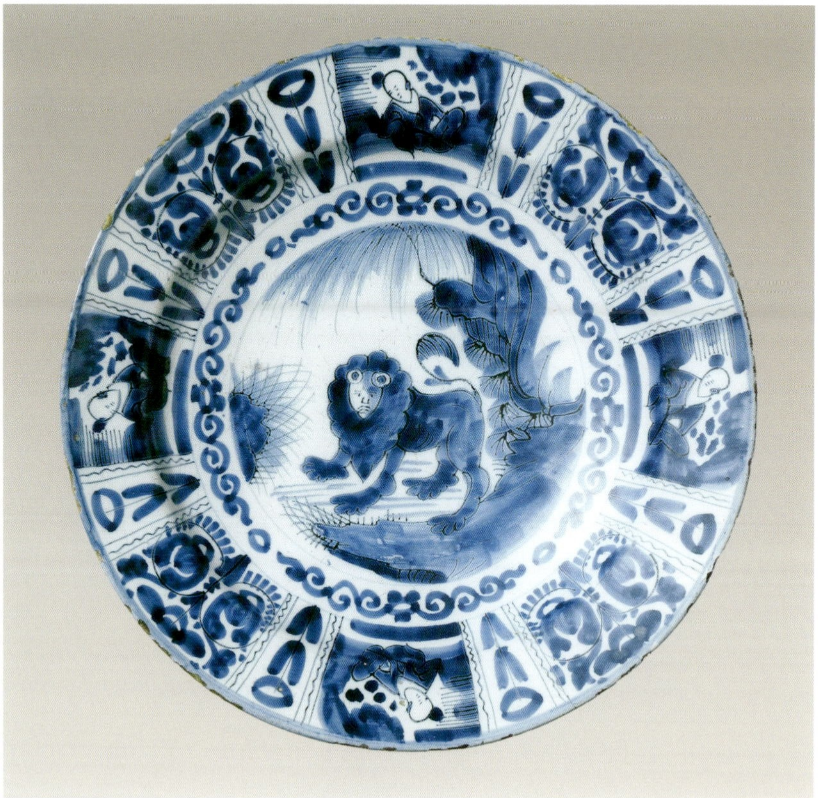

porcelain boasted playful mixes of *kraak* and Ming Transitional patterns such as the dish illustrated in figure 24. Dutch manufactories grew in size and number and began supplying chinoiserie tin-glazed earthenware to the middle-class burghers as well as to customers outside the Netherlands.[14]

Potters in other parts of Europe quickly followed suit by copying the Dutch "delftwares." By the 1650s French potters in Nevers turned their attention to Holland's Asian-inspired designs (fig. 25). In 1666 a French potter set up a pottery in Frankfurt, and the taste for chinoiserie spread to potters in Hamburg as well (fig. 26). All of these European pottery-making centers copied Ming Transitional wares in much the same way. They all featured robed figures with unusual headware wandering in exotic landscapes. Many have distinctively abstracted clouds made from horizontal swaths of alternating color curling up on one end. Europeans also generally copied the Japanese "bottle-brush" tree design.[15]

Immigrants and imports from these Continental production centers took to England the technology and inspiration for tin-glazed earthenware decorated after Ming Transitional patterns. London enjoyed unprecedented expansion in the last quarter of the seventeenth century thanks to England's developing trade networks and its role as a refuge from increasing hostility toward Protestants on the Continent. Among the many skilled artisans who helped make London the largest city in Western Europe were potters from Holland, an influx that contributed to the sudden appearance of chinoiserie in English pottery. Continental potters set up shops in London in the early seventeenth century and quickly copied the *kraak* patterns

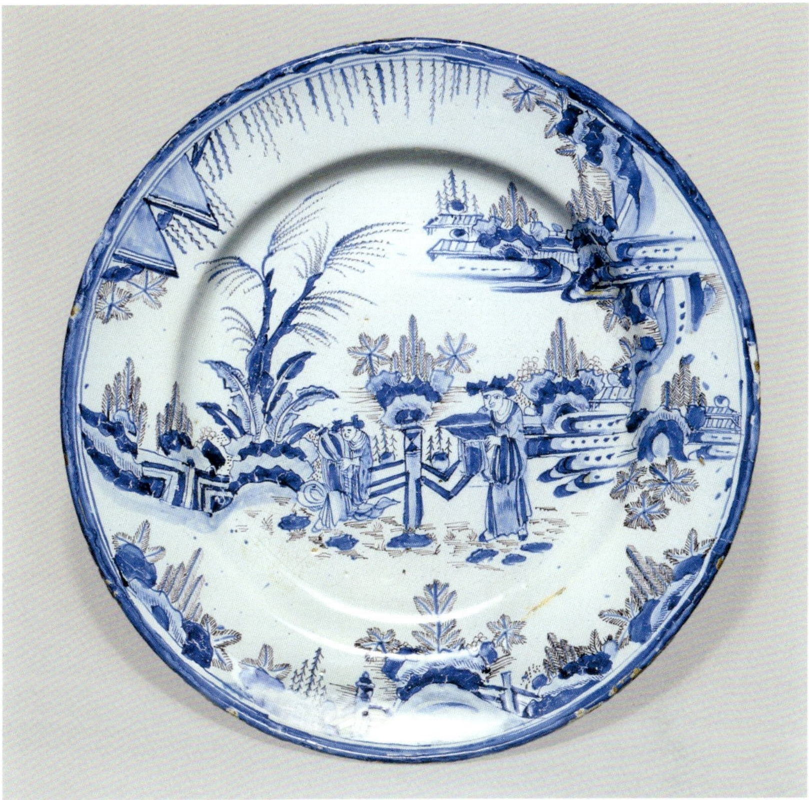

Figure 25 Dish, Nevers, France, 1670–1690. Tin-glazed earthenware. D. 16⅛". (Courtesy, Cambridge University, Fitzwilliam Museum, Glaisher Collection.)

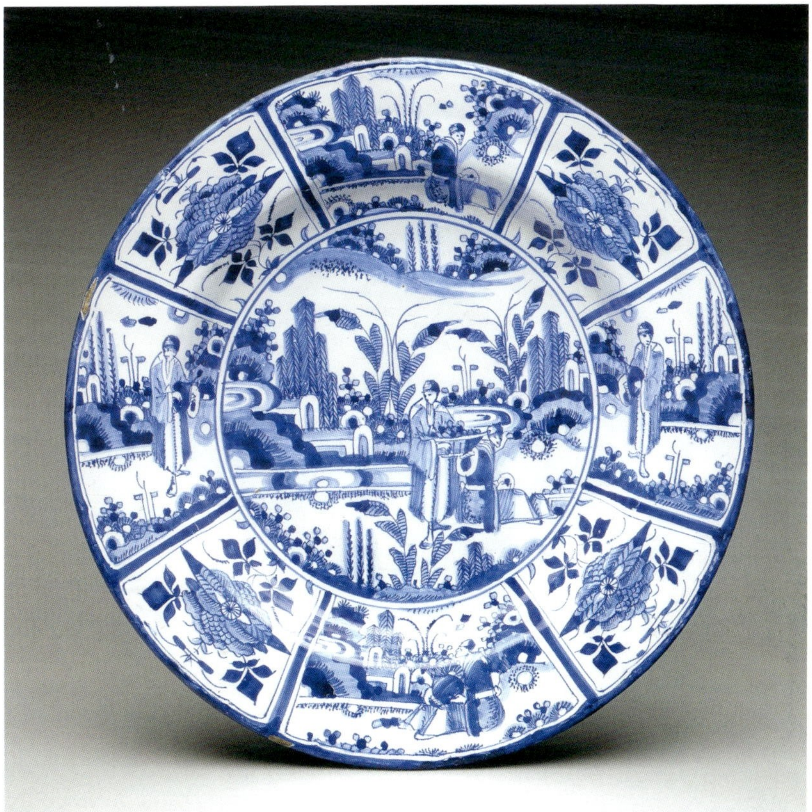

Figure 26 Plate, probably Frankfurt, Germany, 1690–1720. Tin-glazed earthenware. D. 15½". (Courtesy, Milwaukee Art Museum, Gift of Gabriele Flagg Pfeiffer.)

they had been making at home, often onto such traditional European forms as the wine bottles and jugs attributed to immigrant potter Christian Wilhelm. By the late 1670s, potters in Southwark and Lambeth were all making chinoiserie wares using the techniques they learned in Europe and copying the Dutch imports arriving in London. Immigrant potters in Bristol and Brislington were exposed to similar wares and created patterns very similar to those made in London.[16]

Figure 27 Jar, London, England, 1658. Tin-glazed earthenware. H. 12¾". (Courtesy, Pharmaceutical Society of Great Britain, KAC/P1.) The reverse features the date, 1658, and the coat of arms of the Society of Apothecaries.

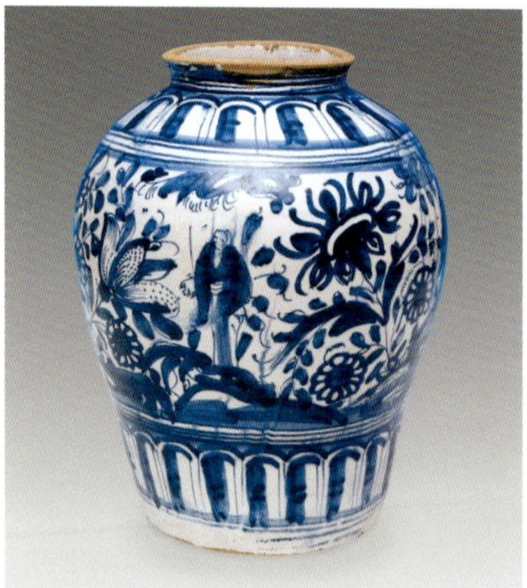

Chinese Scholars in England

Potters in England derived their techniques and decorative schemes from Dutch and Continental sources, but they adapted the Chinese aesthetic quite differently from their European counterparts. They copied the various landscapes and figures taken from Ming Transitional wares, but they also developed a single simple image—the Chinese Scholar pattern—and replicated it over a twenty-year period starting about 1675. The plates and vessels in figures 1–8 demonstrate the remarkable consistency with which English pottery painters reproduced this pattern. The single robed figure appears again and again seated by a stream or rock looking into an abstracted landscape. On many plates, he appears again on the top and bottom of the rim. While Dutch and other Continental potters also simplified Chinese imagery and created figures and compositions similar to the English Chinese Scholar pattern, their compositions did not have the same degree of consistency from piece to piece as the English versions.

In England, Ming Transitional–type decoration was copied, modified, and then codified into the Chinese Scholar image that gained considerable popularity—a process that makes its meaning among English consumers especially potent. The earliest English-made Ming Transitional–type scenes seem tentative and secondary. They appeared first as backdrop decoration for large central armorials or commemorative reserves. A large drug jar dated 1658 features the coat of arms of the Society of Apothecaries, large blue-and-white floral sprays, and several standing Chinese figures on the walls of the jar (fig. 27). A jug dated 1662 features similar Transitional-type

Figure 28 Jug, probably Southwark, London, England, 1662. Tin-glazed earthenware. H. 11½". Mark, inscribed on the shoulder: M / HS 1662. (Courtesy, Longridge Collection.)

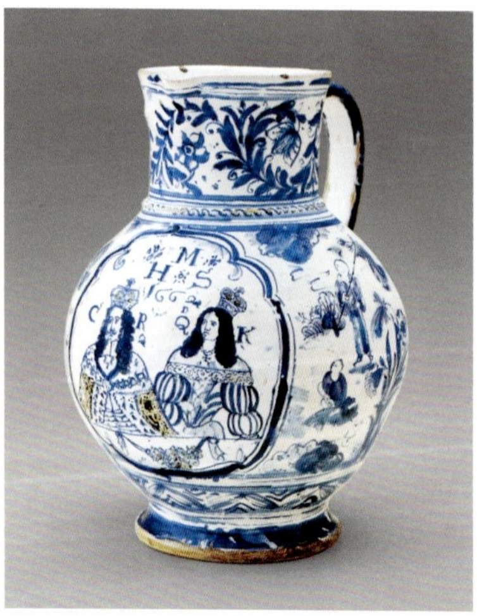
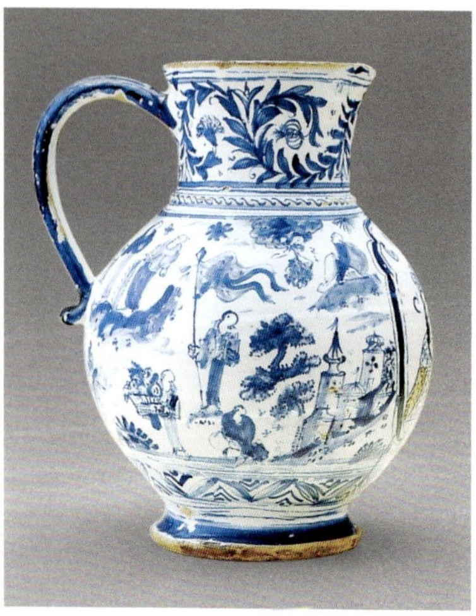
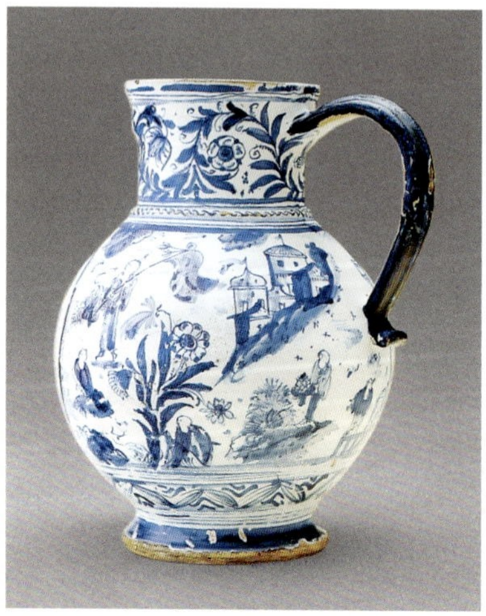

figures surrounding a portrait commemorating the marriage of King Charles I to Catherine of Braganza (fig. 28).[17]

In the 1660s and 1670s the Transitional style became more deliberate and primary, appearing as the dominant decoration on posset pots and caudle cups. Several posset pots survive with Transitional-type figures and the early date of 1669 (fig. 29).[18] Still meant as a backdrop to the commemorative monogram, the chinoiserie decoration dominates these pieces to a greater degree than on the earlier drug jar or jug (see figs. 27, 28). Many similar examples were made in the 1670s and 1680s (fig. 30). This emergence of Transitional-type decoration as a backdrop for commemorative armorials and initials helps identify chinoiserie from its first moments as a style used for public display.

Figure 29 Posset pot, London, England, 1669. Tin-glazed earthenware. H. 4⅝". Mark, inscribed on neck: 16 / S W / 69. (Courtesy, Castle Museum, Norwich.) The image on the right shows the Chinese Scholar motif.

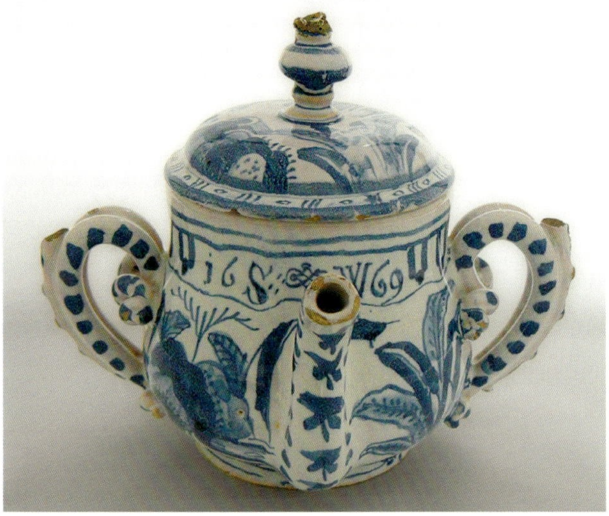

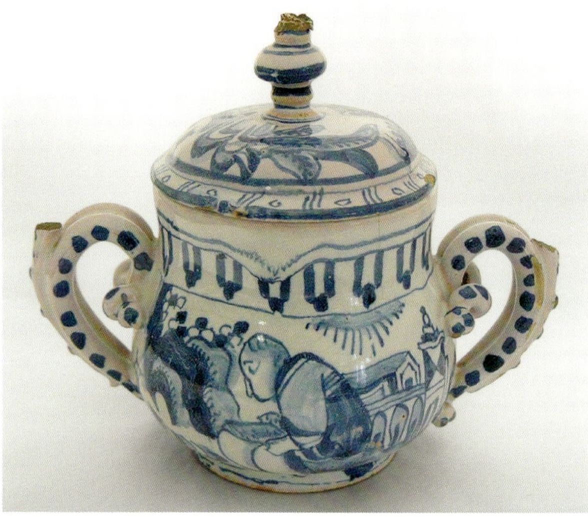

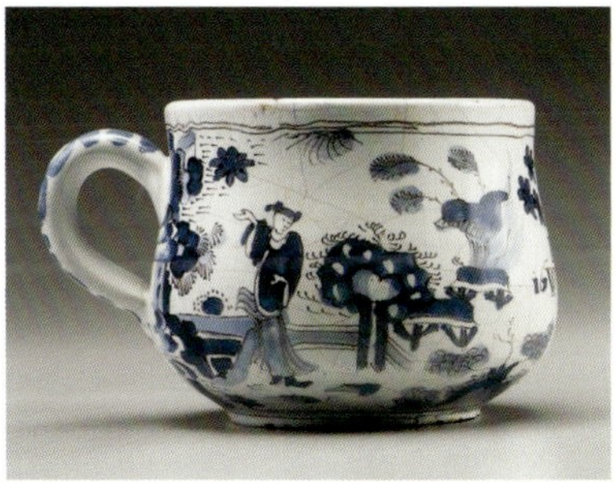

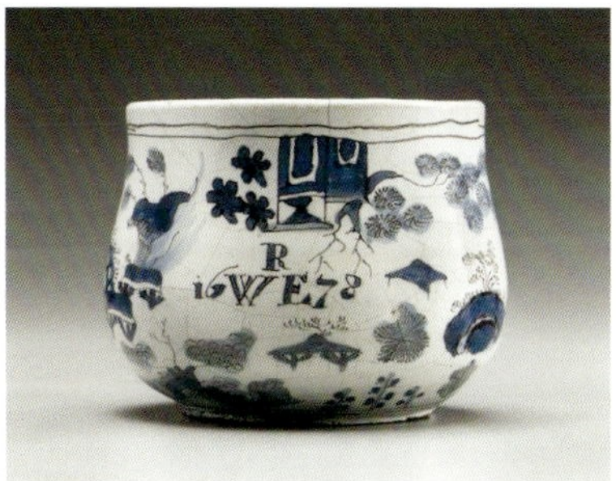

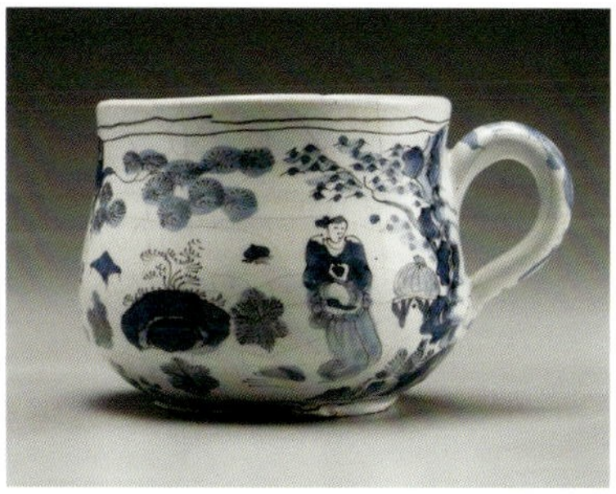

Figure 30 Caudle cup, London, England, 1678. Tin-glazed earthenware. H. 4". (Private collection.)

The solitary seated figure that eventually became the Chinese Scholar pattern first appeared as small details within more elaborate Ming Transitional–style decoration. An enormous dish dated 1673 features the coat of arms of an unidentified London city livery company in the central reserve and a rim with eight oval reserves in a pattern somewhat similar to the earlier *kraak* patterns (fig. 31). Every other oval reserve bears a landscape scene featuring bushy slopes with two-tone highlights, tall, spiky "bottle-brush" trees, and an abstracted Chinese figure who seems to be dipping his toe into a brook.

A very similar Chinese figure was painted on the finial of an enormous covered punch bowl made in London about the same time (figs. 32, 33).

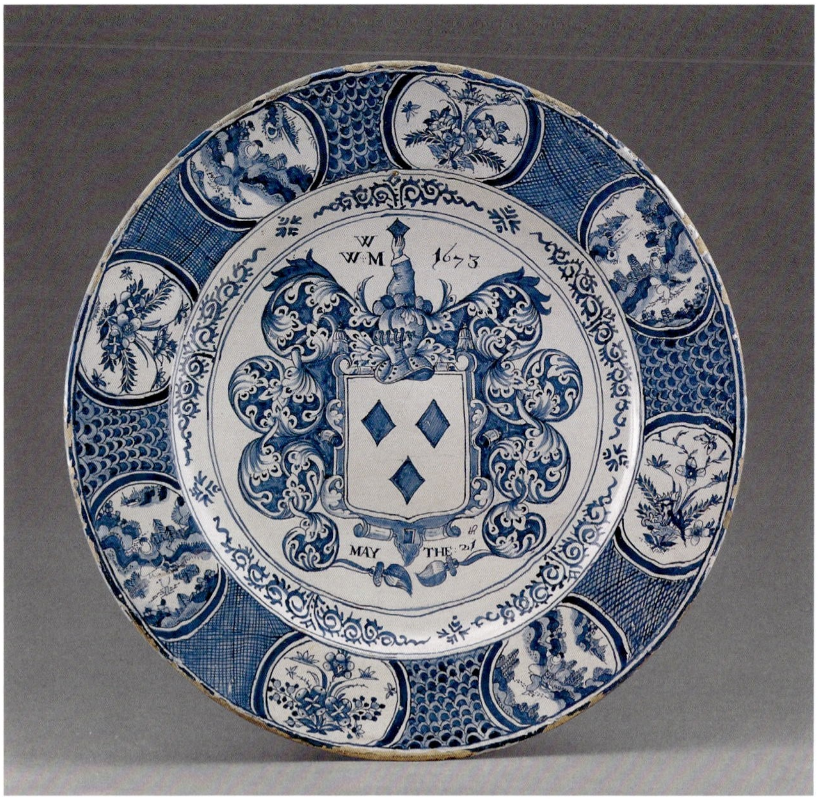

Figure 31 Dish, England, 1673. Tin-glazed earthenware. D. 22⅜". (Courtesy, Longridge Collection.) Note the Chinese Scholar motif in every other reserve around the rim. A sherd that appears to have the same pattern around the rim was excavated in Williamsburg, Virginia, and is illustrated in Ivor Noël Hume, *Early English Delftware from London and Virginia* (Williamsburg, Va.: Colonial Williamsburg Foundation, 1977), p. 94, fig. XV.

While nearly one quadrant of the punch bowl features the heraldry of another London city livery company (possibly the Company of Leatherers), all of its other surfaces feature intricately painted scenes of Ming Transitional–style figures in an exotic landscape. The figure on the finial (see fig. 33) appears to be dipping his toe into a stream in much the same way as the figures on the charger in figure 31. Again, this simple image has a marginal role in the overall decorative scheme and appears more sketchily painted than the more central figures. While the punch bowl is not dated, its painting style and heraldic imagery are similar to those of the 1673 charger. Both objects feature simplified imagery inside the armorial shield, knight masks with the same two-tone twist above, and nearly identical floral spirals on either side of the shield. Further research might determine a link between the painter of these two pieces and those with similar motifs.[19] Regardless, the charger and punch bowl attest that in the 1670s one or several English pottery decorators painted the same image of a Chinese scholar-fisherman dipping his toe into a stream.[20]

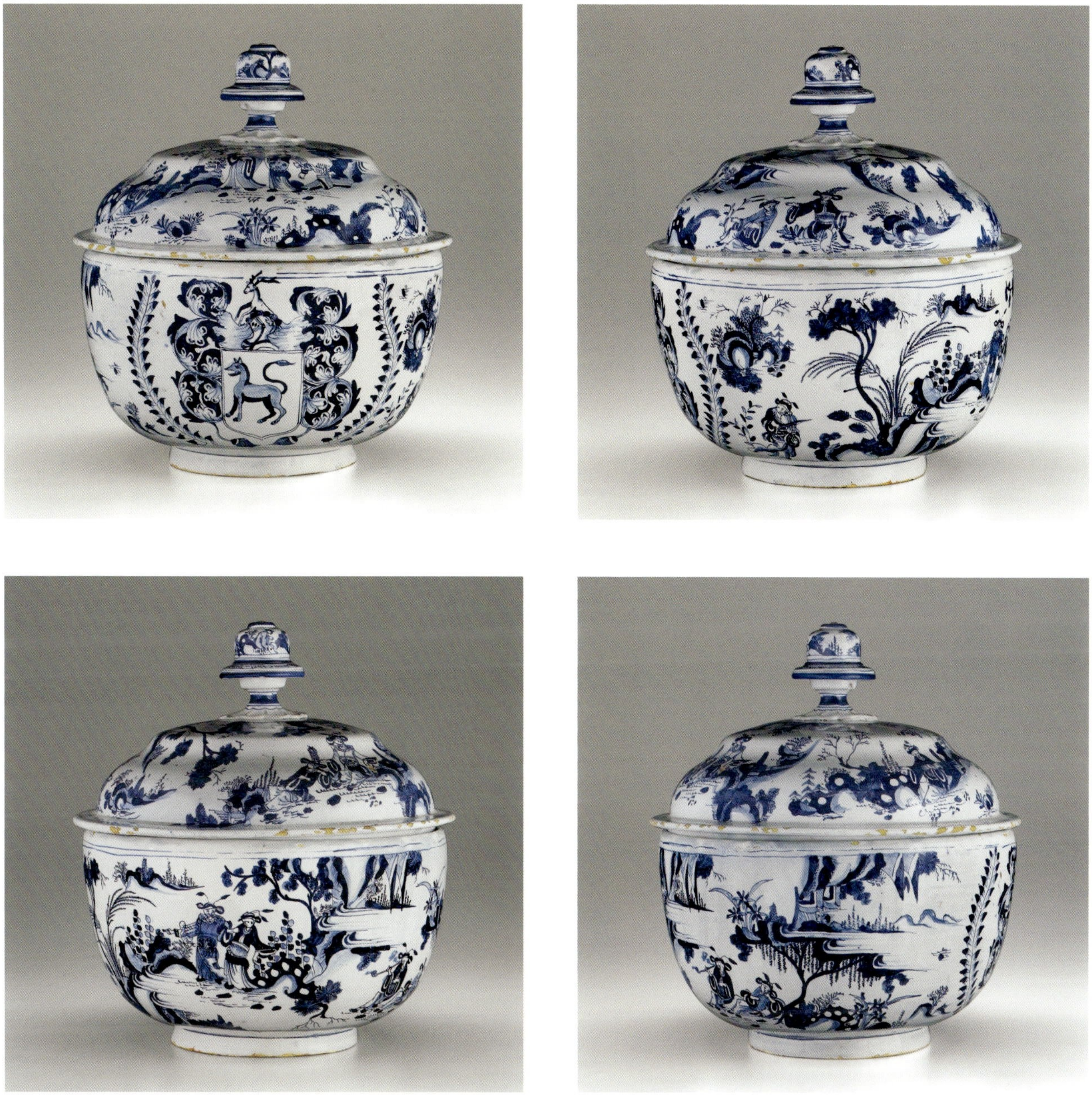

Figure 32 Covered punch bowl, England, 1675–1700. Tin-glazed earthenware. H. 15½". (Private collection.)

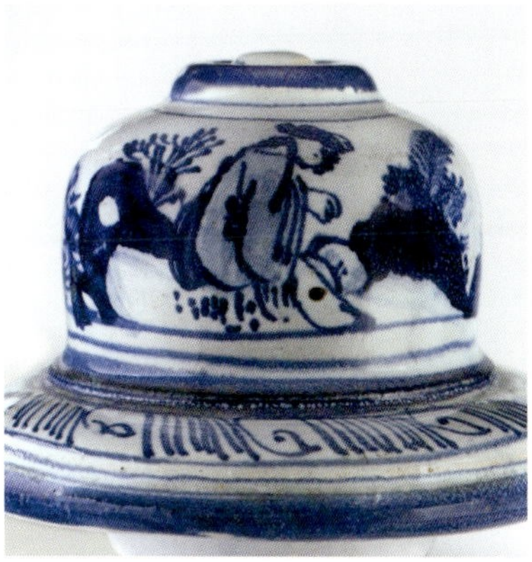

Figure 33 Detail of the Chinese Scholar figure on the finial of the punch bowl illustrated in fig. 32.

Within a few years this toe-dipping Chinese scholar began making solo appearances as the primary decorative image on plates, mugs, and other vessels. The underside of a small plate is inscribed "1675" under the initials "C / T I" (fig. 34). On the plate's face is depicted a seated figure dipping his toe into a stream surrounded by ill-defined rock outcroppings and plants with spiky foliage growing in every direction. This is the earliest known dated English plate featuring the codified Chinese Scholar pattern as the central decoration. The figure is repeated in miniature at the top and bottom of the rim, a composition that appeared regularly in the following twenty-five years except that the year and initials generally appeared on the front of the plate (fig. 35).

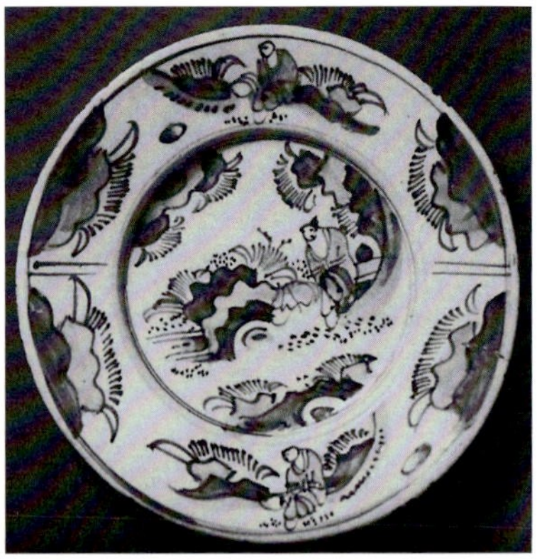

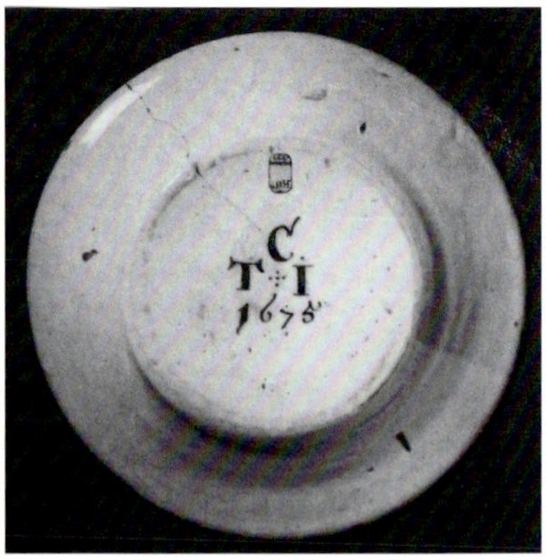

Figure 34 Plate, London, England, 1675. Tin-glazed earthenware. D. 8⅜". Mark, inscribed on the reverse: C / TI / 1675. (Current location unknown.) Illustrated in Louis L. Lipski and Michael Archer, *Dated English Delftware: Tin-glazed Earthenware, 1600–1800* (London: Philip Wilson Publishers for Sotheby's Publications, 1984), p. 34, no. 138.

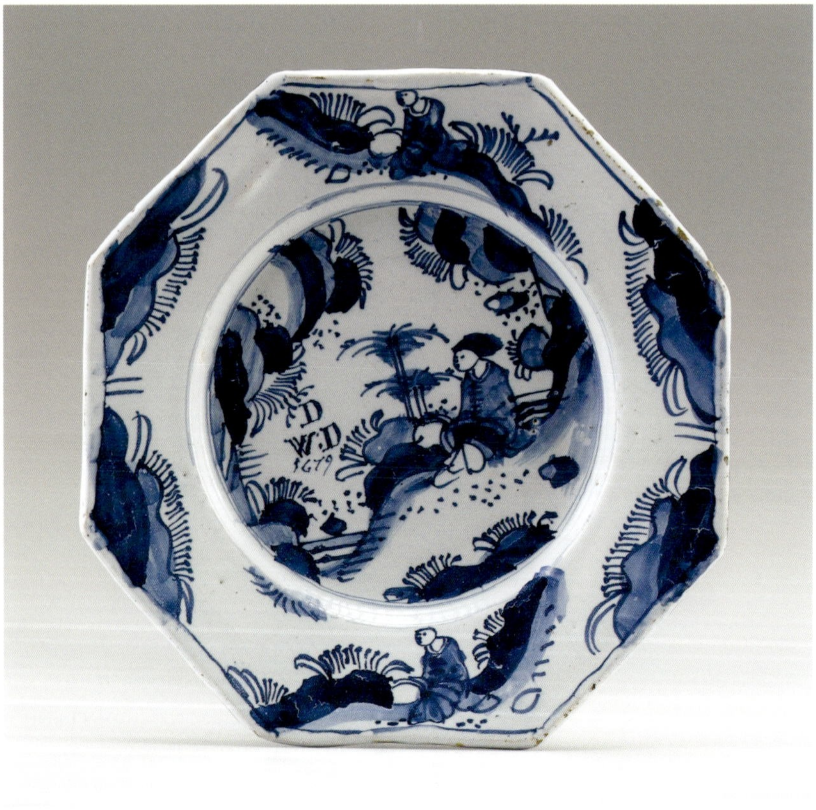

Figure 35 Plate, England, 1679. Tin-glazed earthenware. D. 7⅝". (Courtesy, John H. Bryan.) While the scene inside the central reserve of the Chinese Scholar plates is usually asymmetrical, the rims are generally symmetrical. Often one or two quickly painted lines mark the central horizon where the pattern of spiky foliage and amorphous rocks on the top rim meets the repeated pattern on the bottom rim. This plate is one in a set of three featured in Lipski and Archer, *Dated English Delftware*, p. 51, nos. 143a, b.

From the late 1670s through the 1690s the Chinese Scholar pattern appeared on huge numbers of undated plates and vessels made in London as well as Bristol and Brislington. Many painters simplified the toe-dipping scholar-fisherman even further, into an inactive seated figure in profile. In some cases it looks as if the figure is extending his foot but the painter omitted the stream. In others, the figure makes no overtures toward being a fisherman at all and merely sits cross-legged on a ledge, much like Wang Wei's cloud-viewing scholar. Given the convoluted route by which the imagery came to England, it is likely that any associations with the poet Wang Wei, or even with the academic success and morally upright hermeticism associated with these images in China, would have been lost on English viewers (figs. 36, 37).[21]

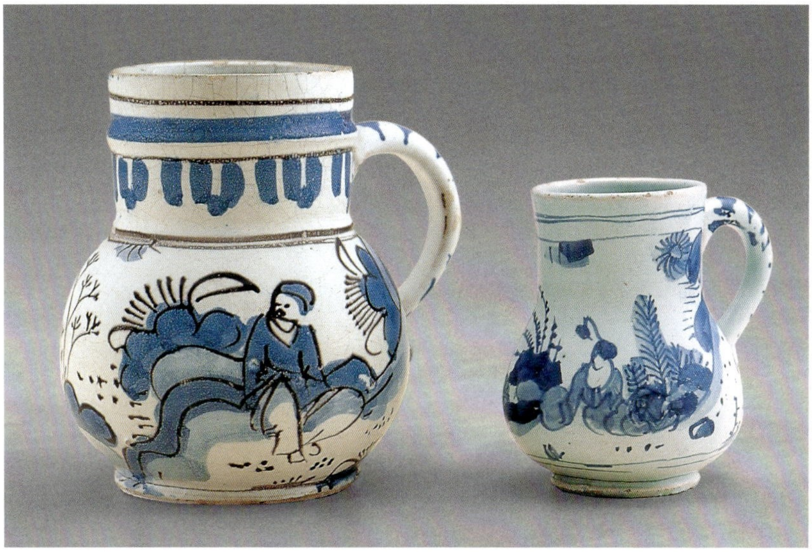

Figure 36 Mugs, London, Brislington, or Bristol, England, 1680–1700. Tin-glazed earthenware. H. of mug (left) 5½". (Courtesy, Longridge Collection.)

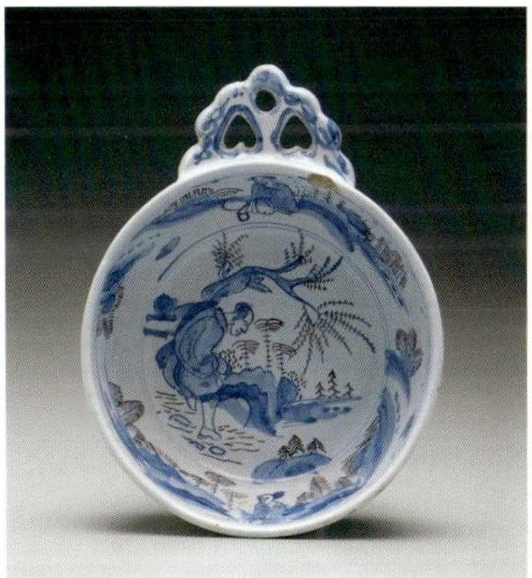

Figure 37 Porringer, England, 1675–1695. Tin-glazed earthenware. H. 7". (Chipstone Foundation.) The Chinese Scholar, dipping his toe in the stream, wears a European-style shoe.

Because of this lack of understanding, the Chinese Scholar pattern as well as the ceramic objects on which it appeared should be considered cultural hybrids. From the first moments that English potters began to copy imported porcelain, they did not understand the symbolism of the Chinese designs and tended to mangle meaning and borrow motifs at random

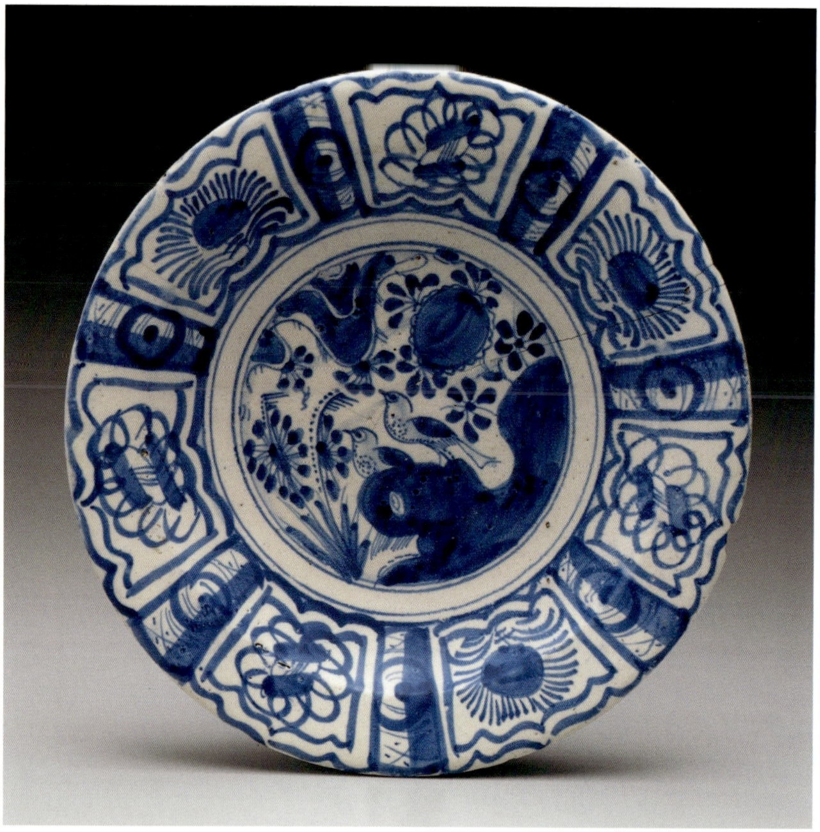

Figure 38 Plate, probably Southwark, London, England, 1630–1640. Tin-glazed earthenware. D. 11¾". (Chipstone Foundation.)

(fig. 38). Likewise, throughout the century Chinese narrative content generally was lost in the English versions, leaving only the decorative details that so fascinated Western consumers: anonymous Chinese people dressed in loose, flowing clothing and unusual headdresses wandering through alien landscapes.[22] In addition to decontextualizing Chinese symbols into generically exotic motifs, potters anglicized them further by incorporating them into familiar conventions of domestic design. On many pieces, potters worked Asian motifs into traditional European systems of symmetry (fig. 39) and often applied these foreign images onto familiar European ceramic forms. Instead of throwing Asian-inspired double-gourd vases or other shapes, potters continued to make conventional caudle cups, tankards, meat pots, and flower vases (figs. 40, 41). By combining exotic imagery (however altered or tamed) with familiar forms and systems of decoration, these potters created hybrid objects palatable to most English consumers. Silversmiths and furniture makers employed similar tactics, and all of their products helped build a market for chinoiserie (figs. 42, 43, 56).[23]

The composition of the Chinese Scholar pattern seems to have been most consistent on plates, remaining remarkably formulaic whether the plates were circular, octagonal, or lobed. The scholar sits in the central reserve and two miniature versions appear at the top and bottom of the rim, at twelve and six o'clock. The posture and details of the seated figures vary from piece to piece, but the miniature versions always replicate the central figure. Rarely are the miniature versions reversed (fig. 44). At least

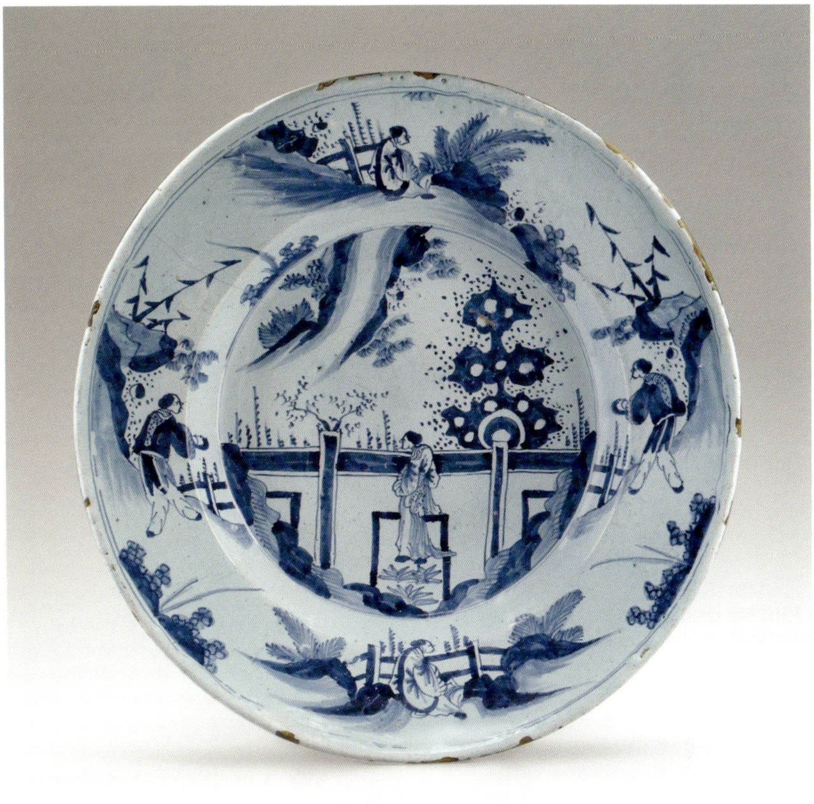

Figure 39 Charger, 1675–1700, England. Tin-glazed earthenware. D. 15¼". (Private collection.) This large Transitional-style charger depicts a scholar's rock, a dramatic geological formation traditionally used by Chinese thinkers to aid their contemplation of the beauty in natural chaos. This anglicized scholar's rock has three symmetrical, graduated tiers with evenly spaced holes; this symmetry would have been an anathema to its original Chinese viewers. The overall composition of the charger is similarly hybrid. It features five sage figures whose details and composition capture the exotic feel of Chinese originals, but the overall arrangement of the scenes in relation to one another conforms to European conventions of symmetry. One figure occupies each of the four cardinal points, and the head of the fifth figure marks the exact center point of the whole plate. By applying imagery that would have been considered foreign and exotic by English observers onto the conventional four-pointed grid of European tradition, this charger and others like it offered newly fashionable chinoiserie in a format familiar to most consumers.

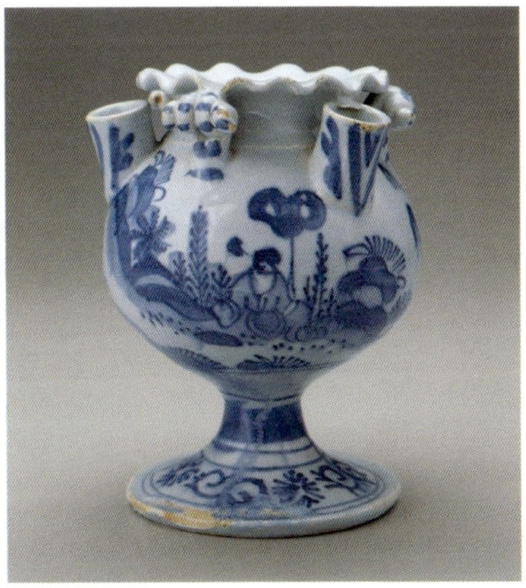

Figure 40 Flower vase, England, 1675–1695. Tin-glazed earthenware. H. 6⅛". (Chipstone Foundation.)

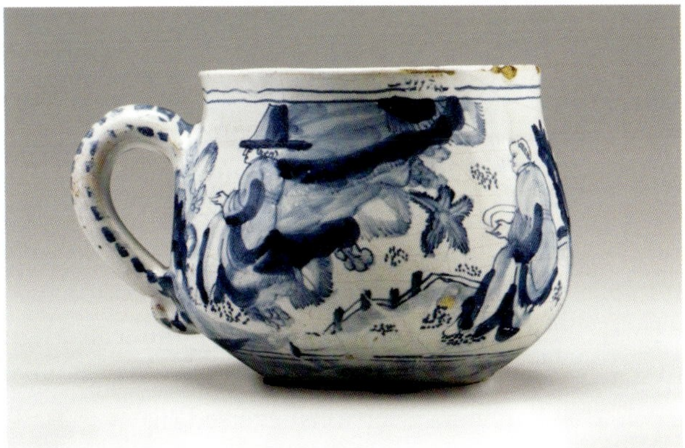

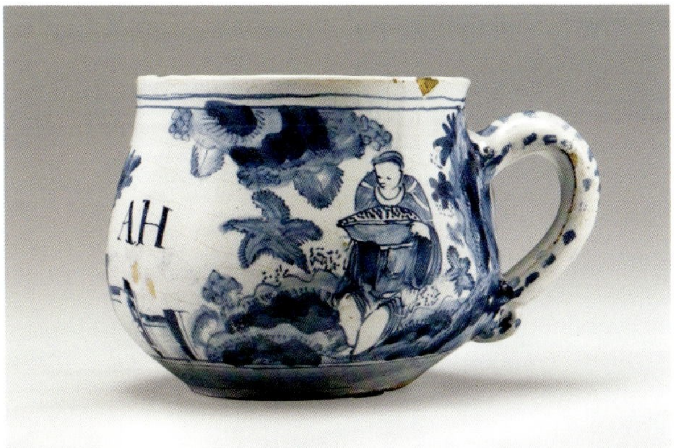

Figure 41 Caudle cup, England, 1670–1690. Tin-glazed earthenware. H. 4", D. 6". (Courtesy, Frederick and Anne H. Vogel.) The decoration on this caudle cup shows curious intercultural mixing. The scene derives from the Ming Transitional style and the figures are seated much like the Chinese Scholar, but they are dressed in English costume.

Figure 42 High chest, Boston, Massachusetts, 1712–1725. Pine with japanning, probably by Nehemiah Partridge (case signed "Park"). H. 61". (Chipstone Foundation.)

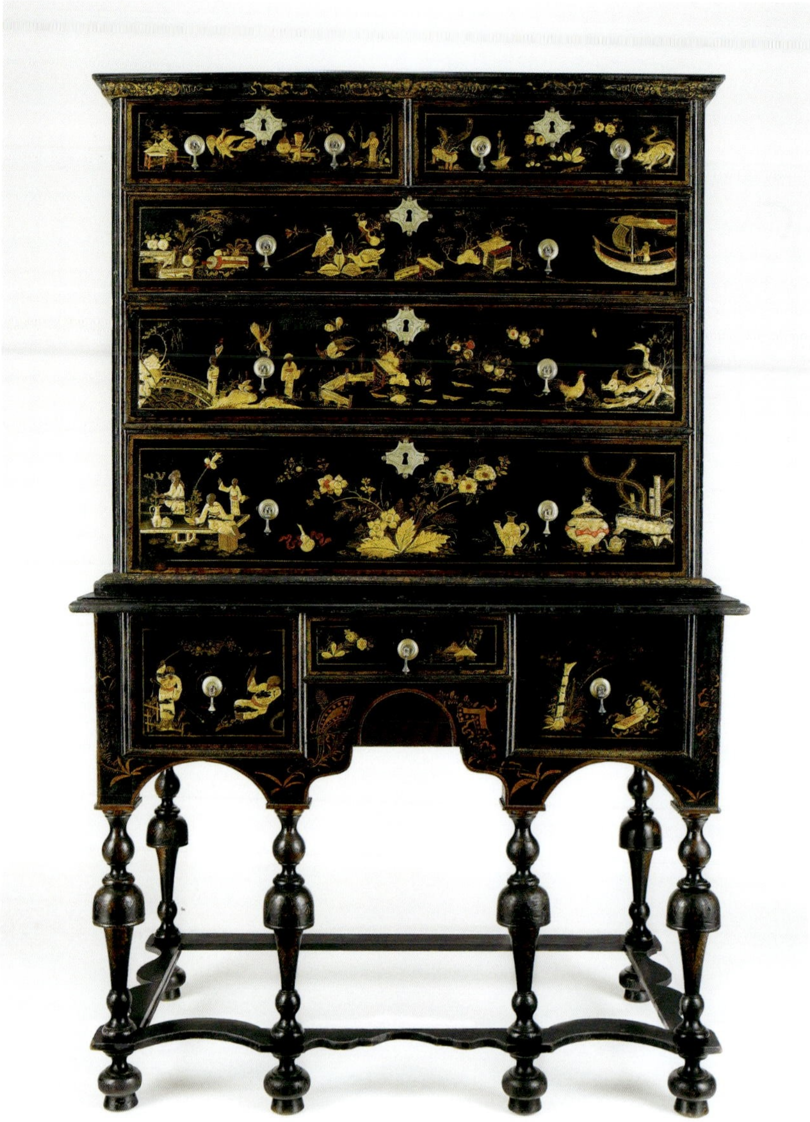

Figure 43 Detail of a drawer on the high chest illustrated in fig. 42.

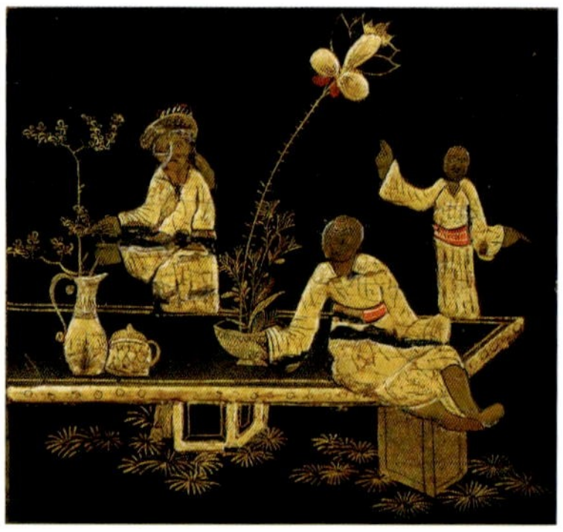

one known plate with four miniature scholars around the rim survives (fig. 45). The first potters to add miniature scholars to the rim decoration may have been looking at the scholar-fishermen inside the oval reserves on large chargers like the 1673 example in figure 31, a suggestion supported by a small plate whose miniature scholars are outlined as if inside Wan Li–type oval reserves (fig. 46).

In the 1690s the Chinese Scholar pattern began to change as the pottery painters' repertoire widened. After the new Qing dynasty was firmly established in China, the Jingdezhen kilns reopened in 1683 and both the Dutch East India Company and the English East India Company began importing vast amounts of Chinese porcelain again. These new wares, with Chinese scholars and popular characters painted in bright green, yellow, and red enamels, offered many new scenes to copy. Robed scholars continued to appear on English ceramics, but the imagery became less formulaic (fig. 47).

Figure 44 Plate, England, 1675–1695. Tin-glazed earthenware. D. 7⅝". (Courtesy, John H. Bryan.)

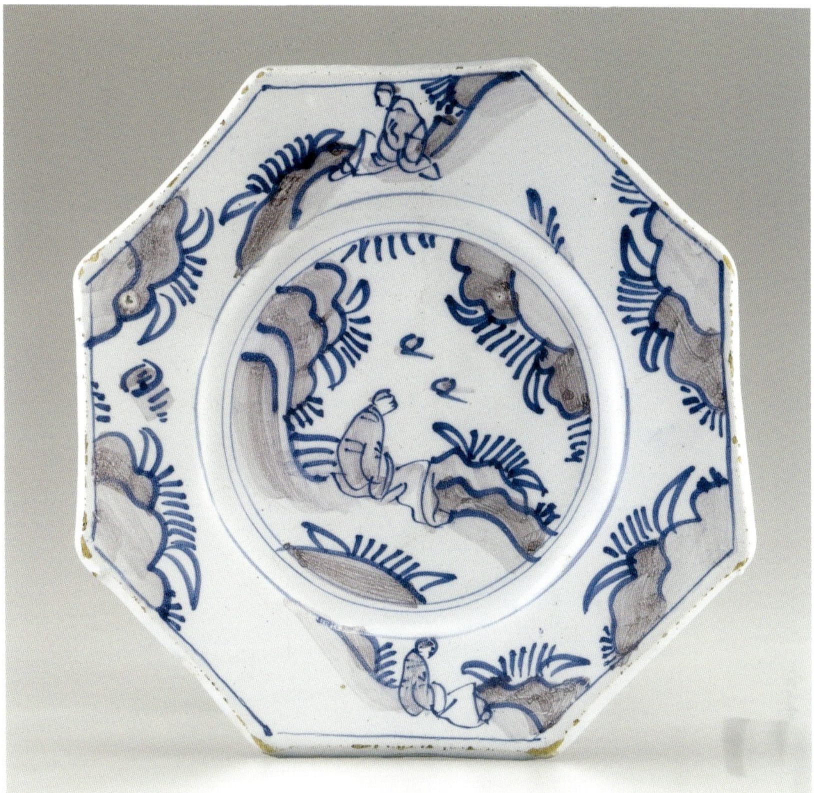

Figure 45 Plate, England, 1675–1695. Tin-glazed earthenware. D. 7½". (Courtesy, Historic Deerfield, Inc.; photo, Amanda Merullo.)

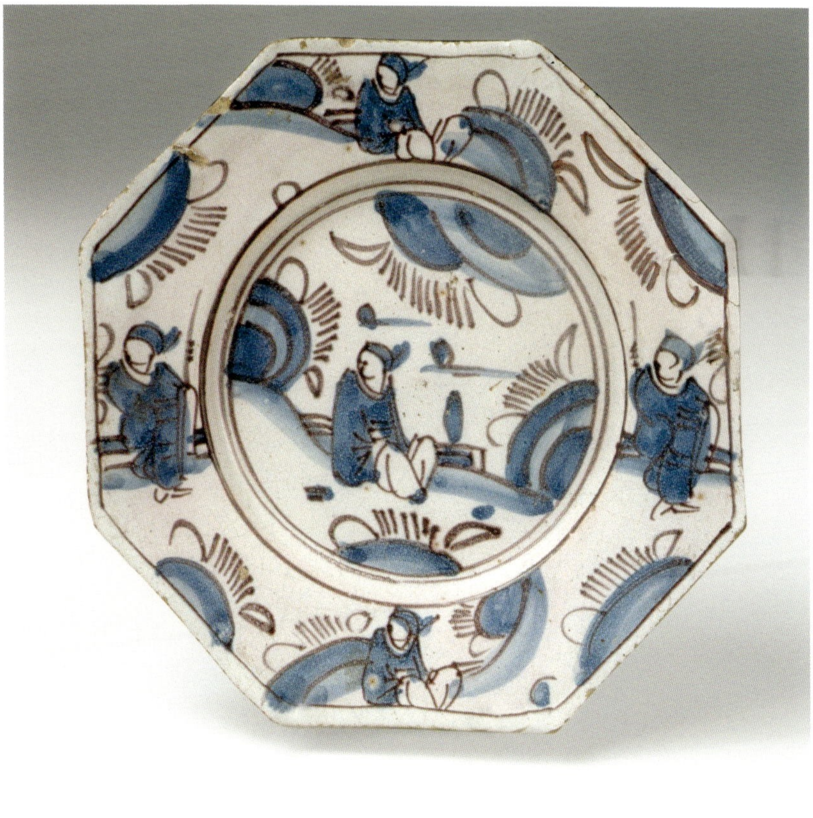

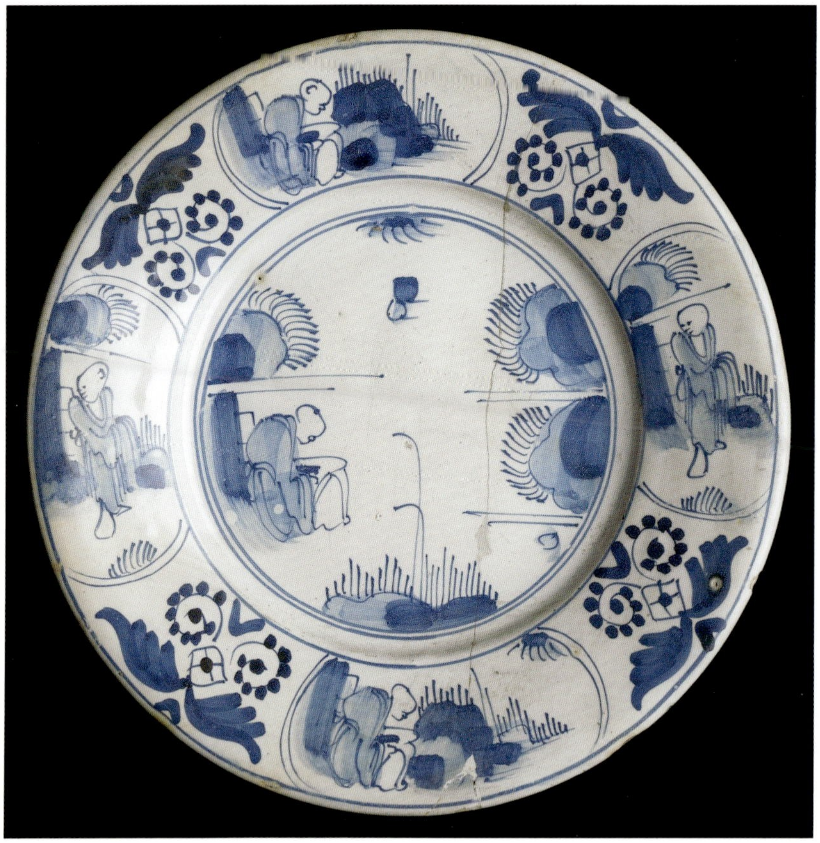

Figure 46 Plate, London, England, ca. 1680. Tin-glazed earthenware. D. 8½". (Courtesy, Museum of London.)

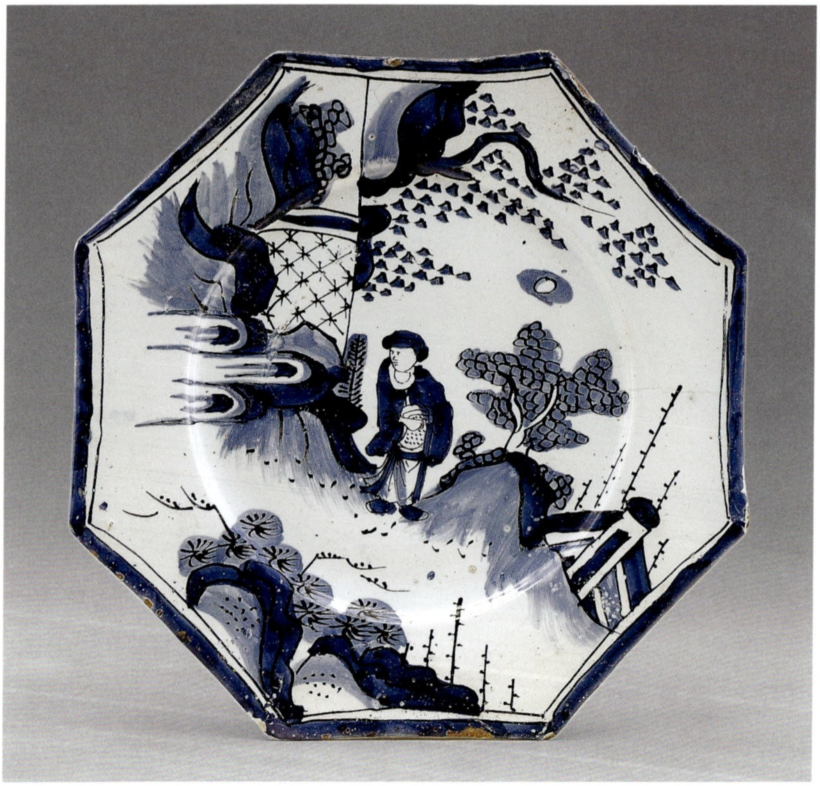

Figure 47 Plate, probably Brislington, England, 1675–1690. Tin-glazed earthenware. W. 10⅜". (Courtesy, Longridge Collection.) Plates with no rim decoration, like this one, tend to feature somewhat more individualized figures painted in greater detail. Allover patterns were relatively unfamiliar to English tastes and may have been influenced by Japanese imports.

This shift was gradual, which makes it difficult to pinpoint when the pattern stopped being produced. Three floral-bordered plates bearing the year 1687 have been cited as the final dated versions known, but several plates with later dates bear variations on the pattern, suggesting its influence continued well into the 1690s. One surviving plate features the miniature scholar figures at the top and bottom of the rim, but the traditional central figure has been replaced with the prominent date of 1696 and the initials "B / I M" (fig. 48). A number of plates dated 1699 feature less animated

Figure 48 Plate, England, 1696. Tin-glazed earthenware. D. 8¾". (Courtesy, John H. Bryan.)

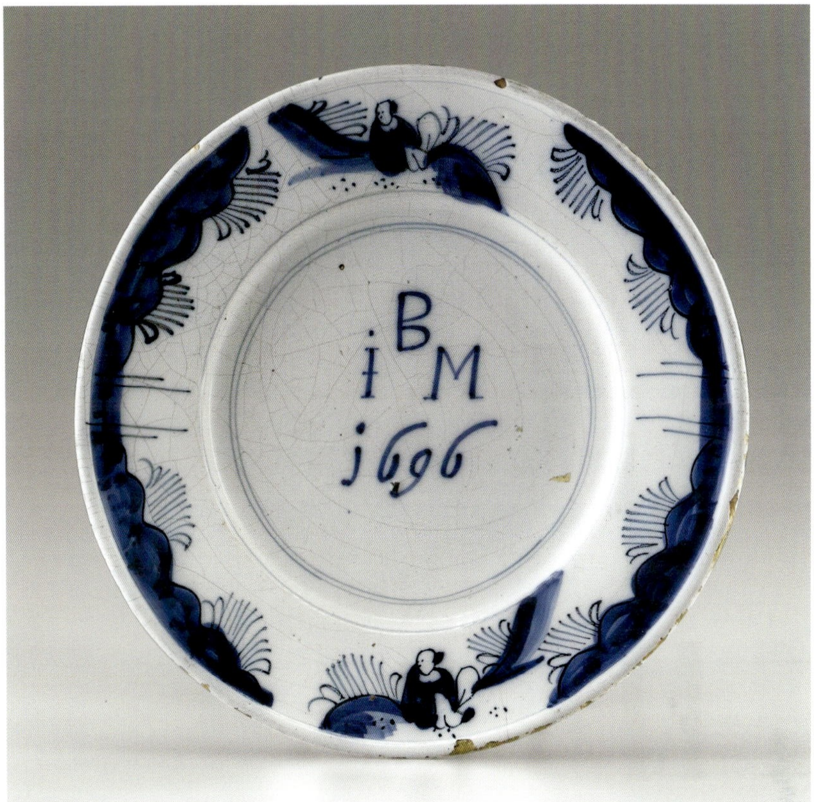

Figure 49 Plate, England, 1699. Tin-glazed earthenware. D. 8⅝". (Courtesy, Sotheby's, New York.)

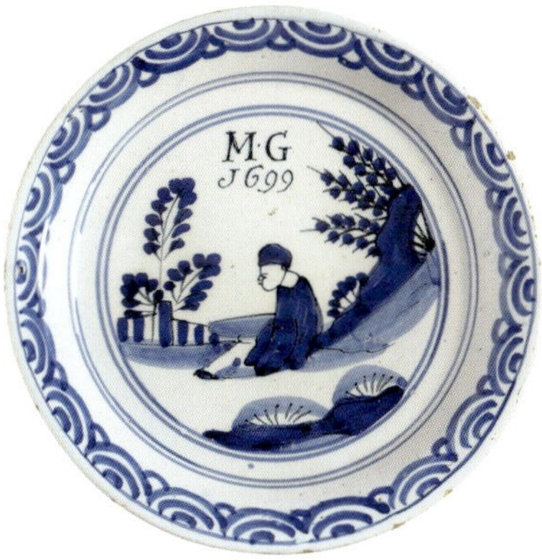

seated figures, geometric borders unrelated to the chinoiserie aesthetic, and plants painted with dabs instead of spiky lines (fig. 49).[24]

Filtered through an elaborate international system of trade and manufacture, the Chinese Scholar pattern that appeared in England barely resembled its original Chinese source. Many of the English potters never saw the Chinese models but were copying Japanese versions or Dutch delftware copies. Their unfamiliarity with Chinese costume, artistic conventions, and symbolism caused many English potters to create Chinese Scholars that are barely decipherable as human figures (figs. 50, 51). These distorted translations of Ming Transitional motifs are examples of what art historian George Kubler called "historical drift." Replications of any object made over a long period of time or in locations distant from the source inevitably become different from the originals. Kubler argued that such drift would cause "diminishing quality" in any series of replicas, and, by conventional aesthetic standards, the Chinese Scholar wares do pale in comparison with

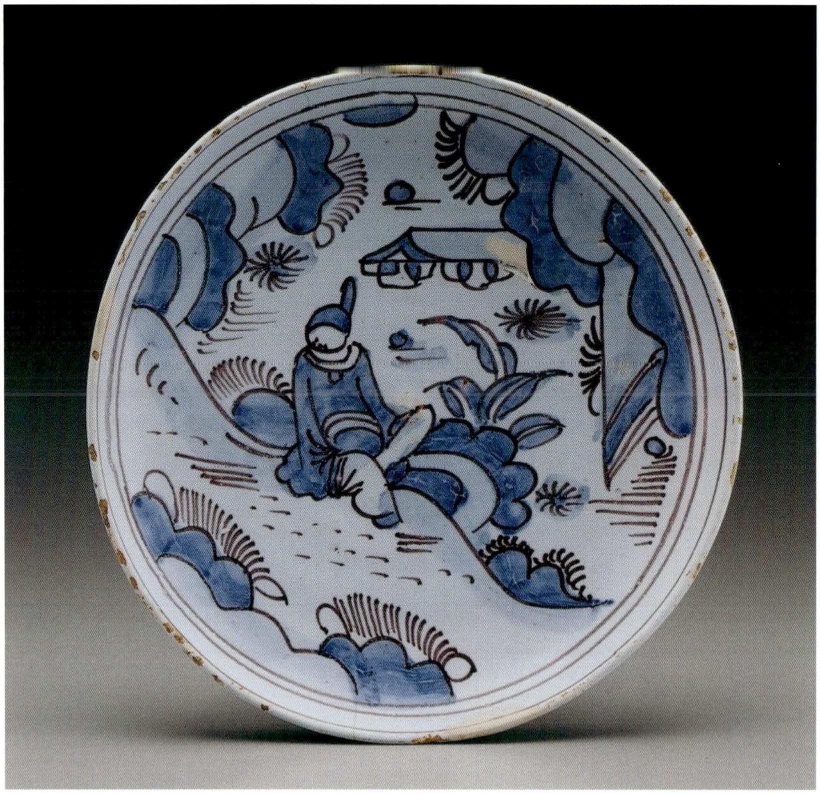

Figure 50 Plate, England, ca. 1685. Tin-glazed earthenware. D. 8½". (Chipstone Foundation.)

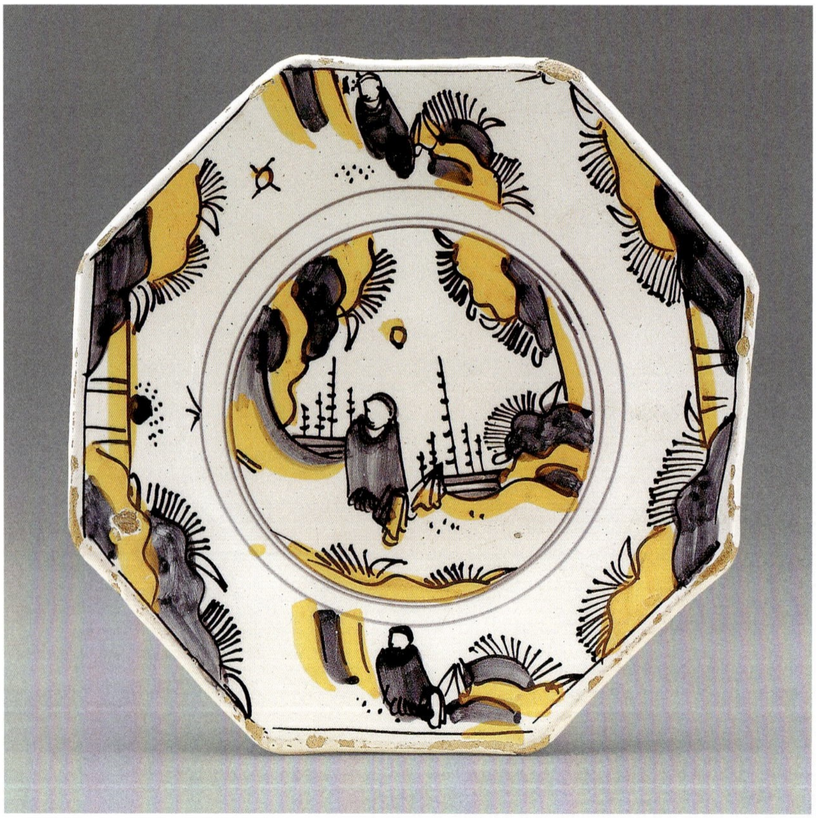

Figure 51 Plate, England, 1675–1695. Tin-glazed earthenware. W. 7⅞". (Courtesy, Longridge Collection.)

the Chinese originals. The clay bodies are thick, the glaze is lumpy, and the imagery less precise. Today's historians, however, can regard the Chinese Scholar wares not as lesser copies of the Chinese originals but rather as manifestations of the frenzied excitement about international discovery and communication that overtook the English public in the late seventeenth century.[25]

The Chinese Scholar and Sharawadgi

In this time of great change the Chinese Scholar pattern can offer cultural insights when studied through three cultural lenses. First, the asymmetry of the Chinese Scholar pattern—however hybrid or tentative—suggests an early role in the aesthetic shift toward the rococo. In the 1670s the rectilinear rigors of Palladian design were just beginning to give way to more playful curvature and whimsy, a shift that eventually led to eighteenth-century rococo styles. Among the first English theorists to write about the allure of asymmetry was Sir William Temple, a retired diplomat and gentleman-botanist who published *Upon the Gardens of Epicurus* in 1685 (fig. 52). While his treatise primarily provided suggestions for organizing the long allées of trees and carefully trimmed parterres in most of the formal gardens in this period, his final paragraphs praised a completely different kind of beauty. He conceded that natural beauty in gardens might be increased by emphasizing the pleasing irregularity inherent in nature. Temple had heard that this agreeable lack of order characterized Chinese gardens and he introduced the Chinese word for it to his English readers:

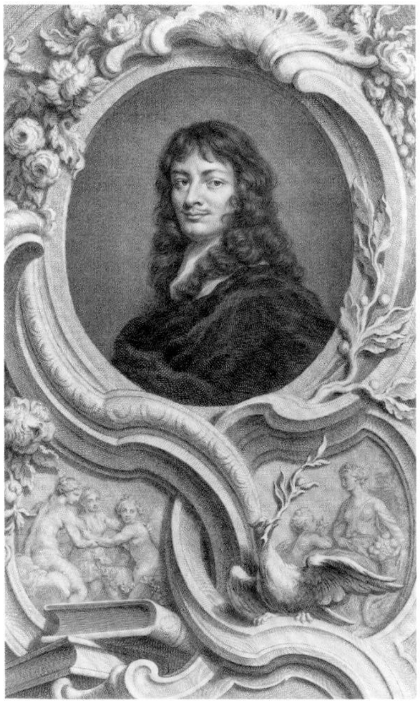

Figure 52 Jacobus Houbraken, after Sir Peter Lely, *Sir William Temple*, line engraving, published by J. and P. Knapton, 1738. Plate 14⅝" x 9⅜". (Courtesy, National Portrait Gallery, London, given by Sir Herbert Henry Raphael, 1st Bt, 1916. NPG D19408.)

> [Their] greatest reach of imagination [among the Chinese] is employed in contriving figures, where the beauty shall be great, and strike the eye, but without any order or disposition of parts that shall be commonly or easily observed: and, though we have hardly any notion of this sort of beauty, yet they have a particular word to express it; and, where they find it hit their eye at first sight, they say the *sharawadgi* is fine or is admirable, or any such expression of esteem.[26]

Sharawadgi, according to Temple, refers to the pleasant surprise enjoyed when the beauty of a natural scene appears unexpectedly and creates a moment of serendipity in the landscape. The word is as mysterious as its effect. Linguists have not convincingly linked *sharawadgi* to any Chinese or Japanese homonym. Temple either anglicized a genuine Asian word beyond recognition or made it up to represent a foreign idea.[27]

Temple's writing is generally regarded among landscape historians as the first philosophical seed of what was to grow into the romantic English garden. The term *sharawadgi* became used widely throughout the eighteenth century to refer to the meandering paths and faux-natural tree groves that characterized the gardens designed by Lancelot "Capability" Brown and others for aristocratic country houses as well as public grounds. A close look at Temple's text reveals that he wrote about *sharawadgi* not only in Chinese gardens but also as a quality in other Asian artifacts. Few design historians have considered the significance of Temple's observations about man-made objects. After defining his foreign-sounding word, Temple continued: "And whoever observes the work upon the best India gowns, or the painting upon their best screens, or porcelains, will find their beauty is all of this kind; that is, without order."[28] According to Temple, *sharawadgi* appeared not only in Chinese gardens but also in the designs on cotton textiles painted or printed in India, lacquer from China and Japan, and porcelain of all kinds. Certainly Ming Transitional porcelain boasted a beauty "without order"; diagonal lines made by tree limbs, craggy mountains, and flowing streams led the viewer's eye wandering over the surface of the vase or bowl in the footsteps of the Chinese sages. This composition placed them wholly outside the regular conventions of European Palladianism.[29]

Like the authentic Chinese porcelains, the English Chinese Scholar pattern also captured *sharawadgi* in its descending cliffs, curving stream beds, and ambiguous swaths of color in the sky. The two scholars sitting at the top and bottom of the rim on many Chinese Scholar plates might not be readily apparent until the viewer's eye stumbles onto them after meandering the plate's surface. This curving and serendipitous gaze is not unlike the "wanton kind of chase" that aesthetic critic William Hogarth praised in his influential 1753 work, *The Analysis of Beauty*. The perfectly swooping "line of beauty" that he advocated in painting as well as objects characterized eighteenth-century rococo in the English taste. Even though Temple had defined *sharawadgi* in 1685, his ideas about asymmetrical garden design were not manifested until Alexander Pope built his gardens at Twickenham in 1714. Likewise, the asymmetry and hybrid exoticism that appeared on the Chinese Scholar wares coincident with Temple's essay predate by several decades the playful expression and exotic content that characterized Hogarth's era. Perhaps the early popularity of the Chinese Scholar pattern opened the door not only to exotic imagery but also to the S-curves, gathered ruffles, and asymmetrical floral sprays found on ceramics, wallpapers, textiles, and other rococo-style objects of the eighteenth century.[30]

The Chinese Scholar and the English Merchant

While the emergence of Chinese Scholar plates and vessels marks an early moment in the shift toward the rococo style, it also suggests a shift in the marketplace. These objects are among the earliest in a group of relatively inexpensive goods newly produced for the growing numbers of successful English merchants. England's greater concentration on shipping and packing raw materials as well as London's population growth allowed a great number of people to support themselves by turning over capital in the cities rather than working the land in the countryside. A new group of merchants created significant wealth for themselves and their nation by moving domestic and imported goods between English ports around the world. These merchants and the professionals who served them constituted a "middle station" or "middling sort" who not only produced profitable goods but also themselves became a new market for fashionable housewares and luxuries previously reserved for the aristocracy.[31]

The growing demand among the middling sort for relatively affordable stylish housewares contributed to growth in tin-glazed earthenware manufacture. Like professional weaving operations and book printers, English potteries had first appeared in the early seventeenth century and then expanded enormously to fulfill the demand from new markets. This increase in tin-glazed earthenware potteries is not remarkable for any particular technical innovation (they relied on Renaissance traditions of maiolica). Their growth instead suggests the shifting attitudes at this time toward ceramics among the nonelite. No longer seen as exclusively utilitarian, ceramic housewares were regarded as items for display and the articulation of taste.[32]

Similarly, chinoiserie also underwent a shift in this period from being available only to the wealthy to being produced for the nonelite. Design historian Dawn Jacobson argues that chinoiserie arose in England primarily outside elite settings in part because of its cool reception within the antiluxury Commonwealth government, which ruled England during the style's early influential years (1649–1660). Even though Charles II was comparatively decadent, the English court did not commission outrageous chinoiserie monuments like the porcelain palace that King Louis XIV of France built for himself in 1670 on the grounds of Versailles. Without strong royal patronage, English chinoiserie at this time offered a distinctly midlevel appearance.[33]

One of the most popular types of chinoiserie tin-glazed earthenware in England, the Chinese Scholar pattern expresses the mercantilist trade system that revolutionized the economy of seventeenth-century Europe. At the core of the mercantilist system was the realization that nonelite desires for luxury goods could be exploited for economic gain. Any nation could succeed by regulating tariffs and exporting more than it imported, a tactic pioneered by the Dutch. Design historian Glenn Adamson has demonstrated that English- and American-made caned chairs, introduced in the 1660s and popular into the eighteenth century, can be read as mercantilist objects because of their novelty, formal emphasis on surface display, relative impermanence, and their similarity to Dutch imports (fig. 53). Much the same can be said of Chinese Scholar plates and vessels.[34]

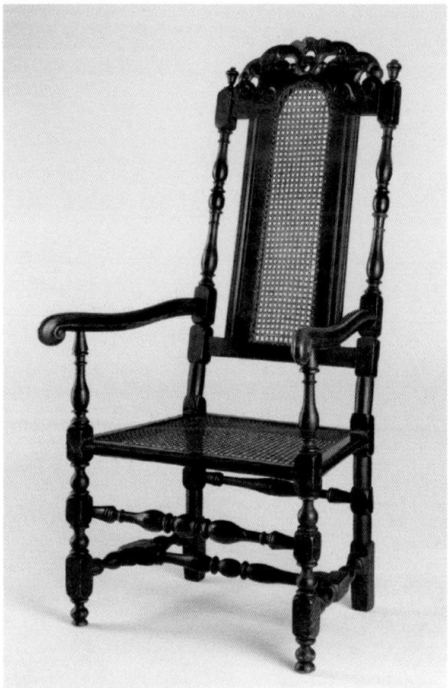

Figure 53 Armchair, Boston, Massachusetts, ca. 1700. Maple and cane. 50¾" x 27". (Chipstone Foundation.)

Similar to the caned-chair makers, potters copying Dutch products thus reduced the number of goods imported from the Continent. Despite calling the tin-glazed earthenwares "delft"—hence emphasizing the Continental source of the tradition—the English may well have seen the success of their burgeoning pottery industry as beating the Dutch at their own game. Another similarity between the Chinese Scholar pattern and caned chairs was their shared emphasis on façade and artificial display. The carved crests of caned chairs could be read clearly only straight-on. The visual power of the Chinese Scholar plates also derived entirely from the frontal views of their surface decoration. While that emphasis was by no means new—decorative plates had been used for interior display for centuries—the quantity of their manufacture and the new distribution within the middling sort suggest an important new emphasis on façade and direct display within a burgeoning consumer culture.[35]

In some ways, the Chinese Scholar pattern may have emblematized mercantilism even more overtly than caned chairs. Its novelty stemmed not just from its similarity to Dutch goods or its formal emphasis on surface but also from its exotic content. By the success of the English East India Company, the prime agent responsible for British mercantilism itself, English eyes were offered intriguing new views of an unfamiliar culture. Like the caned chairs that suddenly began to appear in large numbers in the homes of the nonelite, chinoiserie tin-glazed earthenware—Chinese Scholar items especially—can be considered harbingers of a new era, made in the spirit of England's new identity as a rising mercantilist power.

When considered in the American context, the Chinese Scholar pattern can be seen as part of a common visual language shared among English merchants and their customers throughout their geographically distant settlements. Since the restoration of Charles II to the throne in 1660, the government had renewed its interest in its North American colonies as potential trade centers. With a growing emphasis on maritime trade, the Atlantic economy offered considerable opportunity in these years and attracted investors as well as immigrants. Even New England, a region originally settled for religious idealism rather than trade, had a distinct merchant class by the 1670s. The thousands of English people who settled on Atlantic coastlines in North America and the Caribbean shared a common mercantilist identity with their counterparts back in England. Studied in recent years as members of a transatlantic community tied together by the solidarity of intercoastal trading, English merchants purchased new goods and developed a specific aesthetic that affiliated them with a nation mining the world—from China to Virginia—for fashionable and profitable commodities. Chinese Scholar wares owned in North America, then, can be seen as visually linking merchants not only to English tastes but also to their country's trade ambitions in the East.[36]

These merchant allegiances among colonial settlers in North America become more complex when the archaeological evidence of the Chinese Scholar pattern is considered. Late seventeenth-century colonial sites rarely yield Chinese export porcelain sherds, but they do contain large quantities of Ming Transitional tin-glazed earthenware. Sherds have been found not only from England but also from the Netherlands; it is reasonable to assume that examples from France, Germany, Italy, and Spain also arrived in North America. A number of intact Ming Transitional–style pieces also survive, including a charger probably made in the Netherlands or Germany with an associated history of belonging to Joseph Waterman of Marshfield, Massachusetts. While the large volume of chinoiserie-style tin-glazed earthenware found in the colonies suggests open international exchange of ceramic goods, in reality the English Parliament progressively limited the trade of foreign goods among English-run ports in North America with a series of Navigation Acts in 1650, 1661, and 1672. Despite these regulations, Continental luxuries appeared in American cities, usually smuggled by merchants from all over the Atlantic world. Consumers in the Americas probably did not care about (or even necessarily know) in which country their Chinese Scholar wares were made. Trumping their ties to the English government's rules were their own identities as participants in a growing global network of trade now stretching all the way to Asia. Perhaps many English settlers who bought Chinese Scholar pattern wares from the Netherlands and Germany identified more strongly with their colonial neighbors and their shared commitment to trade than with the Parliamentary protectionism of their own government.[37]

The Chinese Scholar in the English Imagination

Even as the Chinese Scholar pattern contributed to the stylistic shift toward Asian-inspired asymmetry and a shared visual language among merchants, it can also be seen as part of England's widening worldview. The last quarter of the seventeenth century saw a dramatic resurgence in the English economy, rebounding after the Great Fire of London in 1666 and the third Anglo-Dutch War (1672–1674). The English East India Company was now expanding its trade of Indian textiles, successfully securing factories in South Asia. At the same time, more merchants were venturing westward to North America to trade in lumber and furs. As the merchant class became wealthier, more well traveled, and better respected within its own communities, the English public began to see themselves as powerful players in a suddenly expanding world. They envisioned the Dutch, French, and Portuguese as long-standing rivals. The Chinese, by contrast, remained an unknown people about whom the English imagination was quickly piqued. This period of rapid change was the beginning of what historian Joseph Rykwert has called "the domestication of the marvelous." Rising to popularity during this important time of change, the Chinese Scholar pattern helps reveal the English public's complex and ambiguous outlook on Chinese culture.[38]

Throughout the seventeenth century, cultural interactions between Europe and Asia were driven by economic aspirations but also by a certain degree of popular curiosity. Twentieth-century postcolonial theorists led by Edward Said have critiqued the discourse by which the British and French devalued the people in colonized regions in order to bolster their political might with cultural power. More recent writers have suggested that in the sixteenth and seventeenth centuries the early interactions between East and West were far less prejudicial than in later years and involved considerable give-and-take. In the 1670s most British people had little understanding of the geography of China, India, Japan, and other Asian regions let alone the cultural traditions. Most British consumers entertained wildly imaginative (and mostly ignorant) ideas about far-away "Cathay." "India," "China," and "Japan" were often used interchangeably to describe anything even vaguely exotic, whether a genuine import or a locally made imitation. The attitudes of colonialist superiority criticized by Said had not yet fully taken hold, and most people pursued their curiosity more than a sense of imperialism.[39]

Within this early period of relative openness came great admiration among some Europeans for Chinese culture, and especially for Confucius (551–479 B.C.) (fig. 54), the most prominent Chinese personage in the minds of the English people in the late seventeenth century. In London, for example, booksellers in St. Paul's Churchyard advertised Jesuit descriptions

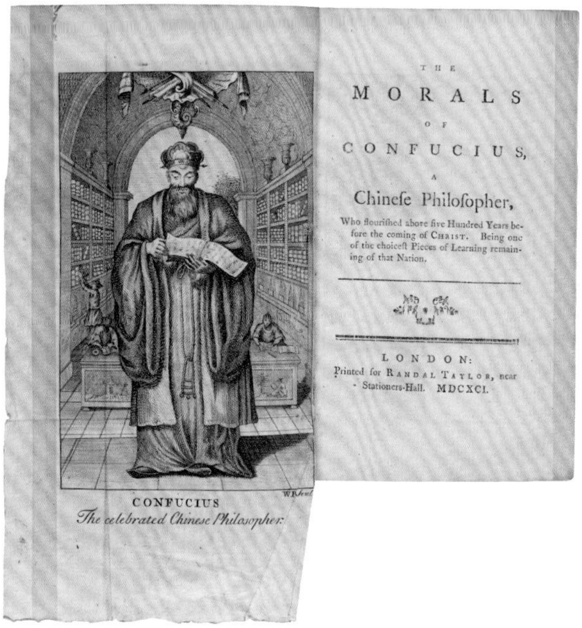
CONFUCIUS
The celebrated Chinese Philosopher.

THE
MORALS
OF
CONFUCIUS,
A
Chineſe Philoſopher,
Who flouriſhed above five Hundred Years before the coming of CHRIST. Being one of the choiceſt Pieces of Learning remaining of that Nation.

LONDON:
Printed for RANDAL TAYLOR, near Stationers-Hall. MDCXCI.

Figure 54 Frontispiece, *The Morals of Confucius* (London: Printed for Randal Taylor, 1691). (Courtesy, Beinecke Rare Book and Manuscript Library, Yale University.)

of China, including Father Couplet's 1691 publication, *The Morals of Confucius, a Chinese Philosopher*, alongside other popular publications such as John Milton's *Paradise Lost* and *Dr. Busby's Greek Grammar*.[40] The robed sage featured in the Chinese Scholar pattern may have evoked associations with this famous figure.

While some viewers might have connected the seated figure on the Chinese Scholar pattern with the religion and philosophy of Confucius, others could have regarded him as a symbol of ancient China's progressive government and ethics. Chief among European proponents of Chinese thinking was German philosopher Gottfried Wilhelm Leibniz, who believed in the shared humanity of all people. Influenced by publications of Confucius's writings in the 1680s, Leibniz praised the "social order" and "public tranquility" in China. Confucian emphasis on the thinking individual, Leibniz argued, enabled the Chinese to "surpass us . . . in practical philosophy" and to set up a society in which "men shall be disrupted in their relations as little as possible."[41] For European observers who followed the progressive thinking of Leibniz and other humanists, China's ancient civilization seemed to offer early modern Europe appealing traditions of equality and order. The hermit-sages in the Chinese Scholar pattern may have recalled for some viewers the disciplined vigor of China's imperial examination system and its orderly, educated society. Seen in any English parlor or tavern, the image of a tranquil Chinese figure dressed in scholar's robes might have signaled an awareness of, and respect for, Chinese ethics and government and, by association, the broad humanist cause.

While the Chinese Scholar pattern gained popularity from this widespread admiration of Confucian morality and society, it also reflected European prejudice and propagated some of the degrading caricatures that came to dominate eighteenth- and nineteenth-century chinoiserie. Conceding that Chinese philosophy was once far ahead of European thinking, only now being selectively mined by Western philosophers, many Europeans saw China as a culture in decline, left to bask in its once and lost glory. The Chinese Scholar image bolstered that view. The figure sits alone, perhaps pondering the natural world. Hanging his head in noble thought or silent despair, the figure could have been seen as a symbol of China's great past captured now by European pottery painters. The hybrid nature of the pattern, which presents an Asian scene through the eyes of European artisans, is an early example of the cross-cultural visual discourse that created the national and racial identities criticized by Said and other postcolonial theorists.[42]

The perception that Chinese people were passive, which became common later in the eighteenth century, may have gained an early foothold with the Chinese Scholar wares. Design historian Stephen Parissien has argued that passive representations of Asian men arose in the 1680s just as the immediate threat of Eastern enemies began to dwindle. The Ottoman Turks, who had posed a threat to landholdings and Christianity in Eastern Europe for generations, began to lose power. Defeated at Vienna in 1683 and again in Hungary in 1687, the Turks finally lost Belgrade in 1688, beginning several decades of military success for the Europeans in the Middle East. As reflected in visual imagery as well as in characters on the London stage, Asian peoples in the English imagination changed from being threatening—if somewhat intriguing—distant aggressors to being subdued, somewhat effeminate inheritors of an ancient civilization in decline. In keeping with this larger trend, the Chinese Scholar sits passively, calmly observing a stream or rock. Missing are the tall mountains and looming tree boughs that lent genuine Ming Transitional scholars their intellectual and moral authority. Instead of resembling a powerful warrior or mythical hermit-sage engaged in active study, the anonymous Chinese Scholar figures suggest no imminent action.[43]

In addition to passivity, eighteenth-century stereotypes about Chinese trading practices and effeminacy also could be seen in the Chinese Scholar pattern. Literary historian David Porter has argued that China's unwillingness to open its borders to international trade led to a widespread "perception among English observers that an unnatural tendency toward blockage and obstructionism was an integral, defining feature of Chinese society as a whole."[44] The inactivity of the Chinese Scholar could have been read as laziness or boredom related to China's seeming disinterest in trade with the West. Less critical observers might have taken the scene simply to suggest that China was a land of leisure, another common assumption in the eighteenth century that promulgated the stereotype of Chinese men being unmanly.

A more complex suggestion about gender can be made by exploring the use and social context of Chinese Scholar wares. Later commentators in the eighteenth century generally associated chinoiserie with women. The delicate porcelain teapots, fantastical imaginary settings, and overall domestic emphasis of the "tea table chat" that took hold in the 1740s were regarded as intrinsically feminine. The link between Asian imports and women in fact had been made as early as 1675, when William Wycherly's play *The Country Wife* first likened the lust for "china" to female lasciviousness. Despite garnering criticism for lewd jokes, the play enjoyed a successful first run on the London stage and influenced such eighteenth-century works as Alexander Pope's *The Rape of the Lock* and William Hogarth's *Rake's Progress* series: both use broken porcelain to symbolize the loss of feminine virginity.[45]

The Chinese Scholar wares, however, do not seem initially to have evoked the same powerful associations with women, female sexuality, or feminized social rituals. Many objects decorated with the pattern were used either for display or drinking liquor and imported hot drinks: plates, chargers, wine bottles, caudle cups, posset pots, gorge mugs, tankards, and monteiths (fig. 55). In the 1680s and 1690s, however, imbibing liquors and tea and coffee still remained an activity associated with men and public teahouses and taverns.[46] This same association between early chinoiserie and

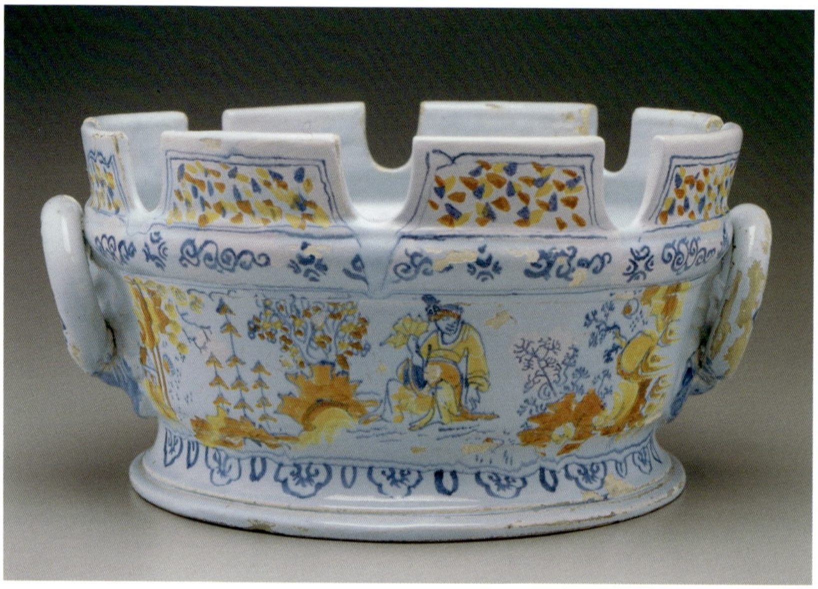

Figure 55 Monteith, England, ca. 1690. Tin-glazed earthenware. D. 12". (Chipstone Foundation.)

male-related objects appears on the flat-chased chinoiserie decoration that appeared on a large number of silver monteiths and tankards made in London in the late seventeenth century (figs. 56, 57).[47] A larger study of early chinoiserie might confirm that the style did not begin as the feminized fashion that it became in the eighteenth century but rather as a style linked to men involved in the burgeoning mercantilist trade.

These speculations about how the Chinese Scholar pattern could have reflected and influenced opinions about China might also suggest some deep conflict in the English imagination. The years between 1675 and 1695 marked a significant turning point in the history of England's empire. With increasing success in India and on the seas, many set their sights on Chinese ports—long dominated by the Dutch—as the next trading frontier, engendering a healthy interest in and admiration of Chinese culture. The excitement of this mercantile ambition was tempered by the fear of rejection from the vast and misunderstood civilizations in Asia. Such insecurities, along with general ignorance about Chinese culture, may have contributed to the condescension seen in the Chinese Scholar pattern and elsewhere. Admiration and condescension often coexist but are more easily projected onto subjects that are especially malleable. Because of its lack of specificity in the eyes of most English people, the Chinese Scholar pattern itself, having been radically simplified and removed from its original context, could absorb and embody different, even contrasting, viewpoints. The faceless, anonymous figure featured in scenes of stylized exoticism could be the great Confucius at one moment and a degrading caricature the next.

Returning to Sir William Temple's treatise can offer a final resonance for the Chinese Scholar pattern. Temple followed Epicurean thought. His biographer wrote that while Temple never publicly embraced the philosophy, "he based his own mature life on the Epicurean precepts that freedom is the most precious fruit of independence and plain living and that the pri-

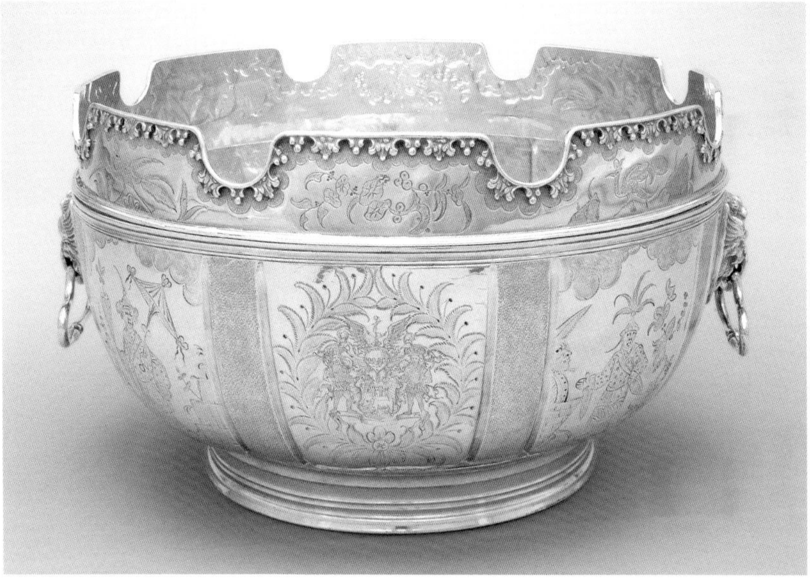

Figure 56 Monteith, George Garthorne, London, England, 1688. Silver with flat-chased decoration. D. 11⅜". (Courtesy, Milwaukee Art Museum, purchase, Virginia Booth Vogel Acquisition Fund.)

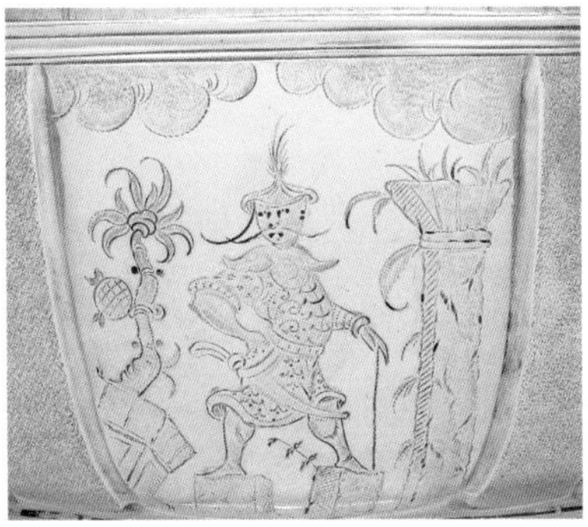

Figure 57 Detail of the monteith illustrated in fig. 56.

vate citizen lives more calmly and safely than the public man."[48] The final paragraphs of Temple's treatise, which celebrate the independent appreciation of nature, may have been an attempt by the aging statesman to encourage his readers, faced with a new question of royal succession, to embrace the sober ideals of Epicurus.

In many ways, the Chinese Scholar pattern might have resonated with Temple and other like-minded thinkers. This robed figure sat in his own personal garden engaged in calm contemplation, much like Epicurus suggested. He seems to embody what literary historian Rose Zimbardo has called the "new feeling about feeling" that arose in the last two decades of the seventeenth century.[49] Writers and theologians began to suggest that they could better understand human behavior through careful self-observation. In 1681 Bishop Samuel Parker explained this new emphasis on individual sentiment: "All men feel a natural Deliciousness consequent upon every exercise of their good-natur'd Passions; and nothing affects the mind with greater Complacency than to reflect upon its own inward Joy and Contentment."[50] When Parker or Temple gathered with companions to drink in a tavern or a private parlor, they might have passed around libations in a cup or posset pot decorated with the Chinese Scholar pattern. Parker might have appreciated the Chinese scholar's inward reflection and speculated about the "natural Deliciousness" he might be enjoying. Temple might have remarked on the pattern's *sharawadgi*. And both men might have noticed that the anonymous seated figure simultaneously evoked Epicurus, Confucius—and also themselves. All were seen as proponents of personal introspection and individual thought, ideas just coming to the forefront of late-seventeenth-century humanist thinking. The visual hybridity of the Chinese Scholar pattern opened itself to multiple readings because it appeared at once foreign and familiar, generic and specific. Indeed, for Englishmen in this era, the Chinese Scholar pattern could have inspired conversation about the curious merits of asymmetrical design, their own prospects for new trade in Asia, Confucian philosophy, and even the ancient garden of Epicurus—a combination of ideas as mixed and provocative as chinoiserie itself.

ACKNOWLEDGMENTS This essay grew out of research undertaken for a 2005 exhibition at the Milwaukee Art Museum called "Enter the Dragon: The Beginnings of English Chinoiserie, 1680–1710." The exhibition can be viewed on the Chipstone Foundation website at www.chipstone.org. I wish to thank Glenn Adamson, the late Tracey Albainy, Ellenor Alcorn, Michael Archer, Gavin Ashworth, Emerson W. Baker, Mary Beaudry, Ellen Berkland, John H. Bryan, Edward S. Cooke Jr., Julia Curtis, Michelle Erickson, George Fayen, Ron Fuchs, Philippa Glanville, Constance Godfrey, the late Dudley Godfrey, Leslie Grigsby, Suzanne Findlen Hood, the late Jonathan Horne, Paul Huey, Robert Hunter, Nigel Jeffries, Amanda Lange, Ethan Lasser, Ann-Eliza Lewis, Ann Smart Martin, Wynne Patterson, William Pittman, David Porter, Jonathan Prown, Haun Saussy, Timothy James Scarlett, and Fred and Anne Vogel.

1. Michael Archer, *Delftware: The Tin-Glazed Earthenware of the British Isles: A Catalogue of the Collection in the Victoria and Albert Museum* (London: The Stationery Office, in association with the Victoria and Albert Museum, 1997), p. 257.

2. For updated names, see Leslie B. Grigsby, *The Longridge Collection of English Slipware and Delftware*, 2 vols. (London: Jonathan Horne Publications, 2000), 2:140, D105; and John C. Austin, *British Delft at Williamsburg* (Williamsburg, Va.: Colonial Williamsburg Foundation, in association with Jonathan Horne Publications, 1994), p. 140.

3 William Temple, "Upon the Gardens of Epicurus; or, of Gardening in the Year 1685," in *Miscellanea, The Second Part, In Four Essays* (London: Printed by J. R. for Ri. and Ra. Simpson at the Sign of the Harp, St. Paul's Church-Yard, 1690), pp. 75–141. A widely available reprint is Albert Forbes Sieveking, *Sir William Temple upon the Gardens of Epicurus, with Other XVIIth Century Garden Essays* (London: Chatto and Windus, 1908).

4. The classic texts on chinoiserie are Hugh Honour, *Chinoiserie: The Vision of Cathay* (New York: Harper & Row, 1961); Oliver R. Impey, *Chinoiserie: The Impact of Oriental Styles on Western Art and Decoration* (London: Oxford University Press, 1977); and Madeleine Jarry, *Chinoiserie: Chinese Influence on European Decorative Art, 17th and 18th Centuries*, trans. Gail Mangold-Vine

(New York: Vendome Press, 1981). More recent noteworthy studies are Dawn Jacobson, *Chinoiserie* (London: Phaidon, 1993); and Anna Jackson and Amin Jaffer, eds., *Encounters: The Meeting of Asia and Europe, 1500–1800* (London: Victoria and Albert Museum, 2004).

5. For references and photographs of sherds found in Maine, see Emerson W. Baker, "The Archaeology of 1690: Status and Material Life on New England's Northern Frontier," paper presented at "Fields of Vision: The Material and Visual Culture of New England, 1600–1830," held at the American Antiquarian Society, Worcester, Massachusetts, November 9, 2007. For photographs of Chinese Scholar dishes from the Hitchcock site, see "Virtual Norumbega: The Northern New England Frontier," online at http://w3.salemstate.edu/%7Eebaker/hitchcock/hitchartifacts.html. For sherds found in New York, see Charlotte Wilcoxen, *Dutch Trade and Ceramics in America in the Seventeenth Century* (Albany, N.Y.: Albany Institute of History and Art, 1987), p. 61, pl. 12, and fig. 12. For sherds found in St. Mary's City, Maryland, and Jamestown, Hampton, and Williamsburg, Virginia, see Austin, *British Delft at Williamsburg*, pp. 144, 146. The St. Mary's City sherd is illustrated in Silas D. Hurry, *"—once the metropolis of Maryland": The History and Archaeology of Maryland's First Capital* (St. Mary's City, Md.: Historic St. Mary's City Commission, 2001), p. 23. A white-on-blue *bleu persan* sherd with the Chinese Scholar pattern also appeared in Williamsburg; see Austin, *British Delft at Williamsburg*, p. 201. The Hampton plate is detailed in Thomas F. Higgins III, Charles M. Downing, and Donald W. Linebaugh, *Traces of Historic Kecoughtan: Archaeology at a Seventeenth-Century Plantation* (Williamsburg, Va.: William and Mary Center for Archaeological Research, 1999), p. 89. For photographs of sherds with the Chinese Scholar pattern found in Port Royal, see Madeleine J. Donachie, "Household Ceramics at Port Royal, Jamaica, 1655–1692: The Building 4/5 Assemblage" (Ph.D. diss., Texas A&M University, College Station, Texas, 2001), pp. 70, 127, 160–61, 164. The findings from excavated sites of this period in New England have not been published. For sherds found in Lambeth, see F. H. Garner, "Lambeth Earthenware," *English Ceramic Circle Transactions* 1, pt. 4 (1937): pls. 12b, 13a; see also Brian J. Bloice, "Norfolk House, Lambeth: Excavations at a Delftware Kiln Site, 1968," *Post-Medieval Archaeology* 5 (1971): 130, drawings 13 and 14. For sherds found in Lambeth and Southwark, see *Excavations in Southwark 1973–76, Lambeth 1973–79*, Joint Publication No. 3, edited by Peter Hinton (London: London and Middlesex Archaeological Society and Surrey Archaeological Society, 1988), p. 326, fig. 139, nos. 1382 and 1383 (maybe), and pp. 331–32, figs. 142–43, nos. 1314, 1316, 1400, 1402, 1403, 1404. For a sherd excavated in Belfast, see Peter Francis, *Irish Delftware: An Illustrated History* (London: Jonathan Horne Publications, 2000), p. 26, fig. 25. For a discussion of the physical similarities between English and Dutch tin-glazed earthenware and the intertrade that makes them difficult to differentiate, see Wilcoxen, *Dutch Trade and Ceramics in America*, pp. 78–79, 92–93.

6. Christiaan J. A. Jörg and Jan van Campen, *Chinese Ceramics in the Collection of the Rijksmuseum, Amsterdam: The Ming and Qing Dynasties* (London: Philip Wilson, in association with the Rijksmuseum, 1997); Christiaan J. A. Jörg, *Fine and Curious: Japanese Export Porcelain in Dutch Collections* (Amsterdam: Hotei Publishing, 2003), pp. 9–10; Christiaan Jörg, "Chinese Porcelain for the Dutch in the Seventeenth Century: Trading Networks and Private Enterprises," in *The Porcelains of Jingdezhen*, Colloquies on Art & Archaeology in Asia No. 16, edited by Rosemary E. Scott (London: Percival David Foundation of Chinese Art, 1993), pp. 183–205; Christiaan J. A. Jörg, *Interactions in Ceramics: Oriental Porcelain & Delftware*, exh. cat. (Hong Kong: Hong Kong Museum of Art, 1984); John Carswell, *Blue and White: Chinese Porcelain and Its Impact on the Western World*, exh. cat. (Chicago: University of Chicago, David and Alfred Smart Gallery, 1985), pp. 45, 165; Michael Archer, "Oriental Influence on English Delftware," in *The Potter's Art: Contributions to the Study of the Koerner Collection of European Ceramics*, edited by Carol E. Mayer (Vancouver: Museum of Anthropology, University of British Columbia, 1997), pp. 52–67.

7. Jörg, *Chinese Ceramics in the Collection of the Rijksmuseum*, pp. 74–87; see also Julia B. Curtis and Stephen Little, *Chinese Porcelains of the Seventeenth Century: Landscapes, Scholars' Motifs, and Narratives*, exh. cat. (New York: China Institute Gallery, 1995); Julia B. Curtis, "Markets, Motifs, and Seventeenth-Century Porcelain from Jingdezhen," in Scott, ed., *Porcelains of Jingdezhen*, pp. 123–49; Stephen Little, *Chinese Ceramics of the Transitional Period, 1620–1683*, exh. cat. (New York: China House Gallery, China Institute in America, 1983); Duncan Macintosh, *Chinese Blue and White Porcelain* (Newton Abbot, Eng.: David & Charles, 1977).

8. For cultural life in late Ming China, see Craig Clunas, *Superfluous Things: Material Culture and Social Status in Early Modern China* (1991; reprint, Honolulu: University of Hawai'i Press, 2004), pp. 141–65, esp. 155–56; Haun Saussy, *Great Walls of Discourse and Other Adven-*

tures in Cultural China, Harvard East Asian Monographs 212 (Cambridge, Mass.: Harvard University Asia Center, 2001), pp. 15–34, esp. 22–23. On Ming painting, see James Cahill, *The Distant Mountains: Chinese Painting of the Late Ming Dynasty, 1570–1644* (New York: Weatherill, 1982), pp. 6–9; and Curtis and Little, *Chinese Porcelains of the Seventeenth Century*, pp. 15–16.

9. Saussy, *Great Walls of Discourse*, p. 23; Chu-tsing Li and James C. Y. Watt, eds., *The Chinese Scholar's Studio: Artistic Life in the Late Ming Period; An Exhibition from the Shanghai Museum*, exh. cat. (New York: Thames and Hudson, in association with the Asia Society Galleries, 1987). For more in-depth readings of the landscape and "scholar's study" imagery, see Curtis and Little, *Chinese Porcelains of the Seventeenth Century*, pp. 15–34, 43–88, and Ireneus László Legeza, *A Descriptive and Illustrated Catalogue of the Malcolm MacDonald Collection of Chinese Ceramics in the Gulbenkian Museum of Oriental Art and Archaeology, School of Oriental Studies, University of Durham* (London: Oxford University Press, 1972), p. 65.

10. For the influence of Wang Wei's poetry on Chinese painting, see Robert E. Harrist Jr., "*Watching Clouds Rise*: A Tang Dynasty Couplet and Its Illustration in Song Painting," *Bulletin of the Cleveland Museum of Art* 78 (1991): 301–23. I would like to thank Haun Saussy and George Fayen for this reference. The translation of Wang Wei's poem "Zhongnan Retreat" comes from Harrist, who cites Pauline Yu, *The Poetry of Wang Wei: New Translations and Commentary* (Bloomington: Indiana University Press, 1980), p. 171.

11. Jörg, "Chinese Porcelain for the Dutch in the Seventeenth Century," pp. 186–87; Jörg, *Fine and Curious*, p. 9. For a specific shipment of Chinese porcelain to the Netherlands, see C. L. van der Pijl-Ketel, ed., *The Ceramic Load of the "Witte Leeuw" (1613)* (Amsterdam: Rijksmuseum, 1982).

12. The most thorough and up-to-date presentation of Japanese imports is Jörg, *Fine and Curious*. Jörg's work builds on John Ayers, Oliver Impey, and J. V. G. Mallet, *Porcelain for Palaces: The Fashion for Japan in Europe, 1650–1750*, exh. cat. (London: Oriental Ceramic Society, 1990), pp. 17–24. The period quote and a discussion of the prototypes sent to Japan for copying can be found in Jörg, *Fine and Curious*, p. 23.

13. For Japanese-made Chinese forms, see Ayers, Impey, and Mallet, *Porcelain for Palaces*, p. 97, no. 38, p. 191, no. 180. For discussion of the V-shaped strokes, ibid., pp. 37, 91; and Jörg, *Fine and Curious*, p. 24. Stephen Little disagrees with Oliver Impey and suggests that V-shaped strokes also appeared on Chinese originals. See Little, *Chinese Ceramics of the Transitional Period*, p. 11.

14. An excellent narrative of Dutch tin-glazed earthenware production has been translated into English. See Jan Daniël van Dam, *Delffse Porceleyne: Dutch Delftware, 1620–1850* (Amsterdam: Rijksmuseum, 2004). The class associations of chinoiserie tin-glazed earthenware in the Netherlands are discussed in Jörg, "Chinese Porcelain for the Dutch in the Seventeenth Century," pp. 193–94. For more on Dutch tin-glazed earthenware in this period, see H[enri]-P[ierre] Fourest, *Delftware: Faience Production at Delft*, trans. Katherine Watson (New York: Rizzoli, 1980).

15. Honour, *Chinoiserie: The Vision of Cathay*, pp. 48–49; Carswell, *Blue and White*, pp. 40–42; and Wilcoxen, *Dutch Trade and Ceramics in America*, pp. 53–54. Michael Archer suggests that the French were copying Ming Transitional patterns before the Dutch. See Archer, "Oriental Influence on English Delftware," p. 57. For a general source on French tin-glazed earthenware, see Danielle Maternati-Baldouy, *Faience et porcelaine de Marseille, XVIIe–XVIIIe siècles: Collections du Musée de la Faïence de Marseille* (Marseille: Musées de Marseille, 1997).

16. While Johannes Nieuhoff's highly influential description of the Dutch embassy trip to China, translated into English and published in London in 1673, may have generated interest in Chinese imagery, his illustrations do not seem to have directly inspired pottery decoration at this time. For extensive studies of the Huguenot migration, see Neil Kamil, *Fortress of the Soul: Violence, Metaphysics, and Material Life in the Huguenots' New World, 1517–1751* (Baltimore, Md.: Johns Hopkins University Press, 2005). An engaging portrayal of English-Dutch relations is Lisa Jardine, *Going Dutch: How England Plundered Holland's Glory* (New York: HarperCollins, 2008). A very good description of the arrival of Asian design into the English pottery industry is found in Archer, "Influences on the Shape and Decoration of English Delftware," in *Delftware*, pp. 29–38; and idem, "Oriental Influence on English Delftware," p. 58. For Chinese Scholar patterns from the Bristol area, see Frank Britton, *English Delftware in the Bristol Collection* (London: Sotheby Publications, 1982), pp. 169–74.

17. For early drug jars with Chinese figures, see Louis L. Lipski and Michael Archer, *Dated English Delftware: Tin-glazed Earthenware, 1600–1800* (London: Philip Wilson Publishers for

Sotheby Publications, 1984), no. 1594 (dated 1658); Garry Atkins, *An Exhibition of English Pottery, March 11th to 22nd, 1997* (London: G. Atkins, 1997), no. 10; and Bernard Rackham, *Catalogue of the Glaisher Collection of Pottery and Porcelain in the Fitzwilliam Museum, Cambridge*, 2 vols. (1935; reprint, Woodbridge, Suffolk, Eng.: Antique Collectors' Club, 1987), vol. 1, no. 2218. Michael Archer (*Delftware*, p. 257) also notes a Transitional-style fragment of probably Dutch but possibly English production excavated at Nonsuch Palace (Ewell, Surrey) dated 1650; see also Martin Biddle, *Nonsuch Palace: The Material Culture of a Noble Restoration Household* (Oxford: Oxbow Books, 2005), p. 72. For the 1662 jug, see Grigsby, *Longridge Collection*, p. 322, D296A.

18. See also the posset pot in the Fitzwilliam Museum, Cambridge (1331-1928), illustrated in Lipski and Archer, *Dated English Delftware*, p. 170, no. 768.

19. The Transitional-style figures, foliage, and heraldic painting on this punch bowl closely resemble the decoration on another punch bowl dated 1681 and now in the Victoria and Albert Museum (c.62-1964). See Lipski and Archer, *Dated English Delftware*, p. 236, no. 1038.

20. While Chinese Ming Transitional wares sometimes portray scholars as fishermen, no definitive source has been identified for the robed figure dipping his toe into a stream who appears on the early English copies. For discussion of the fisherman symbolism, see Curtis and Little, *Chinese Porcelains of the Seventeenth Century*, pp. 23, 134. See also Little, *Chinese Ceramics*, pp. 11, 111.

21. As Christiaan Jörg has argued ("Chinese Porcelain for the Dutch in the Seventeenth Century," p. 187), the Dutch did not have "the faintest idea about the significance of the iconography or ideology of the scenes depicted [on imported porcelain]. For the Dutch it was just 'Chinese figures and landscapes.'"

22. For more on the meanings of Chinese imagery, see Little, *Chinese Ceramics of the Transitional Period*; Curtis and Little, *Chinese Porcelains of the Seventeenth Century*; and Macintosh, *Chinese Blue and White Porcelain*.

23. In the late 1670s and 1680s English japanners applied faux-lacquer finishes on conventional high chests and desks. See Julia Hutt, "Asia in Europe: Lacquer for the West," in Jackson and Jaffer, eds., *Encounters*, pp. 236–38; and Percy Macquoid and Ralph Edwards, *The Dictionary of English Furniture, from the Middle Ages to the Late Georgian Period*, 2nd ed. revised and enlarged, 3 vols. (London: Country Life Ltd., 1954), 3:267–71. Similarly, a few unidentified London engravers used flat-chasing techniques to add chinoiserie decoration to monteiths, tankards, dressing sets, and other luxuries. See Ellenor M. Alcorn, *English Silver in the Museum of Fine Arts, Boston*, 2 vols. (Boston: Museum of Fine Arts, Boston, 1993–), 1:171–76, 183–86; and Carl C. Dauterman, "Chinese Imagery on Restoration Silver," *The Magazine Antiques* 88 (October 1965): 511–15. The impulse to work exotic imagery into conventional European forms was not uniquely English. A large Japanese-for-export charger shares a similar composition with the charger in figure 39 and was made to please the tastes of the Dutch East India Company. See Jörg, *Fine and Curious*, p. 52, no. 28. A similar Dutch tin-glazed earthenware example also survives. See Frits Scholten, *Dutch Majolica and Delftware, 1550–1700: The Edwin van Drecht Collection, Exhibited in the Paleis Lange Voorhout Museum, The Hague* (Amsterdam: E. van Drecht, 1993), p. 130, no. 111.

24. Grigsby, *Longridge Collection*, p. 140, D105. Michael Archer (*Delftware*, p. 257) notes that variations on the Chinese Scholar appear as late as 1711. For the 1699 dated plates, see Lipski and Archer, *Dated English Delftware*, nos. 217, 218; and *Important English Pottery: The Harriet Carlton Goldweitz Collection*, sale cat., Sotheby's, New York, January 20, 2006, lot 34, p. 68. The miniature Chinese Scholar appeared on the rims of plates with unrelated central decoration as late as the 1730s, as illustrated in Grigsby, *Longridge Collection*, p. 169, D142.

25. George Kubler, *The Shape of Time: Remarks on the History of Things* (New Haven, Conn.: Yale University Press, 1962), pp. 71–76.

26. Sieveking, *Sir William Temple upon the Gardens of Epicurus*, p. 54.

27. For more on etymology, see Takau Shimada, "Is Sharawadgi Derived from the Japanese Word *sorowaji*?" *Review of English Studies* 48, no. 191 (August 1997): 350–52.

28. Sieveking, *Sir William Temple upon the Gardens of Epicurus*, p. 54.

29. For a succinct discussion of *sharawadgi* in English gardening, see Yu Liu, "The Importance of the Chinese Connection: The Origin of the English Garden," *Eighteenth-Century Life* 27, no. 3 (Fall 2003): 70–98. Classic texts on the Chinese influence on English romantic gardens include Arthur O. Lovejoy, "The Chinese Origin of Romanticism," in *Essays in the History of Ideas* (Baltimore, Md.: Johns Hopkins Press, 1948), pp. 99–135; Maynard Mack, *The*

Garden and the City: Retirement and Politics in the Later Poetry of Pope, 1731–1743 (Toronto, Ont.: University of Toronto Press, 1969); and Hugh Honour, "The Anglo-Chinois Garden," in Honour, *Chinoiserie: The Vision of Cathay*, pp. 143–74. For a link between *sharawadgi* and design, see Robert Batchelor, "On the Movement of Porcelains: Rethinking the Birth of the Consumer Society as Interactions of Exchange Networks, China and Britain, 1600–1750," in *Consuming Cultures, Global Perspectives: Historical Trajectories, Transnational Exchanges*, edited by John Brewer and Frank Trentmann (Oxford, Eng.: Berg, 2006), pp. 95–123.

30. See William Hogarth, *The Analysis of Beauty*, edited by Joseph Burke (Oxford: Clarendon Press, 1955), p. 42. For more on Alexander Pope's gardening, see Peter Martin, *Pursuing Innocent Pleasures: The Gardening World of Alexander Pope* (Hamden, Conn.: Archon Books, 1984).

31. For more on the growth of the "middling sort" and London's rise to prominence in the European marketplace, see Peter Earle, *The Making of the English Middle Class: Business, Society, and Family Life in London, 1660–1730* (Berkeley: University of California Press, 1989); Mireille Galinou, ed., *City Merchants and the Arts, 1670–1720* (Wetherby, Eng.: Oblong for the Corporation of London, 2004); and David Hancock, *Citizens of the World: London Merchants and the Integration of the British Atlantic Community, 1735–1785* (Cambridge: Cambridge University Press, 1995). For an excellent discussion of market growth in Massachusetts and a definition of *merchant* for this period, see Lorinda B. R. Goodwin, *An Archaeology of Manners: The Polite World of the Merchant Elite of Colonial Massachusetts* (New York: Kluwer Academic/Plenum Publishers, 1999), pp. 49–95.

32. Hinton, ed., *Excavations in Southwark 1973–76, Lambeth 1973–79*, p. 347.

33. Jacobson, *Chinoiserie*, p. 125.

34. Glenn Adamson, "The Politics of the Caned Chair," in *American Furniture*, edited by Luke Beckerdite (Hanover, N.H.: University Press of New England for the Chipstone Foundation, 2002), pp. 174–206.

35. While it remains difficult to generalize about whether Chinese Scholar plates would have been used for decoration or eating, it seems likely that they served a socio-technic function. James Deetz suggests that tin-glazed plates of any decorative scheme were not used for eating until the 1690s. See Deetz, *In Small Things Forgotten: The Archaeology of Early American Life* (Garden City, N.Y.: Anchor Press/Doubleday, 1977), p. 37. The tin-glazed earthenware sherds that Alaric Faulkner excavated in a 1660s context at the French settlement of Pentagoet in northern Maine and southern New Brunswick showed no evidence of use. Faulkner and Gretchen Fearon Faulkner, *The French at Pentagoet, 1635–1674: An Archaeological Portrait of the Acadian Frontier*, Occasional Publications in Maine Archaeology 5 (Augusta: Maine Historic Preservation Commission, 1987), pp. 209–10.

36. John J. McCusker and Russell R. Menard, *The Economy of British America, 1607–1789* (Chapel Hill: University of North Carolina Press, for the Institute of Early American History and Culture, 1985), pp. 54–55. Goodwin, *An Archaeology of Manners*, p. 62. An excellent study of this merchant community in the eighteenth century is David Hancock, *Citizens of the World: London Merchants and the Integration of the British Atlantic Community, 1735–1785* (Cambridge: Cambridge University Press, 1995). Mireille Galinou offers insights into a proposed "merchant aesthetic" in her introduction to *City Merchants and the Arts*, p. xi.

37. A number of studies suggest the paucity of Chinese export porcelain in late-seventeenth-century America. See Julia B. Curtis, "Perceptions of an Artifact: Chinese Porcelain in Colonial Tidewater Virginia," in *Documentary Archaeology in the New World*, edited by Mary C. Beaudry (Cambridge: Cambridge University Press, 1988), p. 20; and Amanda E. Lange, *Delftware at Historic Deerfield, 1600–1800* (Deerfield, Mass.: Historic Deerfield, Inc., 2001), pp. 19–22. For Dutch sherds found in the Caribbean, see Donachie, "Household Ceramics at Port Royal, Jamaica, 1655–1692," pp. 70, 137–38; and for Dutch and English sherds in New York, see Wilcoxen, *Dutch Trade and Ceramics in America*, p. 61. For the Waterman charger, see Jonathan L. Fairbanks and Robert Trent, *New England Begins: The Seventeenth Century*, 3 vols. (Boston: Museum of Fine Arts, 1982), 2:277–78, no. 271. Robert Hunter, "English Delft from Williamsburg's Archaeological Contexts," in Austin, *British Delft at Williamsburg*, pp. 24–29; Ivor Noël Hume, *Pottery and Porcelain in Colonial Williamsburg's Archaeological Collections*, Colonial Williamsburg Archaeological Series No. 2 (Williamsburg, Va.: Colonial Williamsburg Foundation, 1969), pp. 11–12.

38. For more on the expansion of textiles at this time, see John E. Wills Jr., "European Consumption and Asian Production in the Seventeenth and Eighteenth Centuries," in *Consumption and the World of Goods*, edited by John Brewer and Roy Porter (London: Routledge, 1993), p. 136. See also J. P. Losty, "British Settlements and Trading Centres," in Jackson and

Jaffer, eds., *Encounters*, pp. 142–55. Joseph Rykwert, *The First Moderns: The Architects of the Eighteenth Century* (Cambridge, Mass.: MIT Press, 1980), p. 61.

39. Edward W. Said, *Orientalism, 25th Anniversary Edition* (New York: Vintage Books, 1978). For discussions of non-Saidian cultural exchange between England and China in the late seventeenth century, see David L. Porter, "Monstrous Beauty: Eighteenth-Century Fashion and the Aesthetics of the Chinese Taste," *Eighteenth-Century Studies* 35, no. 3 (2002): 400–401. See also Jonathan D. Spence, *The Chan's Great Continent: China in Western Minds* (New York: W. W. Norton, 1998), pp. 41–61; Jackson and Jaffer, eds., *Encounters*, p. 9; Richmond Barbour, *Before Orientalism: London's Theatre of the East, 1576–1626*, Cambridge Studies in Renaissance Literature and Culture 45 (Cambridge: Cambridge University Press, 2003), pp. 3–5. Even if this early period of chinoiserie stemmed more from ignorant curiosity than imperialist enterprises, some cultural historians would still argue, as Homi K. Bhabha has, that "mimicry emerges as one of the most elusive and effective strategies of colonial power and knowledge." See Bhabha, "Of Mimicry and Man: The Ambivalence of Colonial Discourse," in *The Location of Culture* (1994; reprint, London: Routledge Classics, 2004), p. 122. Ralph Edwards (*Dictionary of English Furniture*, 2:267–68) notes the "promiscuous" use of the word *Indian* as early as the sixteenth century to describe anything remotely exotic. He points out that in the seventeenth century two similar probate inventories describe japanned furniture both as an "Indian chest" and a "china gilt cabinet."

40. For more on Jesuit reports about China and the popularity of Confucius in Europe, see Spence, *Chan's Great Continent*, pp. 19–61; Adrian Hsia, ed., *The Vision of China in the English Literature of the Seventeenth and Eighteenth Centuries* (Hong Kong: Chinese University Press, 1998), pp. 10–14; Honour, *Chinoiserie: The Vision of Cathay*, pp. 17–18; and Lovejoy, "The Chinese Origin of Romanticism," pp. 103–7. For a sample advertisement, see the final page of volume 1 of Anthony à Wood, *Athenae Oxoniensis*, 2 vols. (London: Printed for Tho. Bennett, 1691–92).

41. Gottfried Wilhelm Leibniz, *Novissima Sinica or Latest News from China* (1699), as cited in Spence, *Chan's Great Continent*, pp. 84–85; see also John E. Wills Jr., *1688: A Global History* (New York: Norton, 2001), p. 251; and Honour, *Chinoiserie: The Vision of Cathay*, p. 16.

42. Stephen Parissien, "European Fantasies of Asia," in Jackson and Jaffer, eds., *Encounters*, p. 352; and Spence, *Chan's Great Continent*, pp. 62–80, 86.

43. Parissien ("European Fantasies of Asia," pp. 351–52) cites a similar shift toward representations of passive Asian characters in British literature and drama.

44. David Porter, "A Peculiar but Uninteresting Nation: China and the Discourse of Commerce in Eighteenth-Century England," *Eighteenth-Century Studies* 33, no. 2 (1999–2000): 181; and Honour, *Chinoiserie: The Vision of Cathay*, p. 6.

45. David S. Shields, *Civil Tongues & Polite Letters in British America* (Chapel Hill: University of North Carolina Press, for the Institute of Early American Culture, 1997), pp. 99–140; Elizabeth Kowaleski-Wallace, *Consuming Subjects: Women, Shopping, and Business in the Eighteenth Century* (New York: Columbia University Press, 1997), pp. 53–69; and Richard W. Bevis, *English Drama: Restoration and Eighteenth Century, 1660–1789* (London: Longman, 1988), p. 84.

46. For more on the gender associations of drinking tea, coffee, and liquor, see Anne Yentsch, "The Symbolic Divisions of Pottery: Sex-Related Attributes of English and Anglo-American Household Pots," in *The Archaeology of Inequality*, edited by Randall H. McGuire and Robert Paynter (Cambridge, Mass.: Blackwell, 1991), pp. 223–24.

47. Exceptions to the rule, perhaps, are the elaborate silver dressing sets with flat-chased chinoiserie decoration that may have been owned by or were primarily associated with women. See David Mitchell, "A Passion for the Exotic," in Gallinou, ed., *City Merchants and the Arts*, pp. 71–73.

48. Richard Faber, *The Brave Courtier: Sir William Temple* (London: Faber and Faber, 1983), pp. 139–40.

49. Rose A. Zimbardo, *A Mirror to Nature: Transformations in Drama and Aesthetics, 1660–1732* (Lexington: University Press of Kentucky, 1986), p. 13.

50. As quoted in ibid.

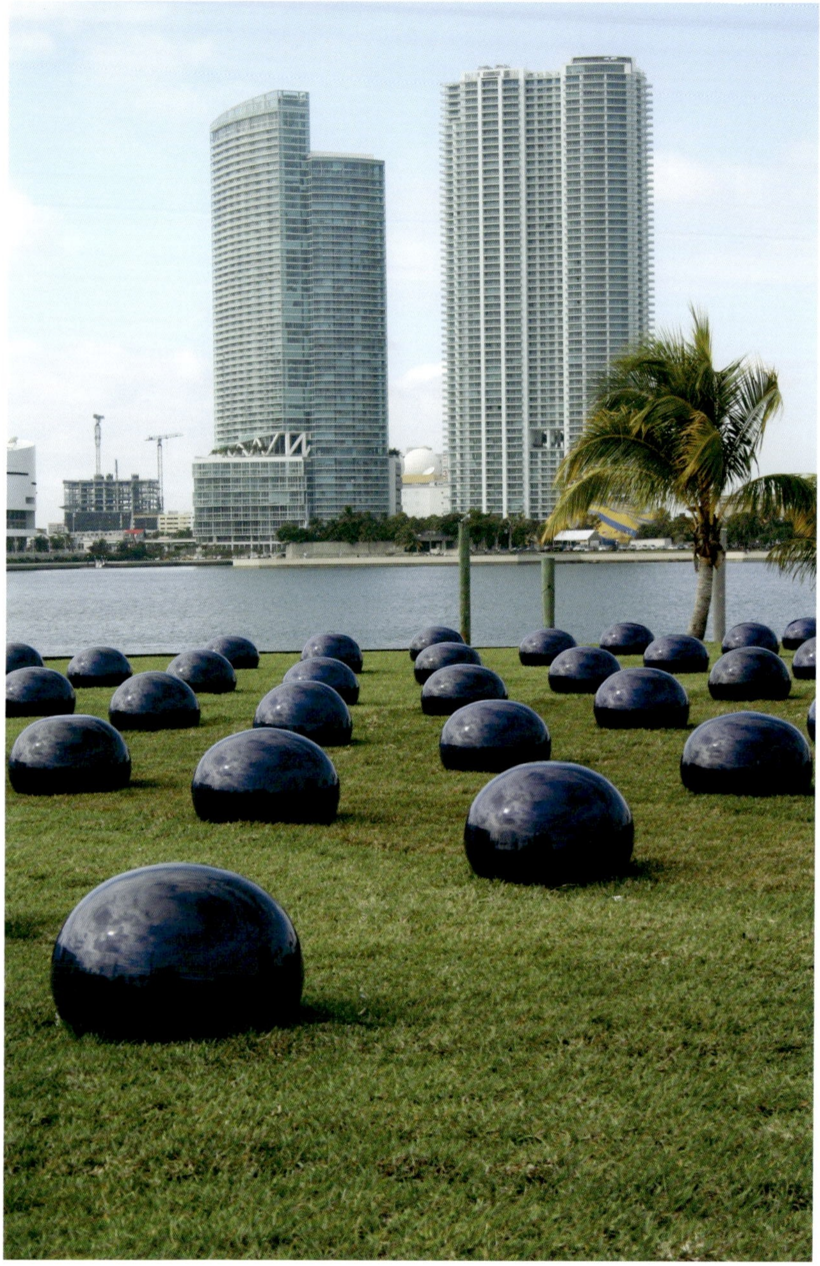

Figure 1 Ai Weiwei, *Bubbles*, installation view, Watson Island, Miami, 2008. Porcelain. (Courtesy, Bigstock.)

Garth Clark

Mind Mud: Ai Weiwei's Conceptual Ceramics

▼ CONTEMPORARY ceramics has never seen a showman like Ai Weiwei. This passionate polymath has taken on the vast history of China's ceramic art, broken it apart (literally), turned it upside down, reassembled it in fresh ways, vandalized "treasures" by Iron Age potters, questioned the authenticity of his country's ceramic patrimony, lampooned the connoisseurship surrounding ancient pottery and porcelain, and used the ceramic tradition to critique his government. Driven by near-demonic ambition, he has produced hundreds of major works comprising from one to one hundred million individual components. And he has mounted the largest contemporary sculptural installations in ceramics to date. Ai has revised the ceramic canon and set the bar at a new and intimidating high.

He has produced an ongoing series of major installations, one more dazzling than the other: *Ghost Gu Descending the Mountain* (2006), several editions (each of ninety-six pots) inspired by the Guiguzi legend; *Bubbles* (2008), one hundred large blue-glazed ceramic orbs each weighing ninety pounds, was installed on Watson Island off Miami during Art Basel Miami Beach; *Field* (2010), also at Art Basel, was a vast grid of porcelain scaffolding incongruously decorated with blue early Ming dynasty (1368–1644) motifs; and there is his blockbuster installation at the Tate Modern, which may have lacked conceptual gravitas but still weighed 150 tons (figs. 1–3).

The floor of the Tate's Turbine Hall was buried under one hundred million sunflower seeds in their husks, handcrafted in porcelain using traditional methods and individually painted by sixteen hundred artisans in Jingdezhen over a period of two years. Three days after the show opened, however, visitors were prevented from walking on the seeds—a significant aspect of the presentation—because the oxide with which they were painted did not "fit" the clay body, and clouds of potentially toxic pigment began to rise in the air. Critics questioned whether, by turning this into a purely retinal experience, the installation had lost both purpose and soul. The debate only added to the work's media coverage, and attendance was huge, placing Ai at the pinnacle of his fame.

How Ai might top the success of this exhibition is, for the present, a moot question. On April 3, 2011, Chinese police at the Beijing airport arrested him just as he was about to board a flight to Hong Kong. He was held incommunicado at an undisclosed location for three months and then released on *qubao houshen*, a legal limbo in which charges are not likely to be pressed as long as Ai pays a fine of about $2 million for unpaid taxes and "behaves" himself, which includes not giving interviews and—implied though not explicit—producing art that criticizes the government. If he breaks any of these rules, formal charges could be brought against him and he could be imprisoned again. So while he is physically free, he has effectively been muzzled, which, together with restrictions on travel, puts a question mark over his future as an artist.[1]

The rise of Ai Weiwei's career has been vertiginous. From 1983 to 1993 he earned his living on the streets of New York City as a quick-sketch artist doing portraits of tourists.[2] He returned to China just as the Western fas-

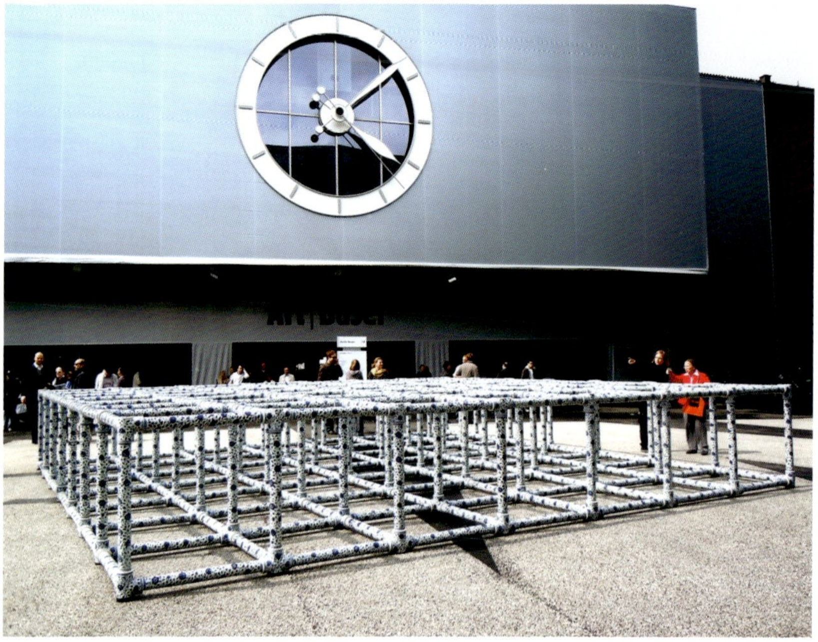

Figure 2 Ai Weiwei, *Field*, installation view, Art Basel, Basel, Germany, 2010. Porcelain, 29 15/16" x 29 15/16". (Courtesy, the artist and Galerie Urs Meile, Beijing-Lucerne.)

cination with contemporary art from his country was beginning to explode. It was only in 2004 that Ai had his first solo museum show, at Kunsthalle Bern, Switzerland, and he has barely taken a breath since. Last year his name was thirteenth on a list of the one hundred most influential figures in contemporary art.[3] In that relatively short time, as Philip Tinari so well summarizes, Ai has moved "from rebel to craftsman, from provocateur to bricoleur" and has shifted "from Duchamp to Joseph Beuys, from readymade to 'social sculpture.'"[4]

While Ai works in architecture, design, photography, and film, and utilizes a wide range of materials (wood, plastic, bronze), judging from the number of repeat visits, ceramics occupies a special place. It was therefore exciting to hear of the cleverly titled "Ai Weiwei: Dropping the Urn (Ceramic Works, 5000 BCE–2010 CE)" organized by the Arcadia University Art Gallery, Glenside, Pennsylvania. The exhibition is currently traveling and will be at the Victoria and Albert Museum in London from October 15, 2011, to March 18, 2012 (fig. 4).

Figure 3 Ai Weiwei, *Sunflower Seeds*, installation view, Turbine Hall, Tate Modern, London, 2011. Porcelain. (Courtesy, Tate.)

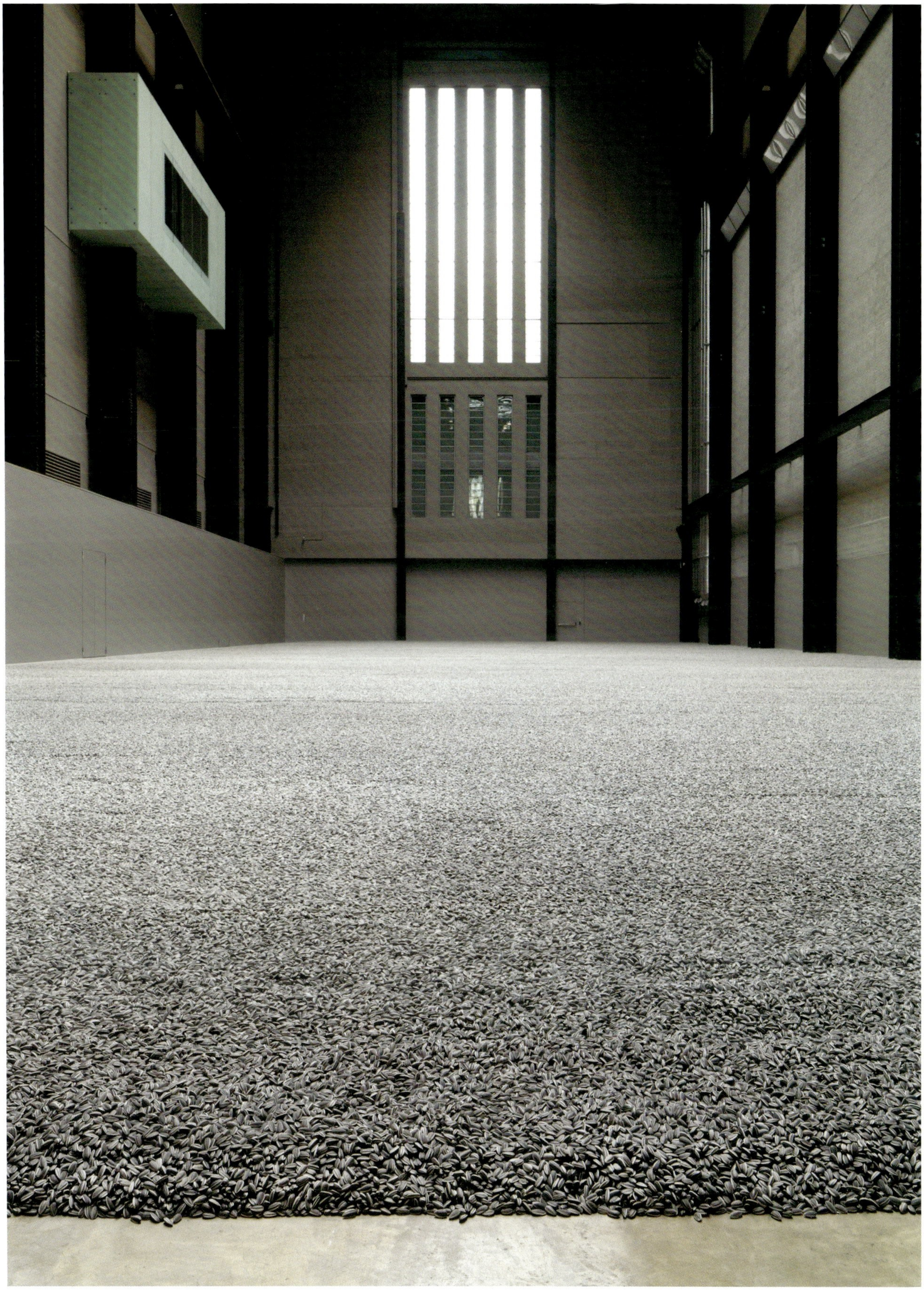

That excitement turned to concern when it was announced that the exhibition would include only twelve works: ten ceramics, one photographic triptych, and a video. Given Ai's larger-than-life presence, his Warholian celebrity, and the circus accompanying his giant ceramic sculptures, could this modest traveling exhibition do justice to his ceramic adventure?

In a David-and-Goliath turnaround, the show reverses this query. While not its direct purpose, it asks whether understanding Ai is better undertaken in an environment in which subtlety and scholarship can be heard above the media clamor. The answer is an emphatic yes. And the exhibition

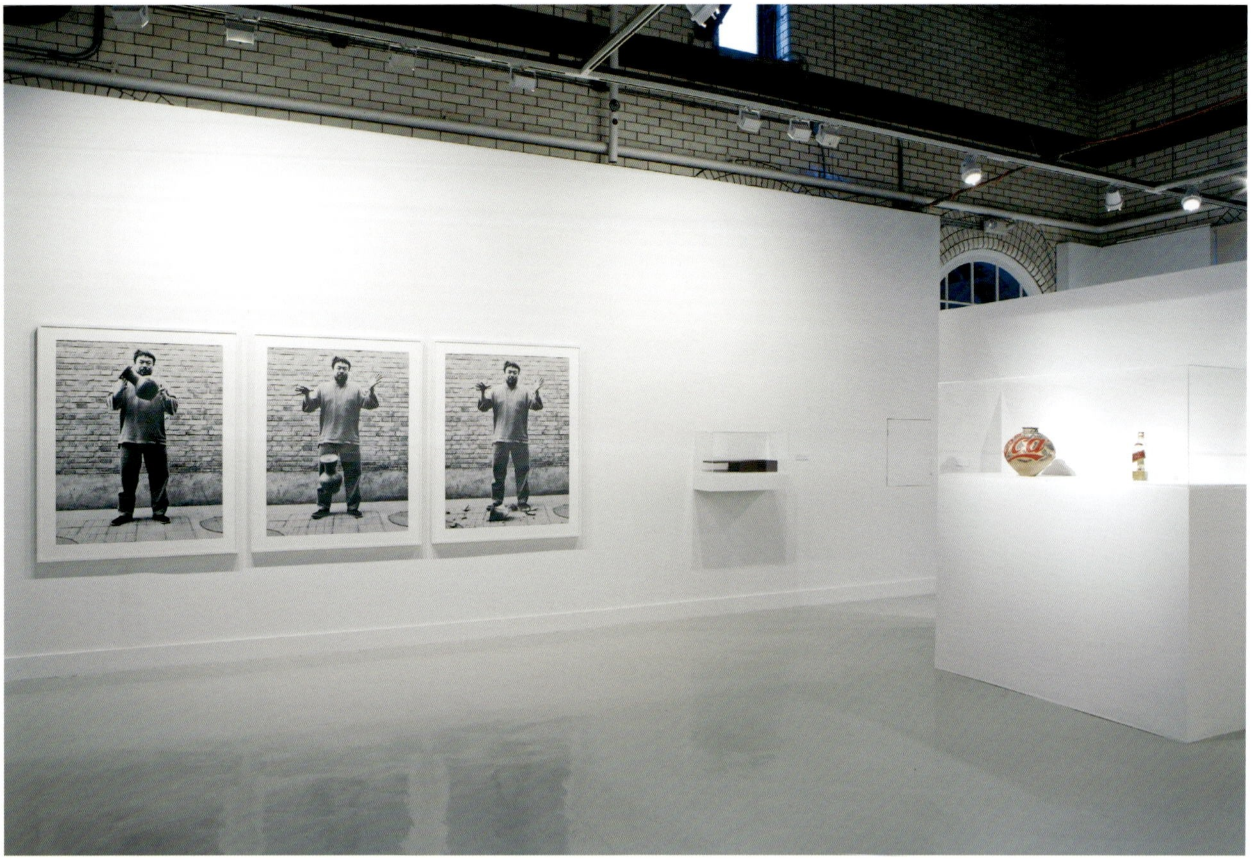

Figure 4 "Ai Weiwei: Dropping the Urn," installation view, Arcadia University Art Gallery, Glenside, Pennsylvania, February 24–April 18, 2010. (Courtesy, Arcadia University Art Gallery.)

succeeds so royally because of exceptional curatorship: Gregg Moore and Richard Torchia have assembled an exhibition that bristles with intelligence, careful analysis, and gentle probing.

The equally modest catalog (124 pages) by Moore, Torchia, Philip Tinari, Glenn Adamson, Dario Gamboni, and Stacey Pierson, plus an interview with Ai conducted by Zhuang Hui, is, essay for essay, a delight to read and an impressive scholarly achievement. Moreover, the spacious installation allows each work the space it needs to breathe and, rare in a fine art project, treats ceramics with respect and historical depth, not just as an accidental material (although knowledge of what has been happening in contemporary ceramics over the last half century seems to be missing). The presentation acknowledges that as much as Ai uses ceramics, ceramics in turn use him. The artist and the medium are locked in a partnership of surprising complicity.

The exhibition takes its name from *Dropping a Han Dynasty Urn* (1995), a series of three photographs that dominates the exhibition and is almost always in sight in the gallery, keeping one aware of the fundamentally transgressive nature of his art. *Urn* is one of Ai's earliest works, made two years after his return to China. For all the art he has produced since, much of it remarkable, *Urn* remains his most iconic. The sheer audacity of the act (dropping a museum-quality urn that had survived, intact, for five thousand years) is perfectly balanced by the look of cool detachment on Ai's face as the vase exploded at his feet into a thousand sherds (fig. 5).

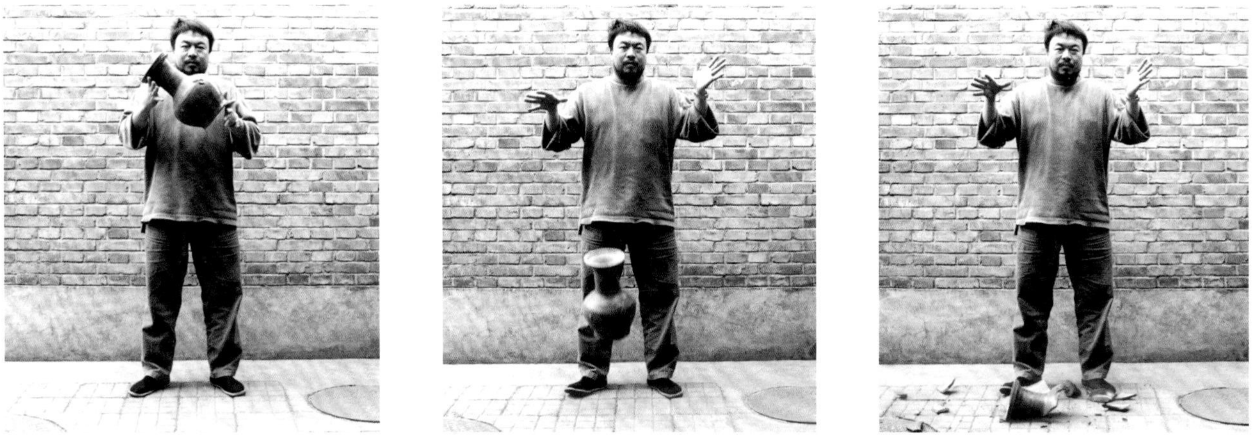

Figure 5 Ai Weiwei, *Dropping a Han Dynasty Urn*, 1995. Triptych of gelatin silver prints, edition of 8, no. 4. Each print 49⅝" x 39¼". (Courtesy, the artist.)

Glenn Adamson, who wrote about this moment in the catalog, takes an interestingly redemptive view:

> Shod in traditional cloth slippers he seems in these three heart-stopping filmic moments to stand astride all of China's culture. Ancientness is made to perform as six frames per second. But this is one big bang we can reverse, by the simple means of reading (as the Chinese sometimes do) from right to left. What could be more touching than his depiction of the artist, gazing implacably as a shattered vessel remakes itself, and floats upward into his hands.[5]

Most read this as sacrilege and accuse Ai the Merciless of creating art by destroying history. The Asian antiquities market was scandalized. But those who knew little about Asian art were even more offended, believing that Ai had demolished an irreplaceable treasure. Urns of this period are relatively common, however, and, depending on condition, can be inexpensive. As Ai pointed out to his critics, the urn was a protoindustrial product, mass-produced in hundreds of thousands, if not millions, of units. If it was industry then, he argued, it remains industry today.

The act was no more barbaric than if an artist five thousand years from now did the same to a vase purchased today from Walmart. Ai has never actually destroyed a work of canonical importance, and yet as late as 2009 the *New York Times* was still describing the dropped urn as "priceless."[6] The urn was worth about $10,000 at most.

At its core, *Urn* is an argument about authenticity. Does the cultural consequence change this common working vessel over time? Is age a valid reason for increasing its value? Does the veneer of antiquity alter an object's original identity; does it shift from artifact to art? Is the reverence that we accord rarity just a Pavlovian response to manipulations of the centuries-old antiquities market? Is rarity merely an accident of survival, a mathematical equation, and, if so, is the honor we give to survivors misplaced?

Together, Marcel Duchamp's porcelain urinal *Fountain* (1917) and *Dropping a Han Dynasty Urn* provide bookends for avant-gardism in the twentieth century.

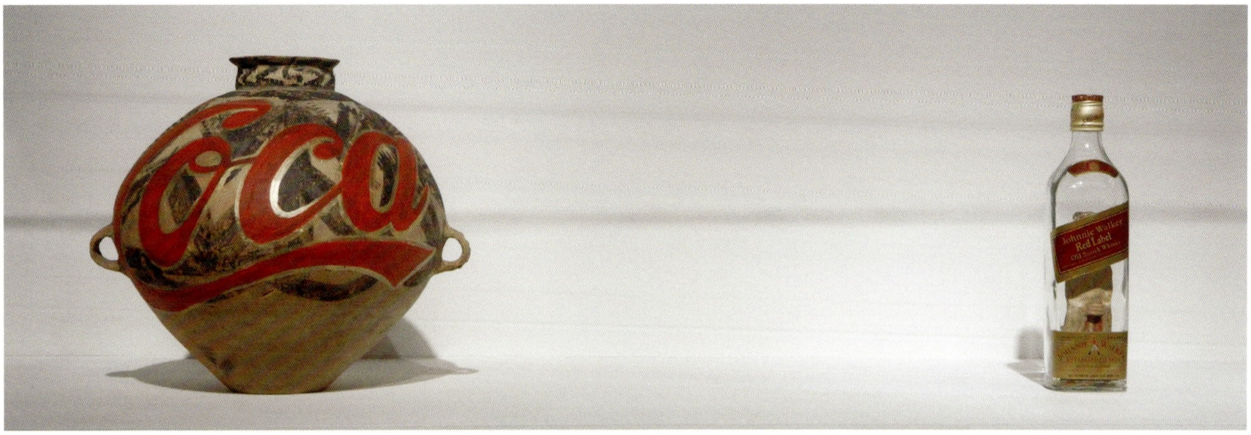

Figure 6 Ai Weiwei, *Untitled* (see fig. 7) and *Coca-Cola Vase* (see fig. 8), installation view. (Courtesy, the artist and Arcadia University Art Gallery.)

tieth century. Both actions, submitting *Fountain* to the Independents Salon in New York and dropping the urn, created a furor. Duchamp's took on early modernism, and Ai's, postmodernism. Both were viewed as offensive and destructive, albeit for different reasons, and the ceramic components met the same fate.

Both were lost (one through indifference and the other by design) and each survives only through photographic documentation. This result was deliberate in Ai's case, but for the *Fountain* it was sheer luck in that, according to legend, Alfred Stieglitz decided to document the urinal's existence before tossing it into a dumpster when he moved to a new studio.

Ai certainly would appreciate the comparison with Duchamp, an artist he reveres. In a conversation with Philip Tinari, Ai commented about violations in his art: "Duchamp had his bicycle seat. Warhol had the image of Mao. I have a totalitarian regime. It is my readymade."[7]

Ai knows how to soften his blows with humor, sometimes sly and witty, other times with obvious one-liners, but beneath is a ruthless revisionism, and the *Urn*'s path of destruction (or deconstruction) continues in four other works of the exhibition: *Untitled* (1993), *Coca-Cola Vase* (1997), *Colored Vases* (2006–2008), and *Dust to Dust* (2009).

Coca-Cola Vase was perceptively balanced by placing it in a spacious cabinet with only one other work, *Untitled*, Ai's first ceramic piece (fig. 6). They clearly enjoy each other's company, almost winking conspiratorially

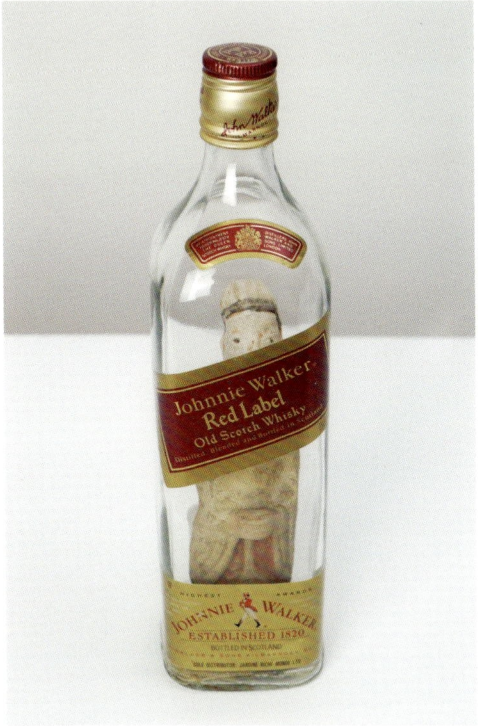

Figure 7 Ai Weiwei, *Untitled*, 1993. Clay sculpture dating from the Song dynasty (960–1279) inserted within a Johnnie Walker Red Label bottle. H. of bottle 9⅝". (Courtesy, Urs Meile Collection.)

at each other, their logos flashing faux gold and alluding to capitalist sumptuary.

Untitled (fig. 7) pays homage to the childhood wonder of seeing a ship model within a bottle; in this case the bottle bears a Johnnie Walker Red Label and contains a ceramic sculpture from the Song dynasty (960–1279). Once one has gotten over the "how," the next question is "why?" Confining this crude figure in the bottle is amusing—its two black dot-eyes peer out of the bottle with wide-eyed surprise, as though wondering how it got into its predicament. The piece forces a chuckle.

But it is also unsettling. The message is political, a sly dig at a regime that had begun to practice communism for profit. And the fact that this brand of whisky is the favorite among China's ruling elite only adds to its resonance. The more one looks at this work, the more uncomfortable it makes one feel. One cannot shake the undertow of imprisonment it suggests: culture held hostage by commerce.

The *Coca-Cola Vase* (fig. 8) visits the same theme but is more complex. It aims to reach the viewer on two levels, both sensually and intellectually. There is a strong, conscious aesthetic component. Ai did not just slap a logo onto the pot but recognized similarities between the swooping lyrical schemas of Neolithic pottery decoration and the Coca-Cola logo's voluptuous cursive scrolling.

The painter (probably not Ai himself but working under his direction) has integrated the logo and the contours of the pot, a perfect fit that is enveloping but elastic. At the same time the fragments of the original decoration, seen underneath the logo, draw the past into a dance with the present. Time is one of Ai's raw materials, and here he gives history an

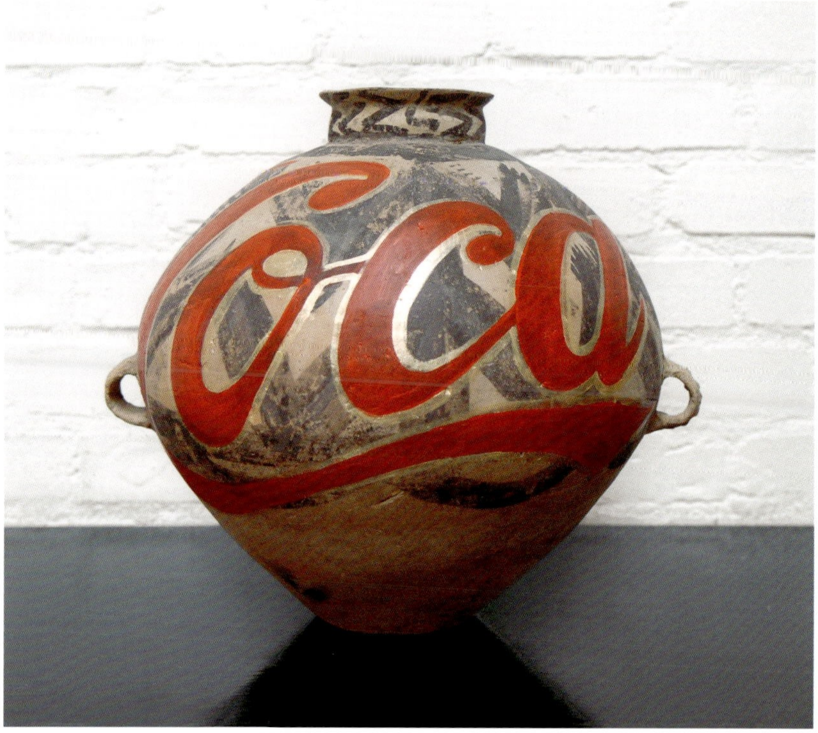

Figure 8 Ai Weiwei, *Coca-Cola Vase*, 1997. Vase from Neolithic Age (5000–3000 BCE), painted. H. 11⅞". (Courtesy, André Stockcamp and Christopher Tsai Collection.)

almost plastic consistency, as though it is malleable, pliant, and can ignore the strictures of chronology.

Ai selected Coca-Cola in part because it is both a beloved, iconic consumer product worldwide and a favorite icon of pop art. Aesthetically blissful, *Coca-Cola Vase* is nonetheless meant to critique unholy alliances, in this case communism and capitalism. In 1979 Coca-Cola became the first Western company permitted to distribute its products in China and today operates more than forty bottling plants and enjoys a 50 percent share of that nation's soft drink market. In Ai's mind Coca-Cola is more than a beverage, it is an enabler of Chinese repression.

Colored Vases is part of a long, ongoing series in which Ai decorates earth-toned Neolithic jars and invades them with color. He has been acquiring these jars inexpensively at flea markets in and around Beijing for years and has amassed a large stock. He dips them in industrial pigment—a household paint—using distinctly contemporary colors, more redolent of Tupperware and Martha Stewart than of art. They are startlingly beautiful in their new clothes. The decorative effect of a cluster of these vases is gorgeously vibrant and cheekily fresh. These forms are clearly historical in origin, but he also paints Neolithic pots that have severe geometric form and appear fully modern (fig. 9).

Moore and Torchia contend that this is not a destructive act but more a "gently compromising" device that neutralizes tradition, allowing the contemporary aesthetic values to dominate: "Masking the luster and bold decoration of their original surfaces, the acrylic paint does not destroy the brushwork, which remains intact beneath a veneer of synthetic pigment. The exact contour of their patterning becomes each vessel's own secret."[8]

Ai's view is that "[t]o have other layers of color and images above the precious one calls into questions both [the] identity and [the] authenticity of the objects. It makes both conditions non-absolute. You can cover something so that it is no longer visible but is still underneath, and what appears on the surface is not supposed to be there but is."[9]

His approach to ceramics is strictly hands-off. He maintained a ceramics studio in Jingdezhen with a staff of twenty and subcontracts work to a small army of more than one thousand independent potters. When asked if he had made any of the sunflower seeds for the Tate project, he admitted, "I made three or four but none of them was any good."[10]

Dipping the vases in paint is the closest that Ai gets to being a ceramist. *(Making of) Colored Vases* (2006–2010), the video of him surfacing the pots, shows that the mechanics are identical to glazing, and the chalky matte pastel paint closely resembles raw glaze. The video also reveals Ai's playfulness, as he moves from vessel to vessel, inventing and reinventing his surfaces on the run; some are dipped in two colors. He sometimes turns the pots upside down so the trickles of colors that flowed upward seem to defy gravity. Clearly Ai relished the simplicity, effectiveness, and spontaneous physicality of this process (fig. 10).

Figure 9 Ai Weiwei, *Colored Vases*, 2006–2008. Nine vases from the Neolithic age (5000–3000 BCE) and household paint. H. 10–14½". (Courtesy, the artist and Arcadia University Art Gallery.)

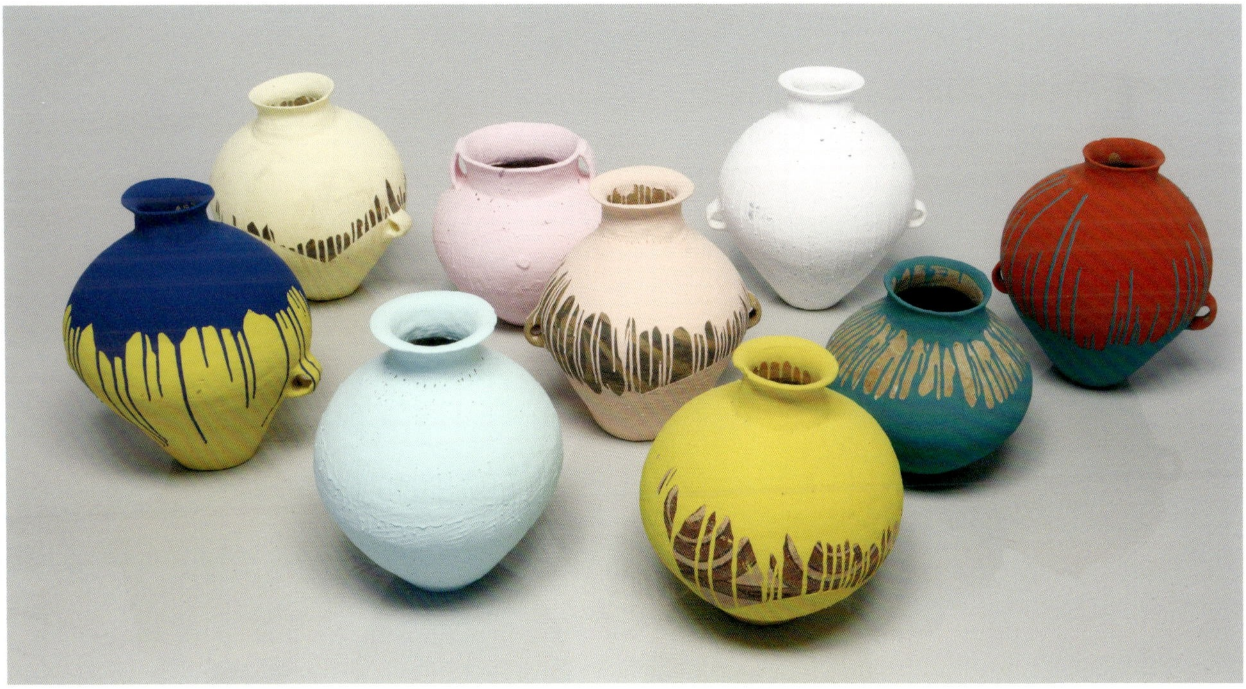

Figure 10 Ai Weiwei, *(Making of) Colored Vases*, 2006–2010. Twelve production stills from single-channel video. (Courtesy, the artist.)

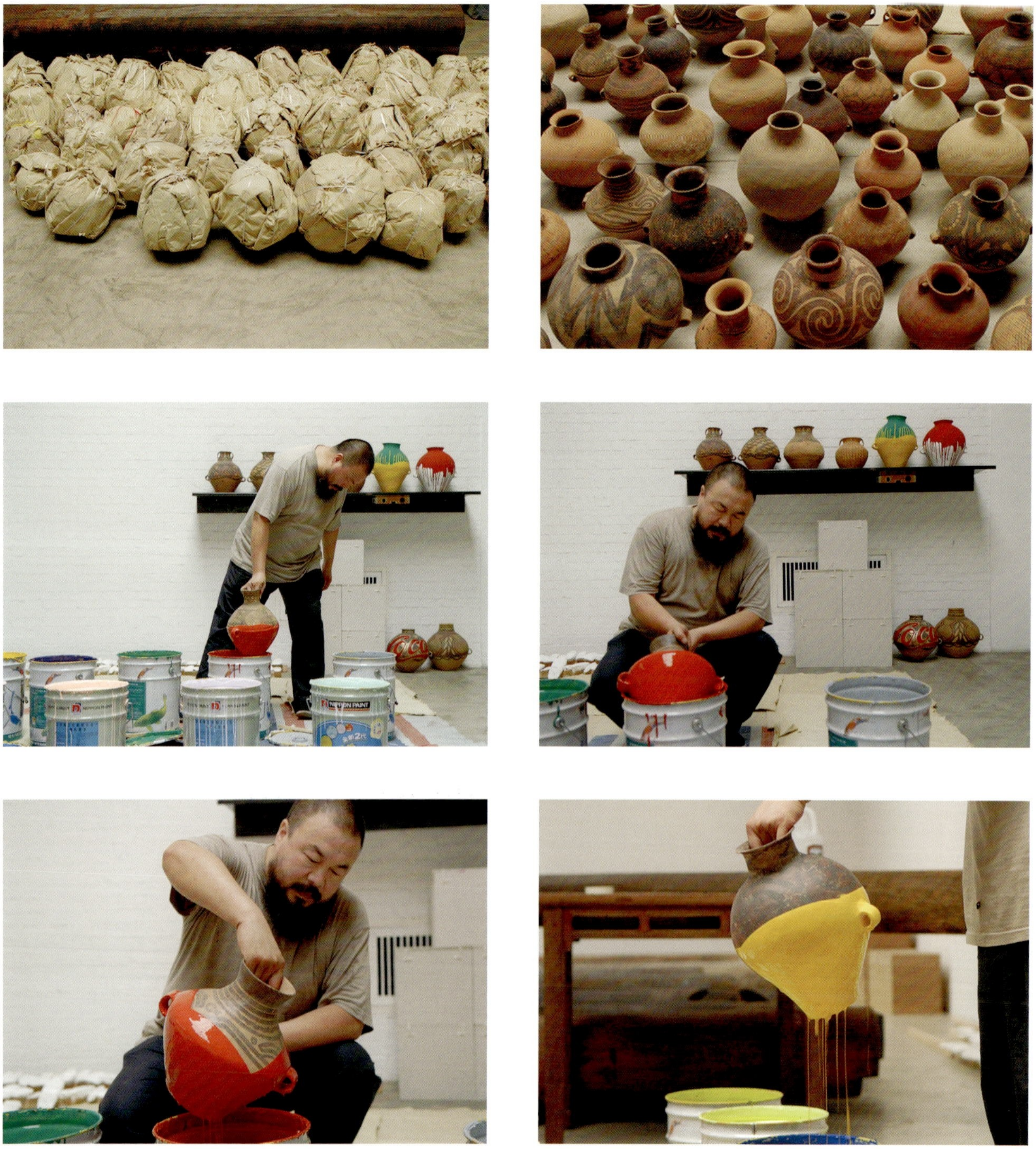

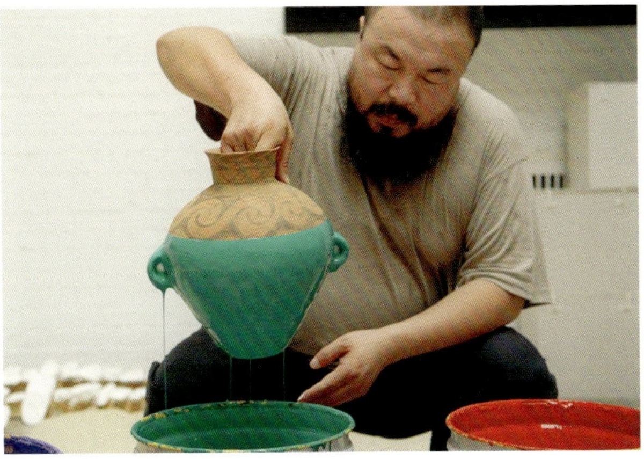

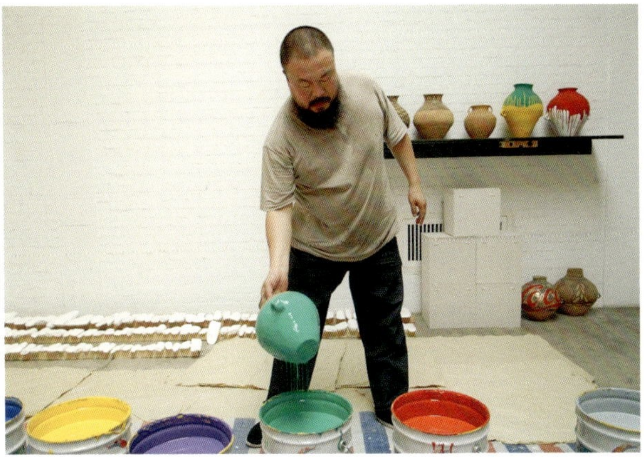

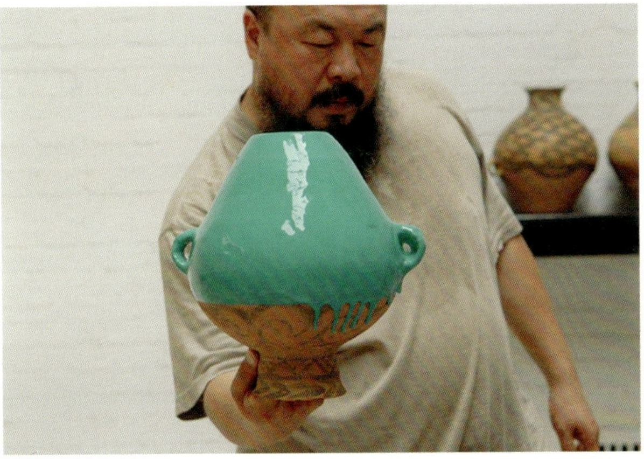

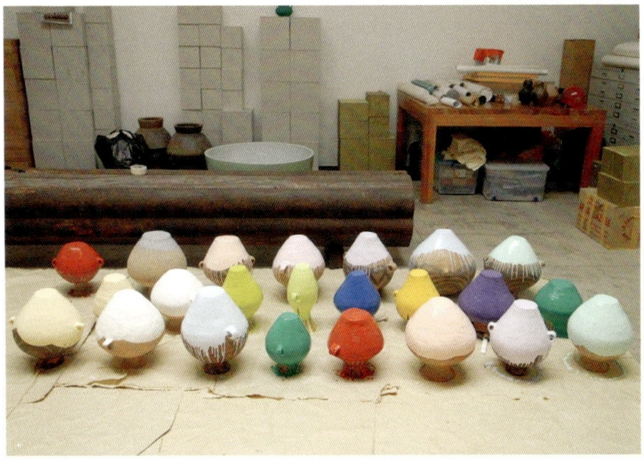

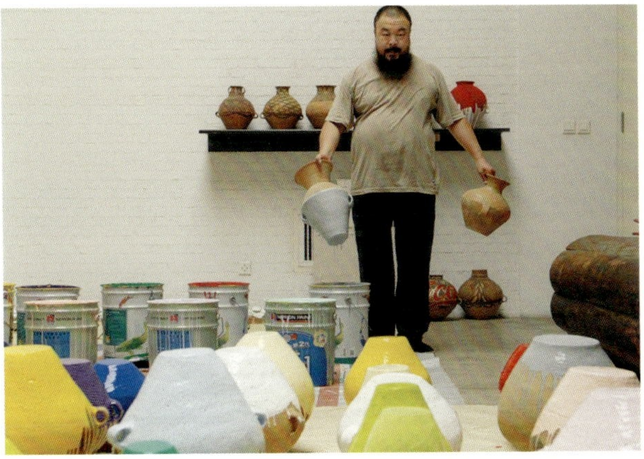

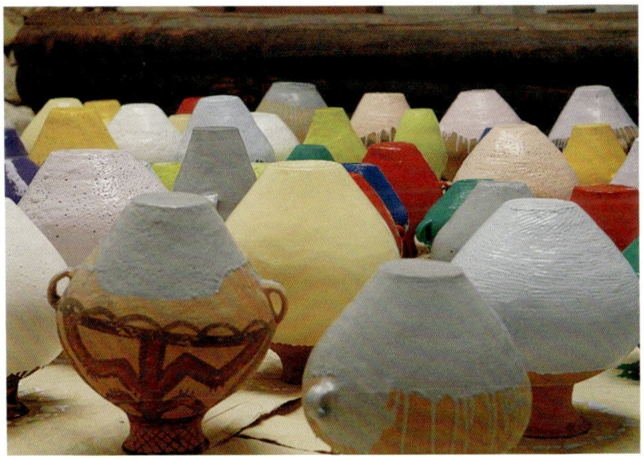

This activity in ceramics is known as "clobbering"—taking an extant ceramic work by someone else and adding surface painting or decoration. It was a popular home industry in the eighteenth century, the purpose of which was to make the object more attractive so it could be sold at a profit. And this is exactly what Ai does, as his groupings are worth many times the cost of the Neolithic hosts. The etymology of the word "clobber"—late-seventeenth-century street slang for beating someone—adds spice, suggesting that even though taking over someone else's object might increase its value, it also, in a sense, does violence.

This is certainly true of *Dust to Dust*, one of the best works in the exhibition. As an object, it perches on that slippery cusp between ordinariness and edgy minimalism. It appears at first glance to be little more than a glass jar containing dust. But in fact it is a funerary urn—the dust is ground ceramic "cremains" of Neolithic jars (fig. 11).[11]

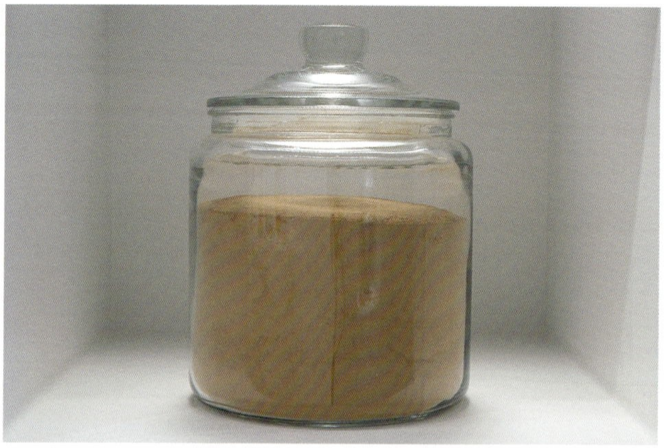

Figure 11 Ai Weiwei, *Dust to Dust*, 2009, installation view. Ground Neolithic pottery (5000–3000 BCE) and glass jar. H. 10¼". (Courtesy, the artist and Arcadia University Art Gallery.)

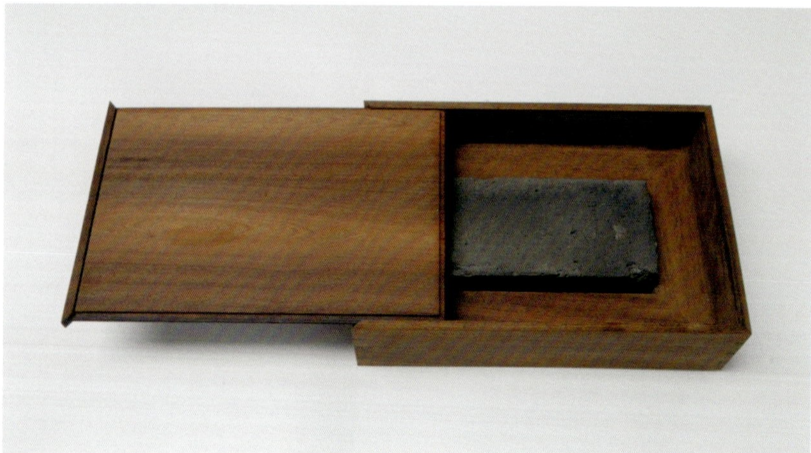

Figure 12 Ai Weiwei, *Souvenir from Beijing*, 2002, installation view. Brick from dismantled *hutong* house, ironwood from dismantled temple from the Qing dynasty (1644–1911). L. 13¾". (Courtesy, the artist and Arcadia University Art Gallery.)

Figure 13 Ai Weiwei, *Template*, installation at "Documenta 12," Kassel, Germany, 2007. Wooden doors and windows taken from destroyed Ming- and Qing-dynasty houses, wooden base. (Courtesy, Galerie Urs Meile, Beijing-Lucerne.)

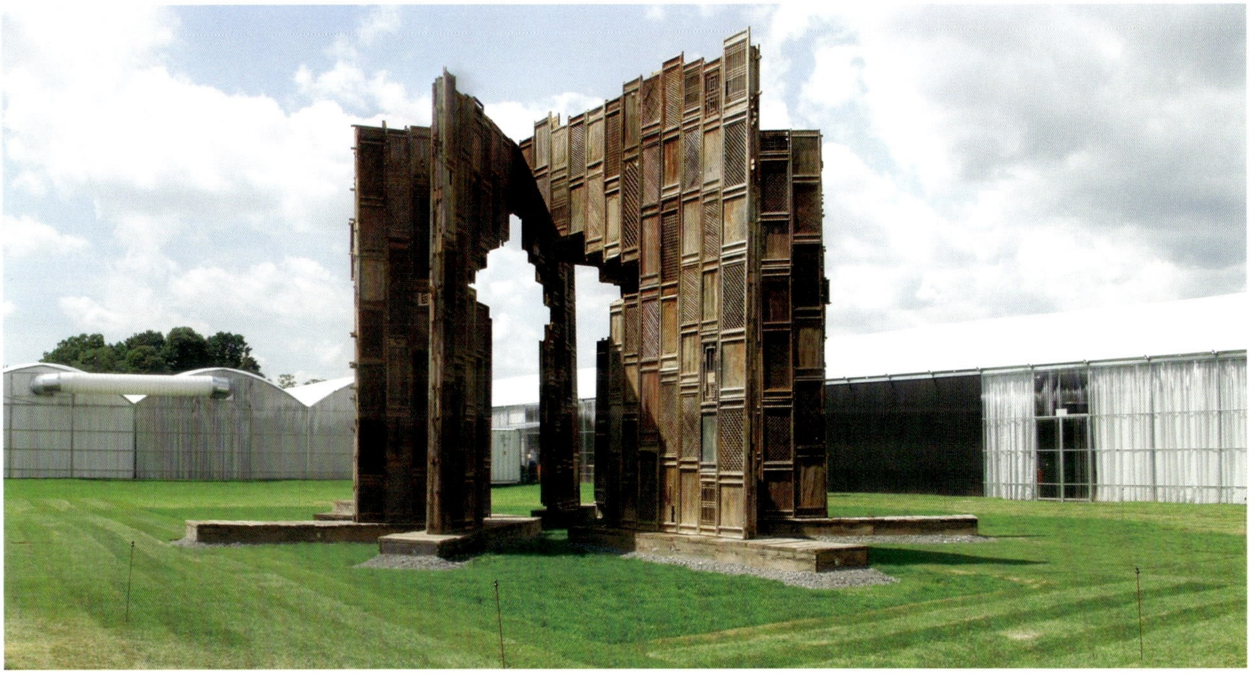

Critics have attached a great number of metaphors to this work, including construction sites, the grim brown pallor of Beijing's polluted air, or the ravaged industrial landscape of China. But it is stronger when viewed simply as cultural mediation between life and death.

The same is true of *Souvenir from Beijing* (2002; fig. 12), a gray brick taken from the ruins of destroyed *hutong* houses (narrow alleys lined with traditional courtyard residences). The brick rests in a coffinlike box made of ironwood salvaged from dismantled temples dating to the Qing dynasty (1644–1911). *Souvenir* offers mournful respite, like the public viewing of a body prior to its burial. It finds its twentieth-century parallel in the somber piles of gray felt that Joseph Beuys showed in the late 1960s.

But with Ai there is always subtext. The title questions the rush of developers to modernize Beijing. Moore and Torchia see this as a link both to *Template* (2007; fig. 13), a massive gate (which collapsed at "Documenta 12" during a windstorm) made of Ming- and Qing-dynasty doors and windows, and to Ai's austere but serene brick-based architecture. Bricks, Ai argues, have a "natural relationship with our hands [that allows] us to build [with] them blindly . . . like using words to write something."[12]

Paradoxically, one of his architectural commissions was designing the Museum of Neolithic Pottery for the Jinhua Architectural Art Park in Zhejiang province, an institution that preserves exactly the same class of wares that he destroyed in *Urn* and painted in *Colored Vases*.

The six blue-and-white pots in the exhibition are handsome but reveal a conceptual weakness in Ai's play with authenticity and provoke the only moment when one misses Ai's large installations. Three of the cobalt painted vases are Ai's inventions: two are from *Ghost Gu*, although they are wrongly described as "replicas" and do not credit Ai's collaborator on this series, the Romanian-born artist Serge Spitzer; the third is *Untitled* (2009).[13] The other three are copies of actual works in the style of the Qing dynasty, Qianlong era (1736–1795).

The genesis for *Ghost Gu* arrived on July 12, 2005, when Christie's sold a fourteenth-century Yuan blue-and-white pot for $27,657,944 (fig. 14). This surprised the Chinese porcelain market because Yuan works of this period had not previously been accorded such importance. Ai was intrigued by the vase's rise in fortune, musing why markets upgrade and downgrade cultural values much like Standard and Poor's bond revisions, but ultimately one suspects he was more amused by market vagaries than perplexed.

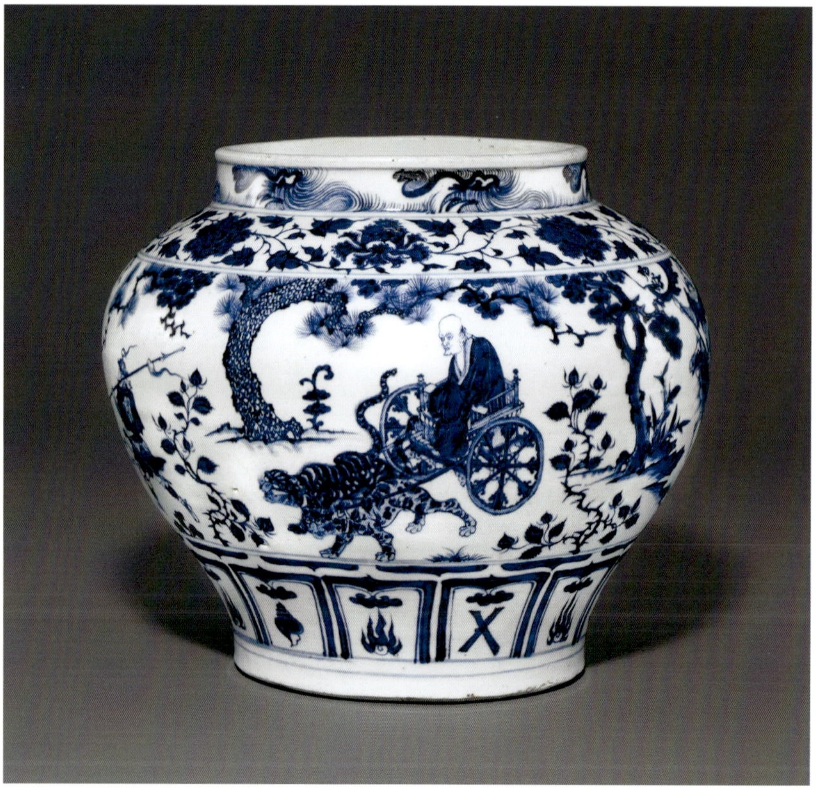

Figure 14 *Guan* with painted decoration of Daoist Immortal, China, Yuan dynasty, mid-14th century. Porcelain painted with cobalt blue under clear glaze. D. 13". (Courtesy, Christie's.) On July 12, 2005, this jar sold at auction for £15,688,000; *Chinese Ceramics and Works of Art, Including Export Art*, sale cat., Christie's, London (King Street), lot 88.

The painting on the vase depicts an ancient morality tale: strategist Gui Gu or Guiguzi (his real name was Wang Xu) races down a mountain into a valley to save a disciple. The complexity of the scene is fulsomely described in Christie's sale catalog:

> Robustly potted standing on a broad, low foot rising to a full rounded shoulder below a short cylindrical neck and a slightly thickened mouth rim, vividly painted in a deep and vibrant cobalt blue around the body with a narrative scene that depicts a robed figure seated in a two-wheeled cart drawn by a tiger and a leopard, following two foot soldiers each carrying a spear, approaching a rustic bridge across a stream beneath a waterfall, the cart followed by two equestrians on either side of dramatically painted rocks, one in military attire and carrying a banner bearing the characters *Gui Gu*, the other in scholar's clothes, on a prancing piebald horse and turning towards the first horseman, the composition punctuated by pine, bamboo, flowering prunus, plantain, rose and willow, all between a classic wave band around the neck, a peony scroll around the shoulder, and a band of upright lappets enclosing emblems around the base. . . .[14]

Photographs taken in the courtyard of Ai's Beijing studio show the original conception of *Ghost Gu Descending the Mountain*, comprising ninety-six pots, exactly as it was installed at the Museum für Moderne Kunst in Frankfurt am Main (figs. 15, 16). It is a stunningly complex vision. When

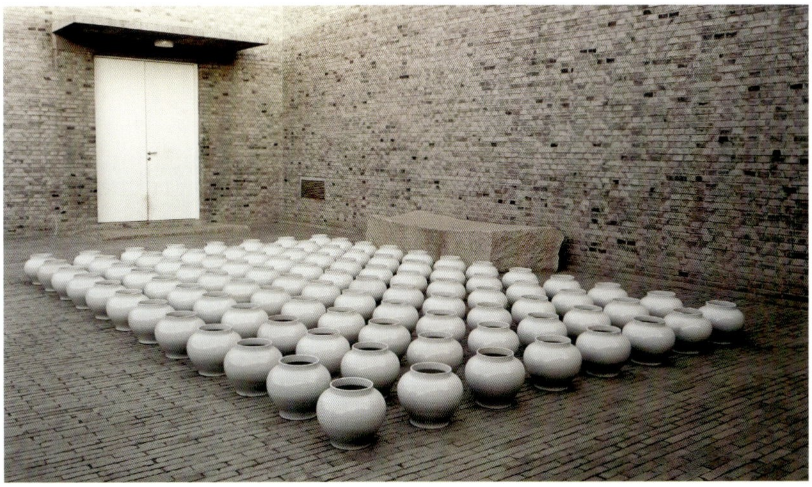

Figure 15 Ai Weiwei and Serge Spitzer, *Ghost Gu Descending the Mountain*, 2006. Blue-and-white porcelain, 96 vases in each group. H. of each vase 10⅝". Ai Weiwei's studio, Beijing, 2006. (Courtesy, Ai Weiwei and Arcadia University Art Gallery; photo, Aaron Igler and Matt Suib.)

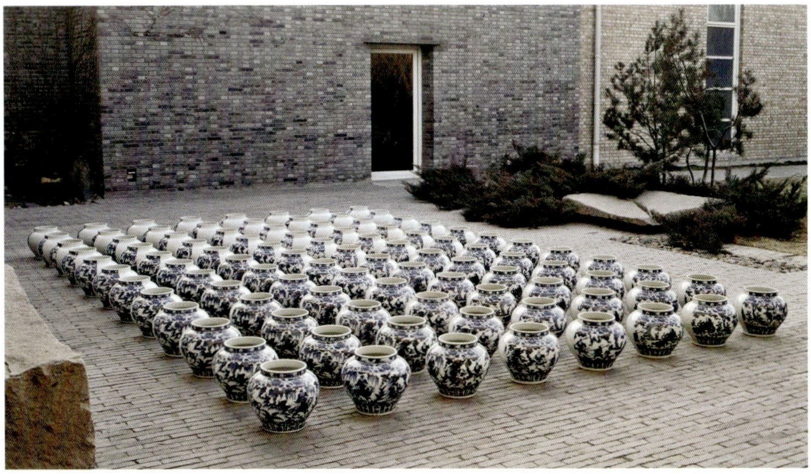

Figure 16 Reverse view of *Ghost Gu Descending the Mountain*, illustrated in fig. 15.

approached from one side, the work appears to be simply white pots, with no sign of decoration. Viewed from the other side, they appear fully decorated. What one sees in the first few rows are pots that have one half of their surface painted. The image is from the original pot but by no means a replica. The painting is heavier, less gossamer, and the pot's shape is not identical, being slightly squatter, so the image is more compressed. Row by row, then, the painted surface becomes a smaller and smaller wedge. But because the pots in front obstruct one's vision, all one can see is the painted area, so one assumes they are all the same.

A close-up view of the pots in the rear, however, reveals that only about 10 percent of their surface is decorated. One can see this only because the photograph is taken at an angle. When this work is shown on a pedestal at close to eye level, it seems as though every pot is fully painted, a wonderful tease, as is walking to the other side, where all the painting disappears. An edition of three of this particular installation was made, as well as other editions with variations in the painting and, in some cases, painted red instead of blue. Ai is nothing if not industrious.

The calming rhythm of the assembled platoons of vases is especially satisfying when viewed from a position in which the lines of pots appear diagonal rather than square. A Zenlike tranquillity rises from this assembly, not unlike the hypnotic brickwork on his buildings. We are reminded that Ai has no difficulty with beauty. He may not actively seek it out, but one rarely sees any work by Ai, no matter what its conceptual purpose, that is not also aesthetically pleasing. He is always touted as an intellectual—and he is—but he is every bit a sensualist, thus his ease with ceramics.

Alone, the two vases that are included in the Arcadia University Art Gallery's show do not quite do justice to *Ghost Gu*, but they are nevertheless fascinating (fig. 17). First, they are a reminder of the dazzling artisanal skills still available in the ancient pottery town of Jingdezhen and how much they contribute to the bravura of Ai's ceramics. The ability of these painters to execute this complex scene with such finesse and felicity on a concave 3-D interior surface is nothing short of breathtaking.

Moreover, by placing the image out of sight unless one is standing above the pot looking down into it connects with the ancient practice of *anhua*,

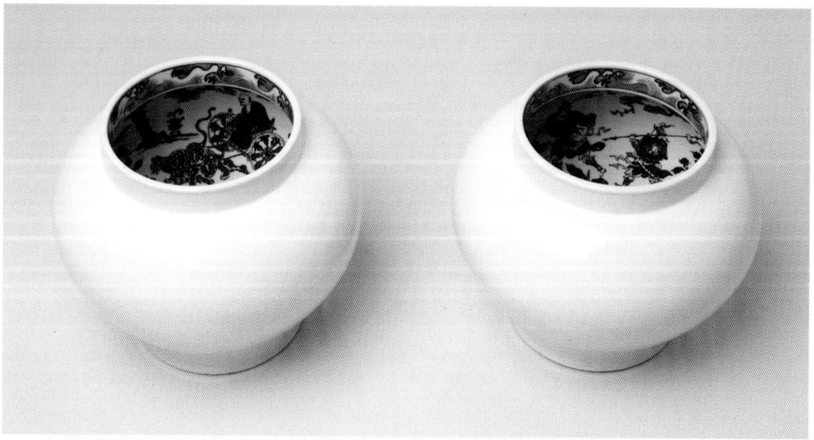

Figure 17 Two vases from *Ghost Gu*, illustrated in figs. 15, 16. (Courtesy, the artist and Arcadia University Art Gallery.)

painting "secret" messages on porcelains with white-on-white so they are discernible only at close view. Such messages typically appeared on high-quality, "imperial" porcelains from early in the Ming period (1368–1644). The purpose was not seditious. One will not find slogans reading "Down with the Emperor." Rather, they impart knowledge about the vase itself, adding a sense of privilege to their owners and, in so doing, flattering their status and the refinement of their connoisseurship.

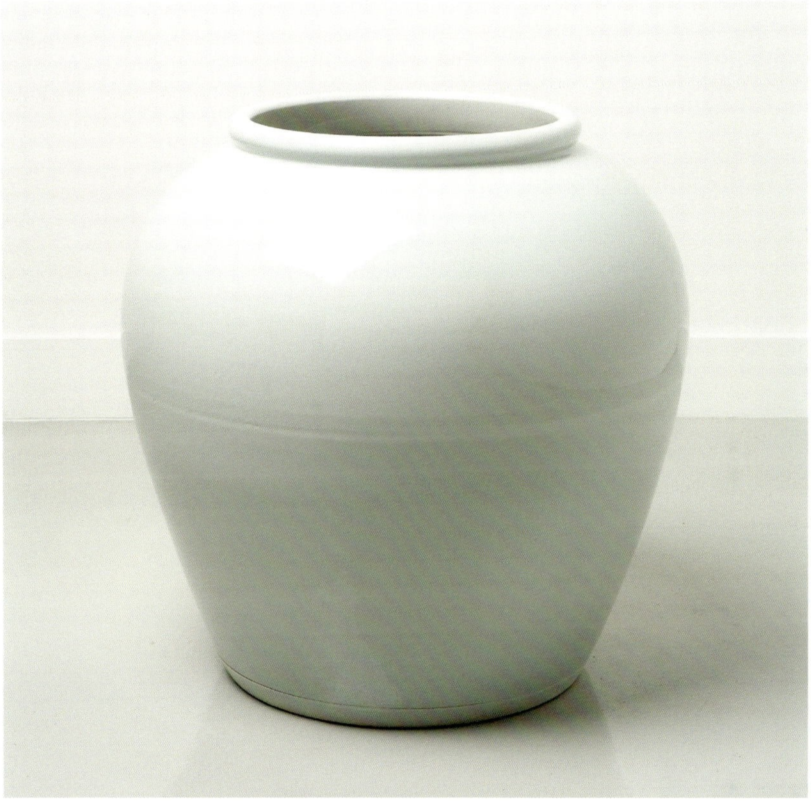

Figure 18 Ai Weiwei, *Untitled*, 2009. Glazed porcelain. H. 30". (Courtesy, the artist and Arcadia University Art Gallery.)

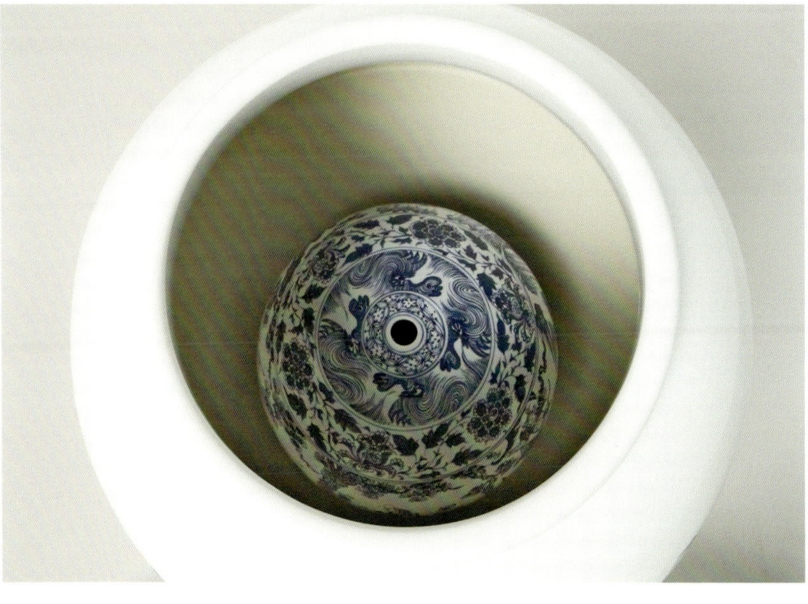

Figure 19 Interior view of the vase illustrated in fig. 18.

The ethereal white exterior of pots connects with the *Ghost Gu* tale. *Gui*, which means ghost, refers to the mysterious valley into which Guiguzi was descending. According to legend it remained shrouded in mist for years at a time, and its swirling vapors, seen as diaphanous ghosts, were feared. The white vases, seen en masse as in the *Ghost Gu* installation, effectively convey the sense of a fog blanket.

Untitled (2009), a large handsome porcelain jar with a satin glaze of celadon, is the most successful as a stand-alone work (figs. 18, 19); indeed, multiplication would not serve this work well. Again, only when standing right next to the pot does one see the orb-shaped, blue-and-white ceramic pot with a small mouth that is nestled inside. It suggests an oculus and causes the same strange sensation. The past is looking at us, maybe judging us. It is a direct piece, obvious in its intent and satisfying without the burden of myriad meanings and metaphors that clutter critical writing about Ai.

Three Qing replicas from his *Blue and White* series—a ginger jar, a moonflask, and a bottle vase—are exquisite and convincing reproductions to the lay eye (fig. 20). This begs Ai's question, "If there is no recognizable

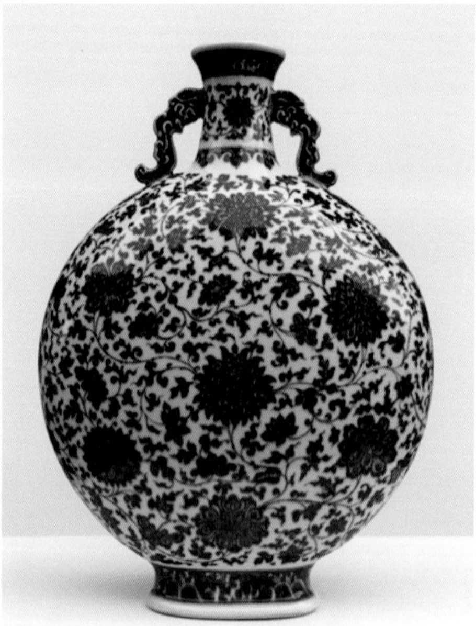

Figure 20 Ai Weiwei, *Blue and White Moonflask*, 1996. Porcelain, cobalt brushwork, glaze. H. 20⅞". (Courtesy, the artist and Arcadia University Art Gallery.) This replica is in the style of the Qing dynasty, Qianlong reign-period (1736–1795).

difference between this piece and an authentic piece, then what does this do to the value of the original period piece, or for that matter a modern replica? Let us say we place an exact copy in a museum setting, would this not completely confuse museum goers? Is it real or is it fake? How can a museum exhibit a modern production as an authentic period piece? Does this not undermine the whole system?"[15]

The issue of authenticity—real or fake?—is almost a fetish for Ai, who even named his studio Fake Design (also because of a confusing closeness in the Chinese characters for *fake* and *fuck*). But there is something about these works that is a touch self-important. The question Ai poses was answered long ago. The thriving forgery industry in China has been doing an exceptional job for centuries and continues today. Forgers now even mix their clay with grog made from older ceramic sherds to provide false positives in thermoluminescent dating tests. The system long ago adjusted its sampling procedures to deal with this. It poses some threat but nothing calamitous. But scholarship and technology have advanced to such a degree that passing fakes off as real has become much more difficult.

The flaw in this series is that it brings nothing new to the table. These vases could easily just be regular forgeries. Having these commissioned by Ai offers no edge, no greater distinction or different perspective. Slipping them quietly into the marketplace, as forgers do, is a much more destabilizing and seditious act. Yes, he is able to exhibit them in art museums, but still as replicas. Where is the tease? Mixing them in among authentic Qing ceramics in museums might be more provocative.

The subject of forgery in ceramics, hardly a Chinese monopoly, has been dealt with better in the past. Rosanjin, a trickster, eccentric potter, and leading epicurean in Tokyo, shared his glaze recipes and taught forgers how to make his costly and highly prized work, so that after his death they could

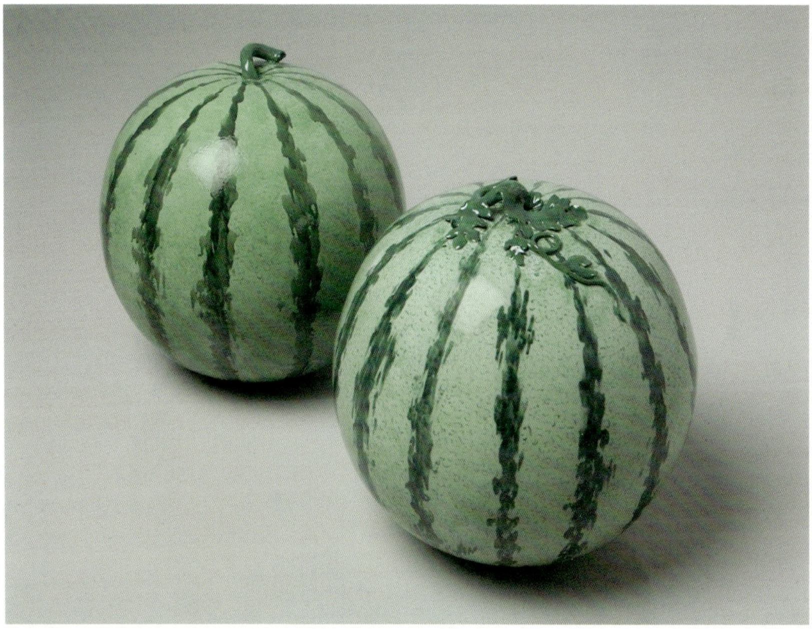

Figure 21 Ai Weiwei, *Watermelons*, 2006, installation view. Glazed porcelain. H. of each 17½". (Courtesy, the artist and Arcadia University Art Gallery.)

make perfect copies, a much more Machiavellian assault on future collecting of his work. Ai's works come nowhere close to this level of transgression.

The exhibition concludes with two snack foods, *Watermelons* (2006) and *Kui Hua Zi* (2009). The watermelons, to continue a dietary theme, can be described as Ai Lite: charming and fun but with no great meaning attached. Nor are they meant to be trompe l'oeil images of the real thing. Ai changed the color scheme from group to group, using different yellow and green glaze combinations (fig. 21).

Watermelons belongs to a series of joke pieces that Ai makes from time to time, the most infamous being *Pee* (2007), disconnected male genitals

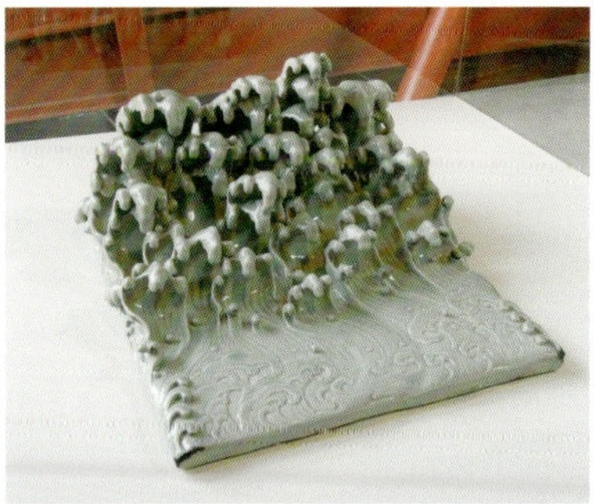

Figure 22 Ai Weiwei, *Wave*, 2005. Celadon-glazed porcelain. L. 16". (Courtesy, the artist and Galerie Urs Meile, Beijing-Lucerne.)

suspended in the air being held up by a stream of urine. Ascribing anything more than a sight gag to this work takes away its unpretentious pleasures. The same can be said of *Wave* (2005), a beautiful homage to the Japanese printmaker Hokusai's image made liquid by a lush viscous celadon glaze (fig. 22).

Kui Hua Zi (Sunflower Seeds) is another matter entirely. It sits on the floor, one ton, more or less, of life-size sunflower seeds in porcelain, painted to mimic the real thing. Its gravitas is both literal and conceptual (fig. 23). When Ai talks about this series it is clear that the sunflower seed is a hugely meaningful thing for him. He recalls that in the labor camp in which his father was held (and where Ai grew up), the detainees would pass seeds to each other as a gift, a rare moment of compassion. And because the act seemed so innocent, it was also a covert way of signifying solidarity. These tales are of course riveting, but they are footnotes. The main subject is Chairman Mao, who during the Cultural Revolution, like Louis XIV before him, was depicted as the sun. Sunflowers are sessile organisms, yet during the day they turn to face the sun, perfect metaphors for Mao's subjects, obediently facing his heliotropic face. The irony is that under Mao's rule hunger was widespread and this seed was often the only abundant food staple.

Writing about the Tate's exhibition of these seeds, Charles Darwent captures the work's essence: "Each seed offers a kind of hope—of food, of kindness—and yet each is like a minute (and inedible) stone monument, so that the work is at once both lively and deathly. In those contradictions of massive numbers and individuality, of humility and grandeur, happiness and pain, Sunflower Seeds is like China itself."[16]

The obvious question that arises is whether the message is communicated more effectively by increasing the tonnage of seeds. Does quantity make the work more or less profound? Does the absence of any distinctive form (the floor of seeds is formless) make a difference?

The pile of seeds at Arcadia is a powerful primordial shape, an early burial mound. Yet it also resembles the traditional conical straw hat of the Chinese laborer, about as iconic an image of China as exists. But perhaps its most spiritual connection is as a symbol of a mountain, an ever-present

Figure 23 Ai Weiwei, *Kui Hua Zi* (Sunflower Seeds), 2009, installation view. Porcelain and ink. Diam. approx. 80", wt. 1 ton. (Courtesy, the artist and Arcadia University Art Gallery.)

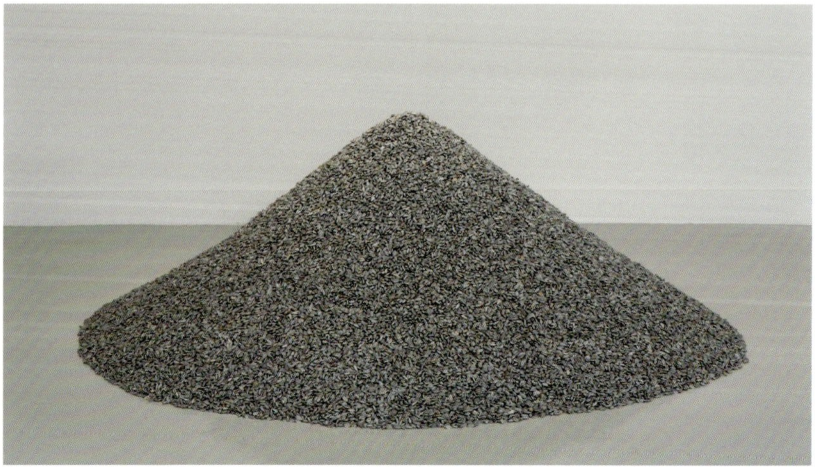

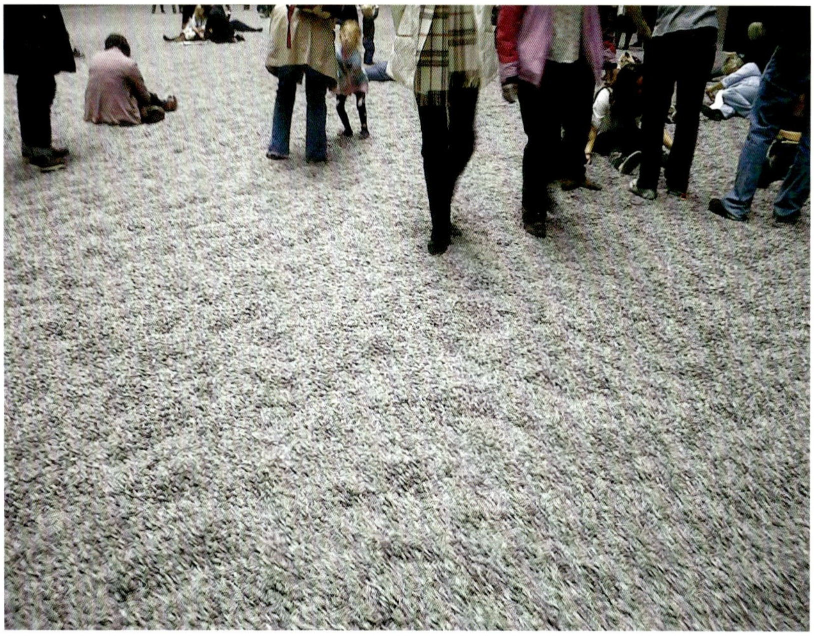

Figure 24 Visitors walking on *Sunflower Seeds*, Turbine Hall, Tate Modern, London, 2010. (Courtesy, Tate.)

and sacred element in Chinese lore (albeit one of rock and dirt), made up, in this case, of humanity—China's citizens. There also is a strong contemporary connection to modern minimalism and abstraction. The pile of seeds is about more than just shape, for the seeds rise closer and closer to one's eye as they approach the crest and, in the process, become increasingly individualistic as we discern that each one is unique. This is masterful, a moving and surprisingly poetic conception.

If humanism is central to this artwork, then that quality is better communicated by *Dropping a Han Dynasty Urn*. The acreage of seeds at the Tate was both dehumanizing and visually monotonous. One had to constantly remind oneself that it consisted of handcrafted seeds and was not just a surface of pebbles or small fragments of gray granite. As for the symbolism of walking on it, when one could still do so, the point totally escapes me (fig. 24). Treading on the populace of China?

Possibly I am making the same mistake that many writers do when discussing Ai, that is, taking his conceptualism too seriously. Richard Dorment takes the argument away from concept and into sensuality:

> Like so much else about China, on paper such figures are almost meaningless. Only by seeing it can you begin to grasp its immensity. Standing before it, we look out over an immeasurable, fathomless grey sea. But the moment . . . you step on it, your relationship to what lies beneath your feet changes. Each crunching footstep merely displaces a thin layer at the top of the pile. Our weight leaves no impression on the millions and millions of seeds beneath our feet. What from afar had been far too immense for the imagination to grasp instantly becomes as worthless as gravel.[17]

Still, had it been a vast mountain of seeds 149 times larger than the piece at Arcadia, towering over the viewers, the experience may have been more transformative both visually and conceptually.

The mantra of the Tate's press office—"150 tons, 100 million seeds, 1,600 workers"—reads more like a hubristic exercise, a warlord bragging about the size of his army rather than having any mathematical logic or even numerological magic. Critic Charles Darwent termed it "almost pervertedly grand."[18] 133,872,485 seeds might have made more sense of this porcelain mass. That would be one seed for every ten people living in China today. Otherwise, the quantity of seeds appears to have no reason but to impress, an exercise of power on the part of the artist and his host.

The Tate project was received with rapturous praise. Even after the original participatory premise was lost and it could only be stared at from above, the press, first having said that the installation worked *because* it could be walked on, reversed its position and claimed it was now better because it increased the "sense of sorrow and stillness . . . reinforced allusions both to ash, with its connotations of cremation, and to the fundamental tensions between the individual and the collective in Chinese society, as the field of seeds, seemingly identical yet each unique, laid dormant."[19]

Weiwei is the darling of the West's art establishment—both good, because he is a great artist, and bad, because it obscures this fact and turns him into a media star. The combination of his Puckish personality, his generally exceptional art, his eloquence and far-ranging intellect, and the frisson of his life-and-death encounters with Chinese authorities has proved irresistible, creating a climate that is adulatory rather than analytical.

But there has been some criticism from the Marxist quarter, questioning whether, in making this piece, Ai has put himself in a contradictory position. He became the boss man, an employer using the low cost of Chinese labor to make his personal monuments. As such, they contend, he is no different from any other entrepreneur in China who exploits the poorly paid workforce for profit.

This is a baseless argument. First, after two millennia of being the leading center for ceramic production, Jingdezhen is failing and has been forced, through competition from other provinces in China, to make low-quality products. In a decade or two, most of the virtuosic skills for which it is known will have disappeared. Factories are closing, unemployment is high, and the pollution casts a dark gray-brown pall over the "Harmonious Porcelain Metropolis," as the town is described on a sign welcoming visitors at the airport.

Jingdezhen is suffering the same agonizing fate we witnessed with the fall of the once great Stoke-on-Trent in Staffordshire, England, whose ceramic production once exceeded that of China. Stoke has lost its historic potteries. Most now operate solely as brand names, outsourcing manufacture to China and other centers in Asia. The "dole" is the town's major industry.

Ai has deep empathy for endangered cultural species, and Jingdezhen falls into that category. He was introduced to the town by Liu Weiwei, a ceramic mentor and the manager of Ai's kilns. They met early in the 1990s at the Beijing flea markets. At that time the numbers of artifacts finding their way to market was staggering, and in a very short period of time Liu and Ai developed an almost scholarly knowledge of Yuan, Ming, and Qing pottery.

About 2001 Liu gave up his market stall and moved to Jingdezhen and introduced Ai to the town. In perfect synchronicity, Liu fires his kilns for only two reasons, for Ai's work and to produce convincing forgeries of Chinese porcelains that are inserted into auction sales. It is the theme of "real and fake" in daily production.

Ai cannot solve Jingdezhen's problems nor reverse its decline. But he does bring employment and admits to paying his crafters slightly more than the normal wage, which is fair. Nor does he run one of the ghastly ceramic sweatshops one finds in this town: walls, workers, and floors (where the workers' younger children play) all caked in red, blue, or other toxic oxides.

Most of Ai's contract artisans work from small home studios; often couples or family groups do most of the work. If anything, Ai is viewed as an angel in this ancient kiln town, filling thousands of rice bowls that would otherwise remain empty. Also, his employment of those with high skills kept alive a practice that the commercial market no longer requires.

It should be pointed out that in the last decade hundreds of artists, both Chinese and Western, have flocked to this town. It is the Disneyland of ceramic appropriation. There is hardly a style of Chinese porcelain they cannot mimic. But not all work is traditionally based. They also produce contemporary ceramic sculpture—usually in vast numbers and at low cost. Ai is just the biggest and most public of these customers.

He recalls that it was a difficult process, bringing the artisans on board. "I tried to explain to [the artisans] what we wanted them for, but they found it very difficult to understand. Everything they usually make is practical, and the painters are used to creating classically beautiful flowers using a high degree of skill. . . . Now they are asking when we can start again. I shall have to think of a new project."[20]

Adamson's essay begins with a good question: "Where is Ai?" Signs asking about Ai's whereabouts hang on the exteriors of several Western museums. Workers in Jingdezhen ask the same question. But when Adamson made this query, in 2010, he was not predicting Ai's arrest but rather trying to get into the artist's head and establish where "Ai situates himself. What is his reality?"[21]

Asking the question with a twist—"Where is Ai in ceramics?"—is instructive. What is his view of ceramics? What are his attachments to the medium? Why does he work so vigorously within this medium? On the surface, Ai does not seem to be much of a fan. In a conversation with the Swiss curator Hans Ulrich Obrist, Ai commented, "Yes, ceramics is kind of crazy. I hate ceramics . . . but I do it. I think if you hate something too much, you have to do it. You have to use that."[22]

If the purpose was to "exorcise" ceramics from his art, as he says later in the discussion, he has failed magnificently.[23] Every year since returning to China, his involvement has increased exponentially. Ambivalence is more likely his position. And this is not because of the pointless art-versus-craft argument that has dogged ceramics through most of the twentieth century. Ai does not recognize boundaries.

Taking their cue perhaps too literally from Ai's remarks, the curators in their introduction write of ceramics as though the medium is a pariah, a "contested, schismatic arena" in which "Ai exorcizes (as well as exercises) his contempt for ceramics by his fearless use of it."[24]

This view of ceramics is outdated. Over the last decade ceramics has become a ubiquitous presence in major galleries. Anthony Caro, Tony Cragg, Jeff Koons, Anish Kapoor, Charles LeDray, Thomas Shutte, and Kiki Smith are among the many artists who have worked in the medium and continue to do so. Even lifelong ceramists such as Arlene Shechert and Ken Price have broken though the clay ceiling into the higher reaches of the canon. Value, too, has adjusted. The medium is no longer "terra worthless" (its nickname in the 1950s and 1960s), with such record sales as the $16.8 million fetched this year by Jeff Koons's ceramic *Pink Panther* (1988) at the May 10 auction at Sotheby's.

Second, these comments impose an early modernist hierarchy and decidedly Western viewpoint of materials on an Asian artist from a country in which ceramics are revered. The rarest porcelain vases are reaching nearly $90 million apiece at auction, largely due to Chinese buyers. What we are witnessing with Ai and his work is a white-hot love/hate affair with all of the ambivalence that generates.

This affair began young. Ceramics was part of Ai's patrimony. His father, the famous modern poet Ai Qing, was a collector. Ai recalls that even when Qing and his family were banished in 1957 and forced to live in a hut half dug into the dirt on a desert labor farm in Xinjiang (a harsh province bordering Russia in which only 4 percent of the land is habitable), rudimentary shelves held a small collection of Chinese ceramics. Qing was reinstated in 1979, and the family returned to Beijing. One of his first acts was to retrieve the more important ceramics in his collection that had been safeguarded during his banishment. As Tinari perceptively states, "ceramics for Ai Qing was a vehicle for memory."[25] His son inherited the same belief.

If anything, one senses that Ai finds ceramics overwhelming, loaded with too much history, symbolism, beauty, romance, and allure. It engulfs, frustrates, and excites him. And bear in mind that when we speak of Ai and ceramics, we are speaking of one aspect of a huge field—porcelain, the "white gold" of China. Ai is a porcelain snob, as are the majority of Chinese ceramophiles. He rarely speaks with aesthetic excitement about the Neolithic earthenware pots that he clobbers. Strangely, he also avoids the obvious reference to his *Sunflower Seeds* installation and the early Ming–era Yixing tradition of making superb trompe l'oeil works of nuts and other foods. But then Yixing wares are stoneware.

However, with porcelain there is rapture. "There were two things in Ancient China that left the outside world incapable of resistance," he said. "One was silk. The other was porcelain."[26] It is also instructive that Ai never clobbers antique porcelain. Possibly this is due to a difference in cost, but it also suggests respect. Only once has Ai smashed old porcelains, during a 1996 performance when he broke two blue-and-white "Dragon" bowls from the Kangxi era (1661–1722) (fig. 25). He is shown with hammer

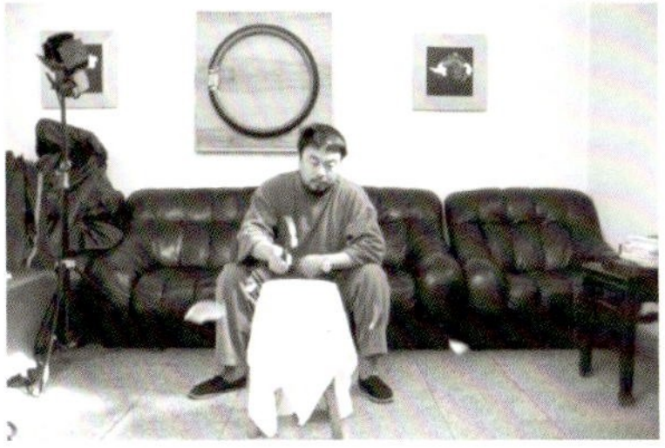

Figure 25 Ai Weiwei, *Breaking of Two Blue and White "Dragon" Bowls*, performance view, 1996. Porcelain bowls from Kangxi era (1661–1722). (Photo, Ai Weiwei.)

in hand, but a white shroud covers the bowls as though he does not want the brutality of the act to be witnessed, an attitude very different from the full frontal destruction of *Dropping a Han Dynasty Urn*.

Ceramics has its own place in the artist's oeuvre. It is less political and more likely to express pure beauty than other mediums. Tinari explains this well:

> Unlike his reconstructed furniture, Ai Weiwei's ceramic work could not be interpreted as riffs on historically Chinese forms, and unlike his New York period of conceptual experiments or his later architectural practice, they could not be explained away as homages to Duchamp or modifications of Mies. Tinier and more cunning than the wry subversions of his work in wood, less sarcastic and provocative than his photographs, Ai Weiwei's ceramic interventions of this brief period are among his most fanciful objects.[27]

He goes on to say, "objects this strange simply did not exist before their creator's urge."[28] Tinari's remark is not accurate, but it is a common misconception about Ai's ceramics. This has caused some antipathy toward Ai in the clay community, and with good reason. His sight-joke sculptures such as *Pee* hark back to the 1960s Californian funk movement and the work of Robert Arneson, among others. Smashing pots, forests of seven-foot pots, oil spills with clay and glaze, Coca-Cola logos on seemingly ancient forms, trompe l'oeil dresses and Chinese snack foods, and satiric play with the blue-and-white tradition have been staples, even clichés, on the ceramic scene for decades (fig. 26).

Figure 26 Ai Weiwei, *Pillars*, installation view, Ai Weiwei's studio, Beijing, 2006. Sixteen glazed porcelain vessels. H. 70–86". (Courtesy, the artist.)

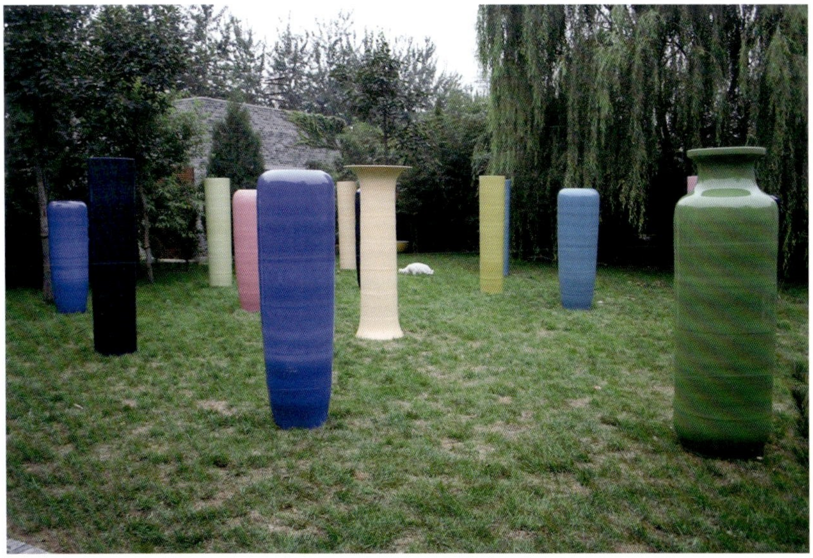

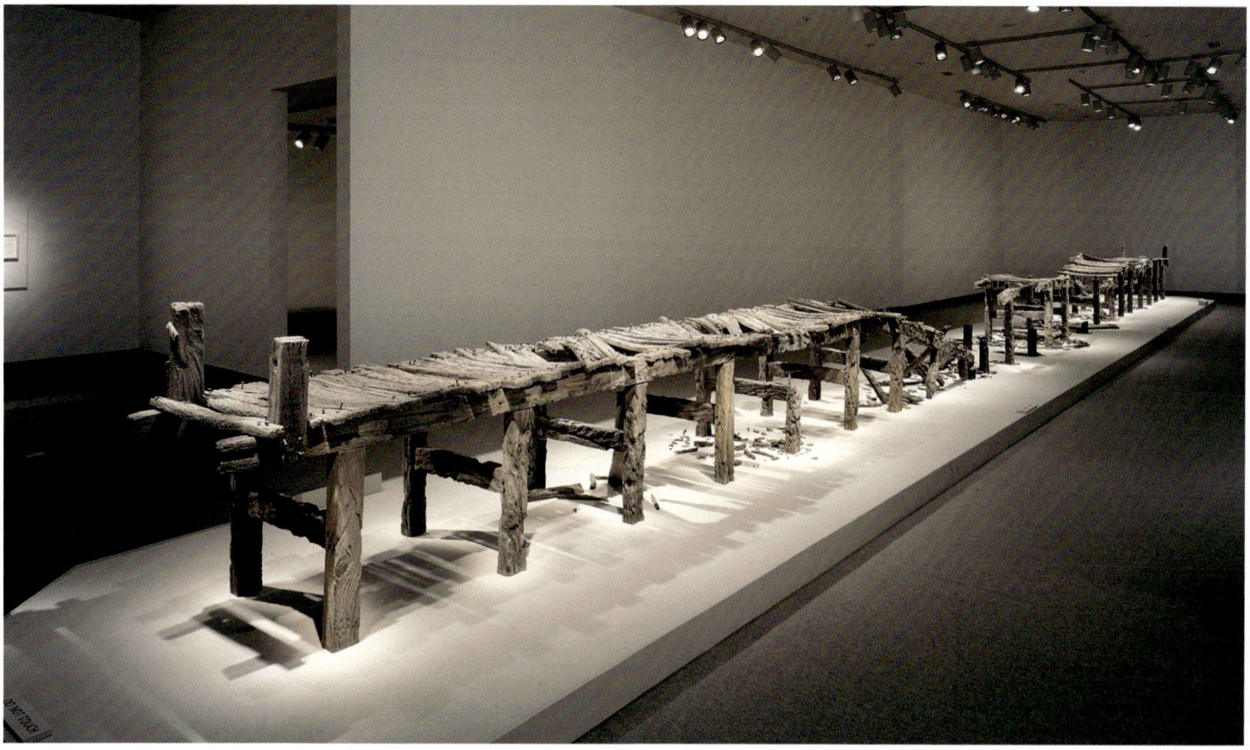

Figure 27 Ah Leon, *Bridge*, installation view, Freer Sackler Galleries, Smithsonian Institution, Washington, D.C., 1997. Stoneware. L. 60'. (Courtesy, Freer Sackler Galleries.)

And when it comes to scale and trompe l'oeil magic, there are other artists who have traveled this avenue. Taiwan's Ah Leon is one example. *The Bridge* (1997) is a sixty-foot-long gnarly weathered "wood" bridge made entirely of stoneware with seemingly real metal spikes of the same material holding it all together (fig. 27). Ceramists have long followed roads parallel to *Sunflower Seeds* even if to different destinations. There is little that Ai has done in ceramics that has not been done before him within the contemporary ceramics community.

This a good moment to reverse the question and ask, Where is ceramics in terms of Ai? Where does ceramics place him? Is he hero or villain? The rub for contemporary studio ceramists is less their love of his art—they get it better than most—but their invisibility in these discussions. It would be refreshing if, when ceramics and art are discussed, critics would cross the street, seek some contextual correspondence with the clay community, and give credit for innovation where it is due. But this is an in-house art historical argument. It matters little to the quality of Ai's art.

Art is not a race in which the first horse out of the gate is declared the winner. The prize, as such, goes to whoever best exploits an idea, whose work is the most transformative in the long haul. While not always the innovator, Ai wins hands down in almost every arena he invades. He has taken ceramic concepts that were at best interesting and made them profound. He has embraced and shed the limitations of the decorative, overturned scale limits, and, to a large degree, reinvented what he has inherited from Chinese clay. It has been a breathtaking, optimistic ride for those of us who a decade ago wondered whether ceramics still had a future in art.

No contemporary ceramics artist has so forcefully and repeatedly thrust clay into the heart of the fine arts, raising the bar at every outing. It is fair to say that because of Ai the arts are more aware of the history, power, and seductiveness of ceramics than at any time in the last one hundred years.

What many have missed is that Ai is the first ceramics star to emerge from China in nearly two centuries. The great irony is that China, the greatest ceramics culture on earth, never made it into modernism. In Europe, America, and Japan, ceramics were gradually transformed, beginning with the Reform Movement in 1850, into an adventurous art form. Were one to list the top one hundred twentieth-century ceramists, there would not be a single artist from the People's Republic of China. Similarly, if one lists the most important visitors to ceramics, artists who are not ceramists but who worked in ceramics and transformed the art status of the medium in the twentieth century, such as Paul Gauguin, Pablo Picasso, Lucio Fontana, and Isamu Noguchi, again one will not find a single Chinese artist. Until now.

Ai has pushed the gates of Fortress Ceramica—a conservative, cautious, status-obsessed, tradition-bound community—wide open. In this decade we have already seen his influence on the progressive fringe of the ceramics community. He has made it possible for fine arts to travel deeper into the medium and impossible for the medium to ignore its potential and its commercial success. And with modernism's spell broken, Chinese artists can once again begin to contribute great ceramic art to world culture. The doors are open to everyone in the arts who wants to follow this artist's example. Everyone, that is, but Ai, trapped, at least for the moment, in legal limbo.

1. Edward Wong, "Dissident Chinese Artist Is Released," *New York Times*, June 22, 2011, available online at http://www.nytimes.com/2011/06/23/world/asia/23artist.html (accessed July 10, 2011).

2. See Corey Kilgannon, "Before Fame, or Jail, Ai Weiwei Was a New York 'Starving Artist,'" *New York Times*, May 13, 2011, available online at http://cityroom.blogs.nytimes.com/

2011/05/13/before-fame-or-jail-ai-weiwei-was-a-new-york-starving-artist/?scp=2&sq=%22ai%20weiwei%22&st=cse (accessed June 12, 2011).

3. "2010 Power 100: Annual List" at www.artreview100.com/2010-artreview-power-100/ (accessed June 12, 2011).

4. Philip Tinari, Peter Pakesch, and Charles Merewether, *Ai Weiwei: Works 2004–2007* (Lucerne, Switzerland: Galerie Urs Meile, 2008), p. 121.

5. Glenn Adamson, "The Real Thing," in *Ai Weiwei: Dropping the Urn (Ceramic Works, 5000 BCE–2010 CE)*, edited by Gregg Moore and Richard Torchia, exh. cat. (Glenside, Pa.: Arcadia University Art Gallery, 2010), p. 49.

6. Michael Wines, "China's Impolitic Artist, Still Waiting to Be Silenced," *New York Times*, November 28, 2009, p. 30, available online at www.nytimes.com/2009/11/28/world/asia/28weiwei.html?scp=1&sq=China%27s%20Impolitic%20Artist&st=cse (accessed June 12, 2011).

7. Quoted in Philip Tinari, "Postures in Clay: The Vessels of Ai Weiwei," in *Ai Weiwei: Dropping the Urn*, p. 39.

8. See the exhibition's "Illustrated/Annotated Checklist," available online at http://gargoyle.arcadia.edu/gallery/09-10/Ai-Annotated-Checklist.pdf.

9. "Interview with Ai Weiwei," in *Shooting Back*, edited by Gabrielle Cram and Daniela Zymans, exh. cat. (Vienna: Thyssen-Bornemisza Art Contemporary, 2007), p. 36.

10. Charlotte Higgins, "People Power Comes to the Turbine Hall: Ai Weiwei's Sunflower Seeds," *The Guardian*, October 11, 2010, available online at www.guardian.co.uk/artanddesign/2010/oct/11/tate-modern-sunflower-seeds-turbine (accessed July 10, 2011).

11. The title, which references the phrase "earth to earth, ashes to ashes, dust to dust" from *The Book of Common Prayer* and the Anglican burial service, derives from Genesis 3:19 (King James Version): "In the sweat of thy face shalt thou eat bread, till thou return unto the ground; for out of it wast thou taken: for dust thou art, and unto dust shalt thou return."

12. Ai Weiwei, quoted in "Conversation: Hans Ulrich Obrist and Ai Weiwei," in *Ways beyond Art: Ai Weiwei*, edited by Elena Ochoa Foster and Hans Ulrich Obrist (London: Ivorypress, 2009), p. 25.

13. There was a tradition of painting the insides of vessels, but long after the fourteenth century, when the Yuan pot was made, and usually for smaller objects such as snuffboxes. Also the collaborator on this work, the Romanian-born American artist Serge Spitzer, is not credited. Even more important, the pots were not conceived as stand-alone works.

14. For a succinct masterpiece of scholarship that tracks all six of the vases in this series and explains exactly why this particular vase is unique and important, see www.christies.com/LotFinder/lot_details.aspx?intObjectID=4549655 (accessed June 12, 2011).

15. Ai Weiwei, interview with Robert Bernell, in *Chinese Art at the End of the Millennium: Chinese-art.com 1998–1999*, edited by John Clark (Hong Kong: New Art Media, 2000), p. 176.

16. Charles Darwent, "Why Should Millions of Specks of Porcelain Upset the Chinese? Because Ai Weiwei Never Treads Carefully and Neither, It Seems, Do the Patrons of Tate Modern," *The Independent on Sunday*, October 7, 2010, p. 29, available online at www.independent.co.uk/arts-entertainment/art/reviews/ai-weiwei-tate-modern-london-2108669.html (accessed June 12, 2011).

17. Richard Dorment, quoted in Isabelle Busnel, "We Can't Walk on Ai Weiwei Sunflower Seeds Anymore. . . ," November 21, 2010, http://thinkingthroughthings.blogspot.com/2010/11/we-cant-walk-on-ai-weiwei-sunflower.html (accessed July 10, 2011).

18. Darwent, "Why Should Millions of Specks of Porcelain Upset the Chinese?"

19. Higgins, "People Power Comes to the Turbine Hall."

20. Quoted in ibid.

21. Adamson, "The Real Thing," p. 49.

22. Quoted in *Hans Ulrich Obrist: The China Interviews*, edited by Philip Tinari and Angie Baecker (Hong Kong: Office for Discourse Engineering, 2009), pp. 24–26.

23. Ibid.

24. Moore and Torchia, "Doing Ceramics," in *Ai Weiwei: Dropping the Urn*, p. 12.

25. Qing brought back ceramics from countries he visited and added them to his collection of Chinese works. "He would spend no money on his trips abroad," Ai recalls, "and then on last day splurge on very nice [ceramics]." Tinari, "Postures in Clay," p. 32.

26. Quoted in ibid., p. 39.

27. Ibid., p. 32.

28. Ibid.

George Schwartz

Digging Up Salem's Golden Age: Ceramic Use Among the Merchant Class

▼ IN THE FALL OF 1974, while archaeologists were excavating in Massachusetts at the Narbonne House on the property of the Salem Maritime National Historic Site, another dig was conducted on historic Chestnut Street in Salem.[1] Maritime artist Racket Shreve initiated repairs to the foundation of his studio barn attached to number 17, which was built in 1804–1805 for Captain Stephen Phillips, a successful sea captain and merchant during Salem's golden age. While working around the foundation area, Shreve discovered fragments of Chinese porcelain sticking out of the ground near the southwest corner of the barn. A quick survey of the surface soil yielded many more pieces of high-quality ceramics. His curiosity piqued by these finds, Shreve halted his foundation work and over the next three summers diligently excavated what turned out to be a stone-walled trash pit measuring 6 feet by 5 feet (fig. 1). In the process, he uncovered a treasure trove.

In all, several thousand artifacts were unearthed, including a plethora of Chinese export wares: creamware, pearlware, redware, stoneware, and glassware. After mending a number of the artifacts, Shreve exhibited a few of them at the 1976 Salem Antiques Show at Hamilton Hall (also on Chestnut Street) and subsequently donated the collection to the Peabody Essex Museum. The only analysis of the collection conducted at the time

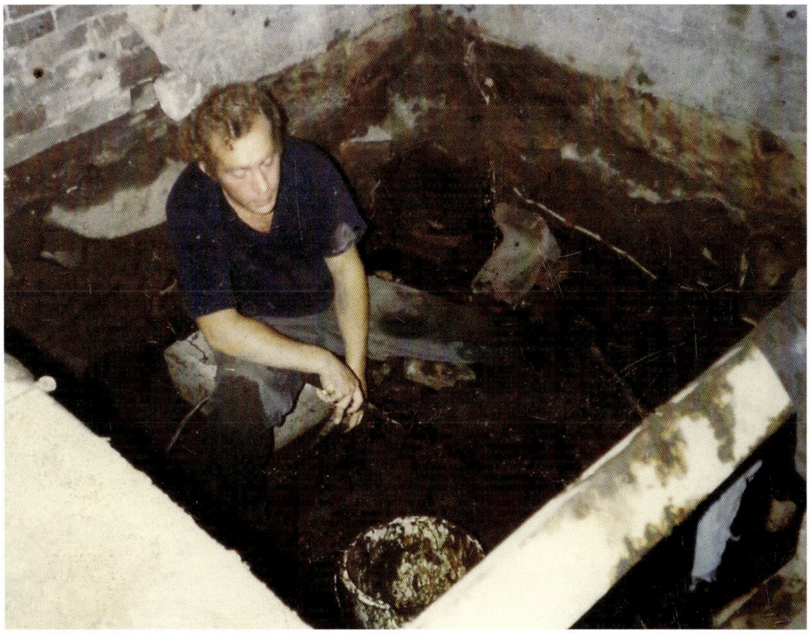

Figure 1 Racket Shreve excavating the Captain Stephen Phillips trash pit, Salem, Massachusetts, 1974–1976. (Courtesy, Racket Shreve.)

was a four-page article in the show's publication, which focused primarily on the physical attributes of the objects and the identification of certain wares without contextualizing them into the larger scope of Salem's mercantile world.[2]

The archaeological history of Salem, one of the most important colonial and early American seaports, has a significant gap. Though work has taken place in the city since the late 1930s—including several undertakings at the Salem Maritime National Historic Site—no excavations of the dwellings owned by the prosperous merchant class during Salem's golden age have been initiated.[3] Thus, the majority of what we know about Salem's mercantile class and the goods they owned and used during this time comes from only a few contemporary sources, most notably shipping papers, diaries, and letters. The remainder of our knowledge has been based on the finest ceramics owned by merchant families (many in museum collections), historic homes that have been renovated throughout the years, and history books on Salem that were written nearly one hundred years after the city's

Figure 2 James Frothingham, *Captain Stephen Phillips*, Salem, Mass., ca. 1818–1826. Oil on canvas, 34¾" x 29¾". (Courtesy, Peabody Essex Museum.)

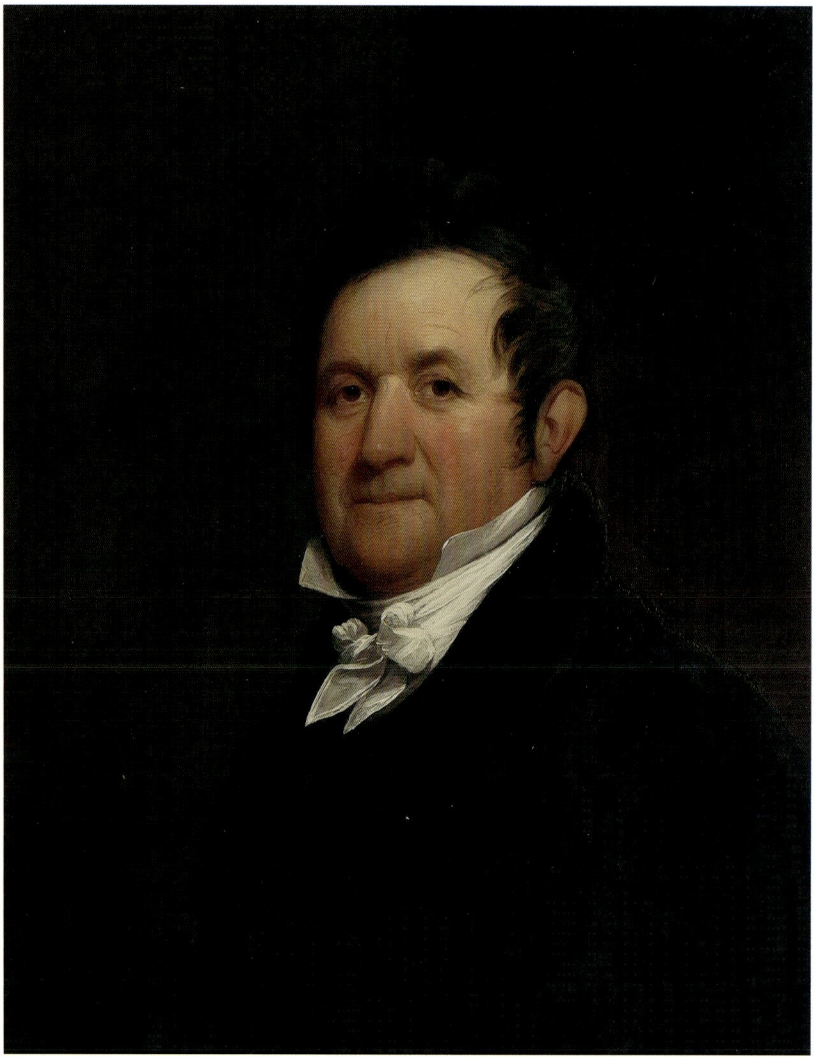

heyday, some of which are poorly documented and often loosely based on archival material. Fortunately, the collection of artifacts from the Phillips house now fills this niche for Salem archaeology and opens a window into the ceramic tastes of the mercantile class of the time.[4]

The artifact assemblage revolves primarily around Captain Stephen Phillips (1764–1838), whose life was an embodiment of Salem's rise as a global port in America (fig. 2).[5] Phillips, who grew up during a tumultuous time in Revolutionary New England, spent his entire career aboard ships on the pioneering voyages to the Far East that helped Salem transition from a fishing port to an entrepôt (fig. 3).[6] At the end of the eighteenth century, several prominent merchants in Salem who wanted to retire to "a quieter and more pleasant place to live" than the crowded and noisy wharves on Derby Street relocated to Chestnut Street, an area between Essex and Broad Streets that was laid out in 1796.[7] Captain Phillips, now strictly a merchant shipowner and businessman, bought a portion of land in this recently developed area for himself, his second wife, Elizabeth, and

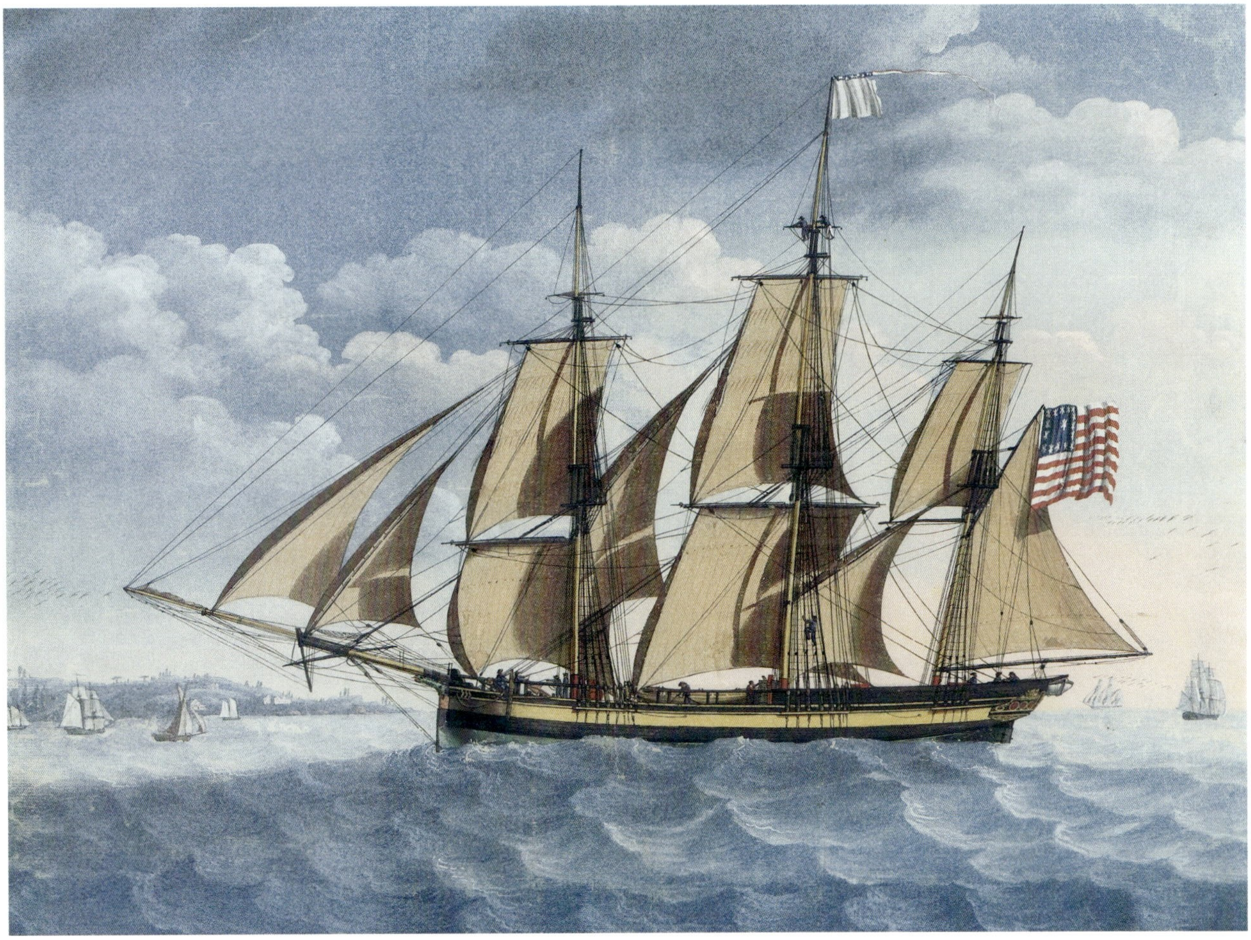

Figure 3 Michele Felice Cornè, Ship *John*, Salem, Massachusetts, 1803. Watercolor, 17¼" x 23¼". (Courtesy, Peabody Essex Museum.)

his son, Stephen Clarendon, and had a three-story, hipped-roof house built in 1804 1805 (fig. 4).

The trash pit discovered by Racket Shreve was found under the foundation of a two-story, hipped-roof wooden barn attached to the west end of the house by a one-story shed that had been constructed between 1837 and 1851.[8] The pit was used primarily as a refuse dump for ceramics and miscellaneous items such as toothbrushes and pipes. The majority of the assemblage consists of nearly whole and a few intact vessels (fig. 5) and organic items, like animal and fishbones, which make up only a minuscule amount of the material deposited. These characteristics are typical of artifacts found in urban archaeological sites where rapid filling occurred, usually after the death of the head of the household or when the house changed hands or occupants, and often many ceramic and glass vessels can be almost fully reconstructed.[9]

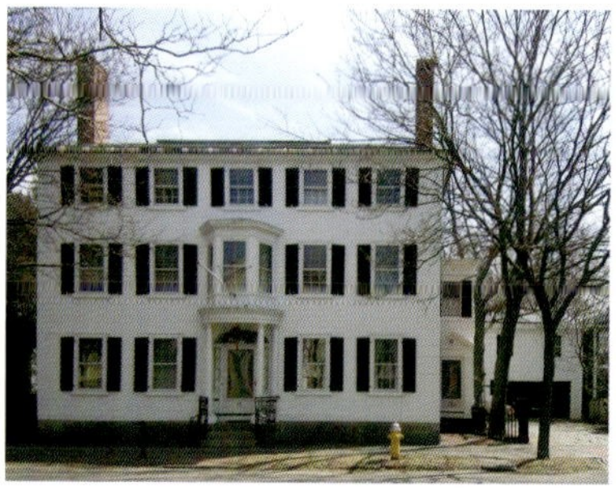

Figure 4 Number 17 Chestnut Street, the Captain Stephen Phillips House, 2005. (Photo, George Schwartz.)

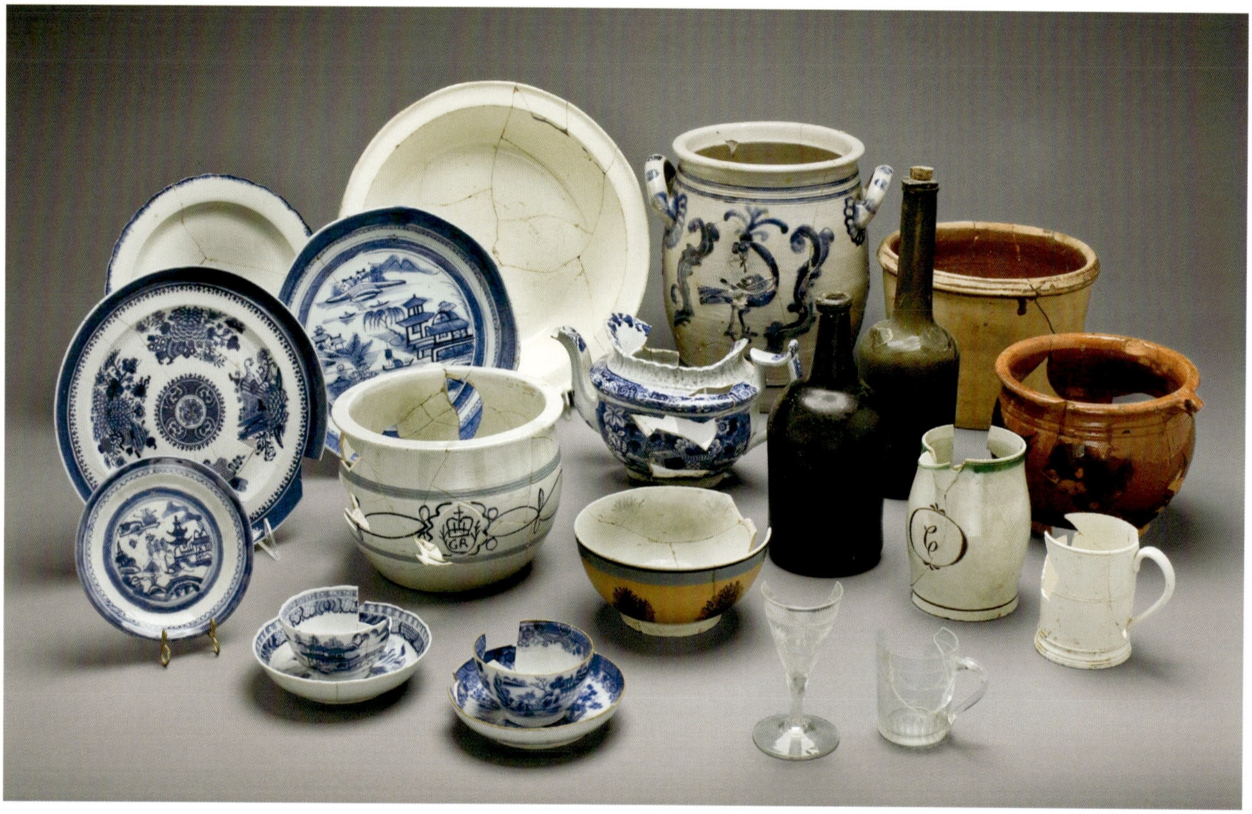

Figure 5 Selected examples of the ceramic and glass assemblage excavated from the Phillips trash pit. (Courtesy, Peabody Essex Museum; unless otherwise noted, all artifact photographs by Walter Silver.)

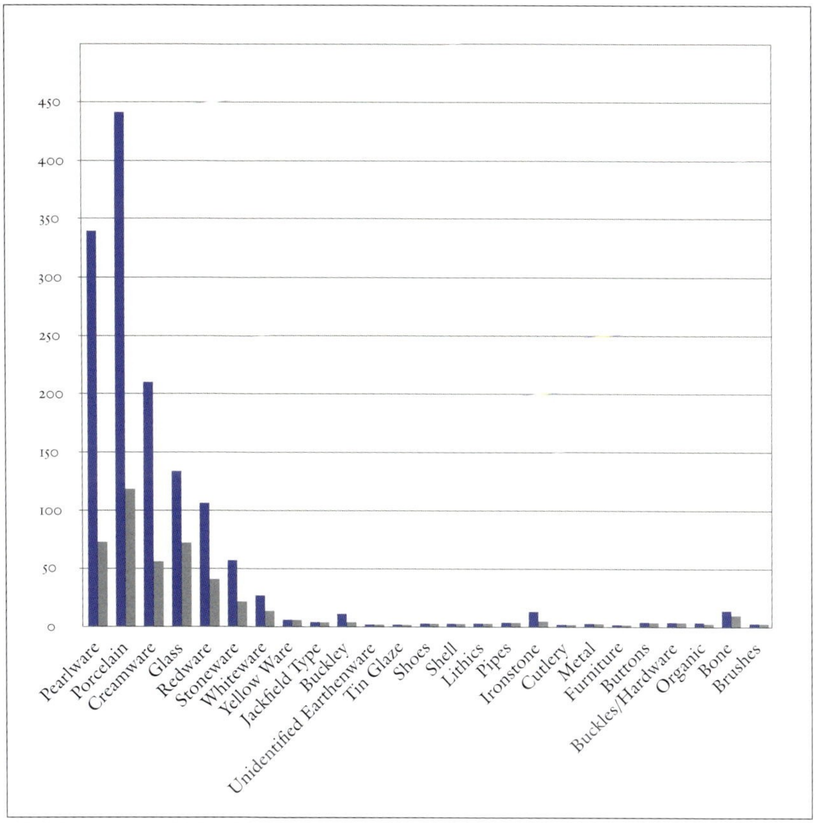

Figure 6 Distribution of artifacts recovered from the Phillips trash pit.

Number of sherds

Minimum number of vessels

Shreve pulled more than thirteen hundred artifacts from the trash pit, representing well over four hundred vessels (fig. 6). Most of the artifacts are ceramics and glass that date between 1790 and 1830. The forms are primarily ceramic plates, teabowls, and saucers (figs. 7, 8) as well as numerous fragments of wine bottles and Bohemian table glass. Most of them—the focus of this article—are Chinese export wares, pearlware, and creamware.[10]

Figure 7 Distribution by ceramic type of plates recovered from the Phillips trash pit.

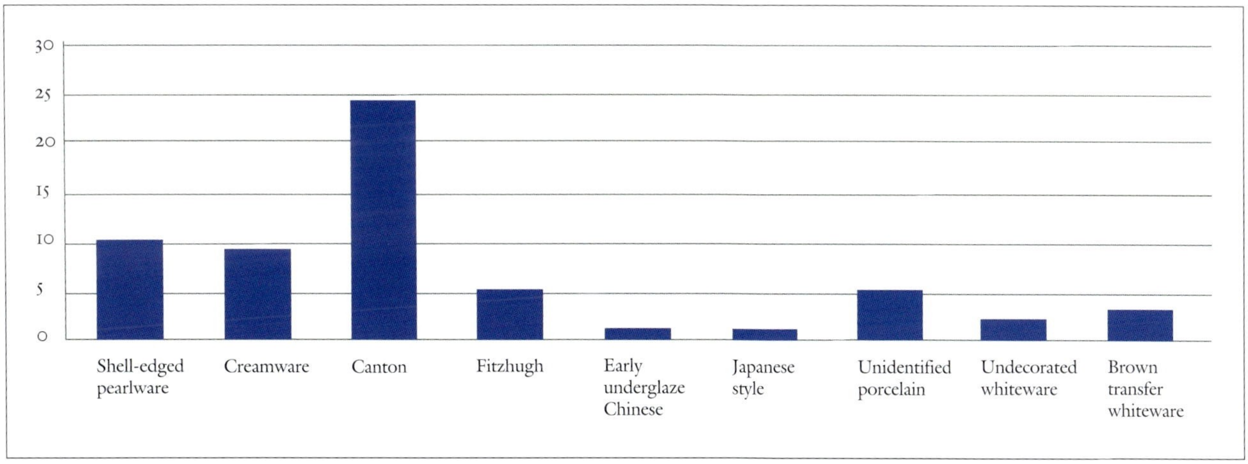

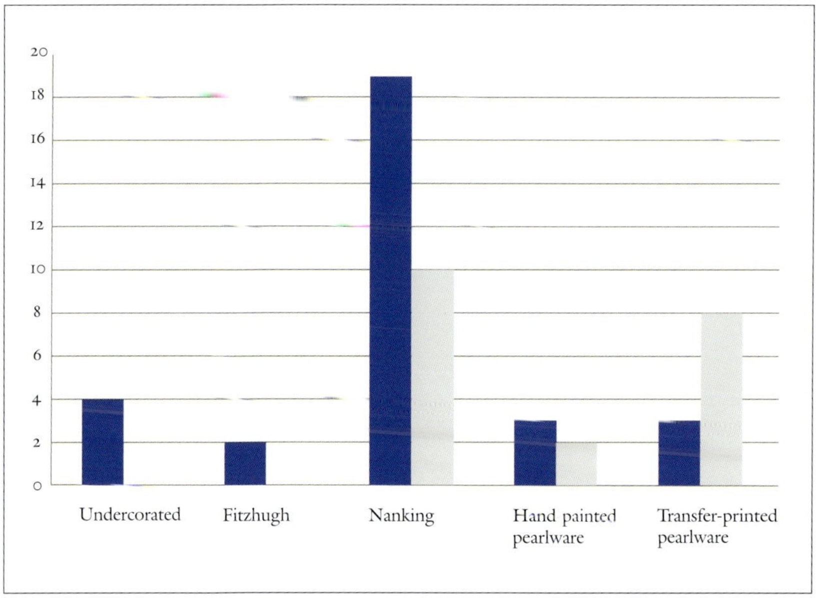

Figure 8 Teabowls and saucers recovered from the Phillips trash pit.

■ Teabowls
■ Saucers

Chinese Porcelain

Porcelain, the most expensive of all the ceramic types available to the American market in the seventeenth and early eighteenth centuries, was not usually purchased by people of marginal wealth. It is not often recovered from North American archaeological sites that predate the 1730s, after which it became more popular and available.[11] Between 1780 and 1830, before the Industrial Revolution, Chinese blue-and-white porcelain was the most widely produced and exported item in the world.[12] As a result, Chinese

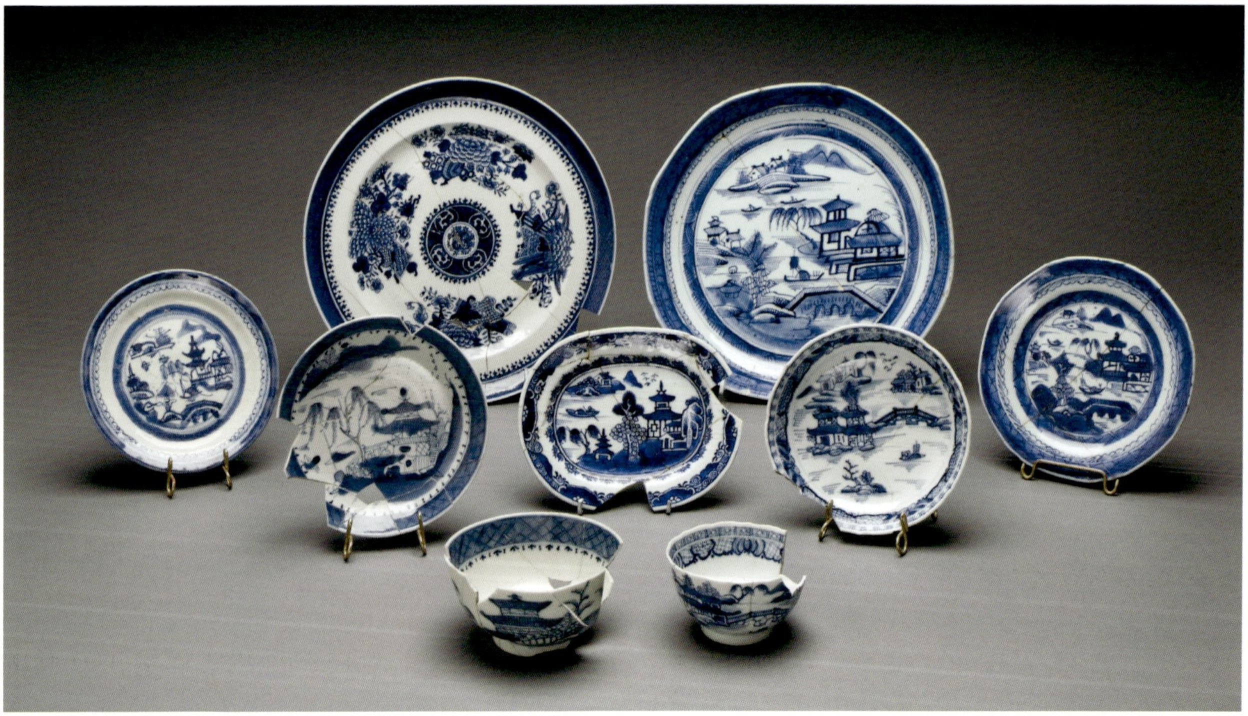

Figure 9 Selected examples of the Chinese porcelain recovered from the Phillips trash pit. (Courtesy, Peabody Essex Museum.)

Figure 10 Types of Chinese porcelain recovered from the Phillips trash pit.

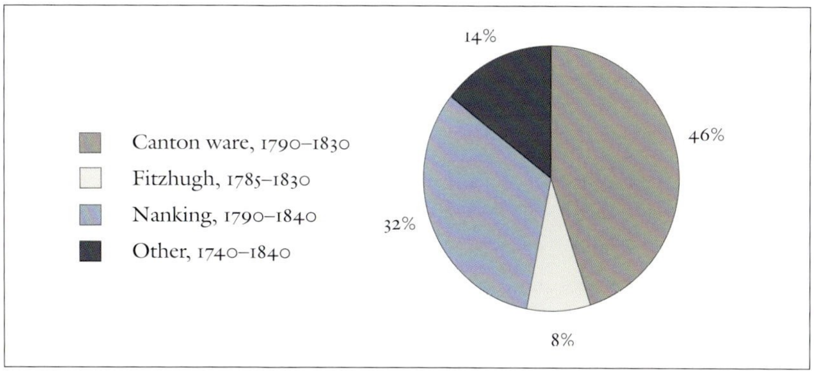

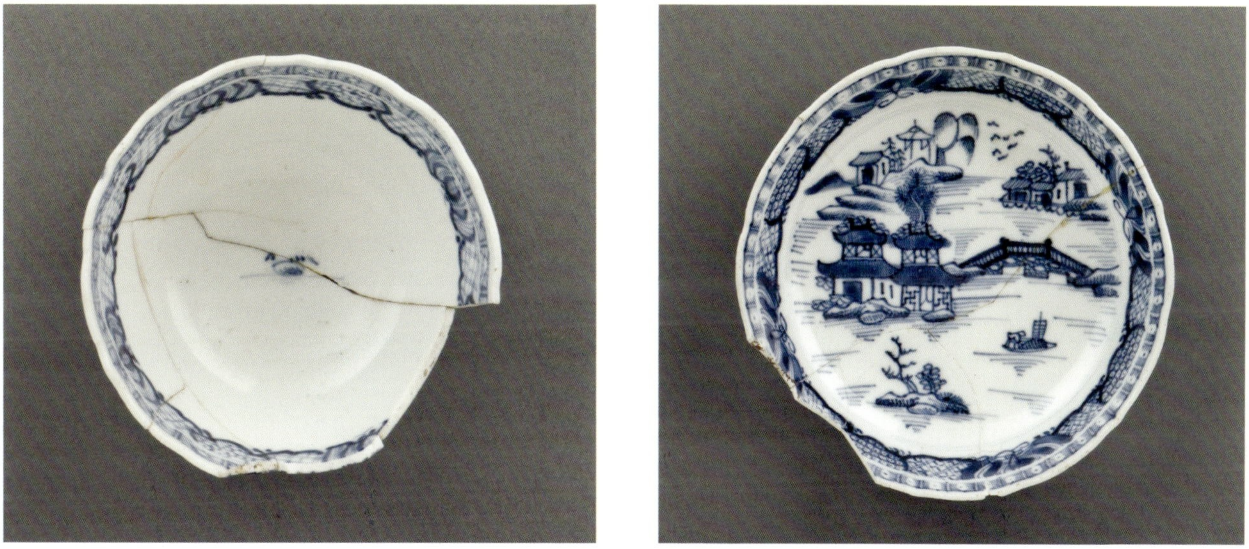

Figure 11 Teabowl and saucer, China, 1760–1780, recovered from the Phillips trash pit. Hard-paste porcelain with underglaze-blue decoration in the Nanking style. D. of bowl 3½", D. of saucer 5½".

porcelain is often present in the archaeological assemblages of upper-class households, particularly in the coastal areas of the United States.

One of the main reasons porcelain became immensely popular by the late eighteenth century was its elegant appearance, a strong contrast to the pewter and wood utensils more commonly used in the colonies. Chinese porcelain was particularly popular because it had become more durable than European earthenware and its Eastern origin and decorative motifs fascinated and pleased American buyers.[13] Once the American colonies achieved independence, Chinese export ware could be purchased directly from China, making it far more accessible.[14] Like their European counterparts, wealthier Americans often ordered pieces with patriotic or personal motifs such as flags and family crests.[15]

More than one-third of the ceramics uncovered in the Phillips trash pit (117 vessels) were Chinese export porcelain (fig. 9), which is remarkable given what has been found in other historic sites excavated in New England. Of the blue-and-white porcelain from the trash pit, 46 percent is what is commonly referred to as Canton ware. Nanking pieces account for 32 percent (fig. 10) and are almost exclusively in the form of teabowls and saucers (fig. 11).[16] These Nanking objects point to the continued importance of tea drinking in America after colonial times and suggest that more refined ceramics had become a requirement for this social practice among the merchant class.

In comparison to the finer Nanking porcelain, Canton ware was often marred by a variety of imperfections: the use of unrefined materials; the rigorous treatment of vessels during firing, which sometimes led to warping or "sooting"; and the overuse of glaze coating, which caused it to crawl, bubble, and become pitted under high kiln temperatures.[17] The combination of these factors often produced either a curdled "potato soup" or "orange peel" surface.[18] The relative poor quality of most Canton ware was understood by merchants. In 1815 the owners of the Salem brig *New Hazard* made it clear to master Benjamin Shreve that the porcelain was to be

> smooth, the cups & saucers in particular not thick & clumsy. Let the patterns of the Enamelled Tea Sets, Cups & Saucers & Bowls be delicate, of lively patterns and shades. Those of a thick heavy daub will not bring so good prices by a good deal & cost as much. The blue & white Dining sets should be of uniform shade & same pattern. They are often put up without attention to these particulars.[19]

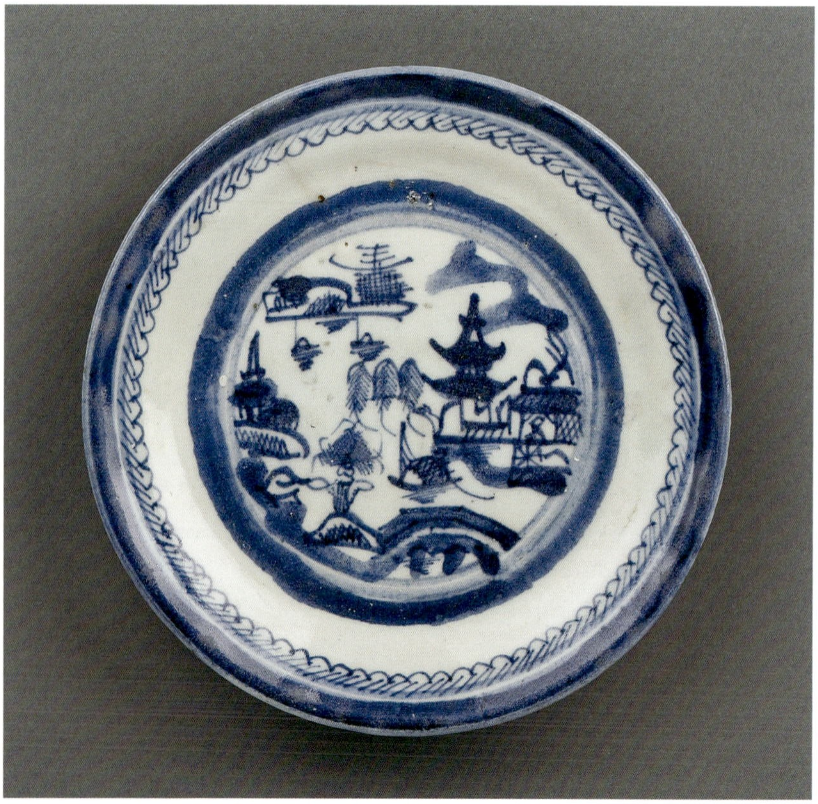

Figure 12 Plate, China, 1790–1820, recovered from the Phillips trash pit. Hard-paste porcelain. D. 5¾". The rim of this low-quality Canton-style porcelain is heavily pitted.

China trade historian Jean Mudge believes that the abundance of poor-quality Canton ware was a major factor in its low cost. Much of the Canton ware in the assemblage from the Phillips trash pit is coarse, pitted, and carelessly decorated and displays many imperfections (fig. 12).

While the vast majority of Chinese export porcelain found in the trash pit was underglaze blue-and-white ware, a few notable examples of overglaze pieces were uncovered as well. Fragments of at least three drum

Figure 13 Plate fragments, China, ca. 1750, recovered from the Phillips trash pit. Hard-paste porcelain.

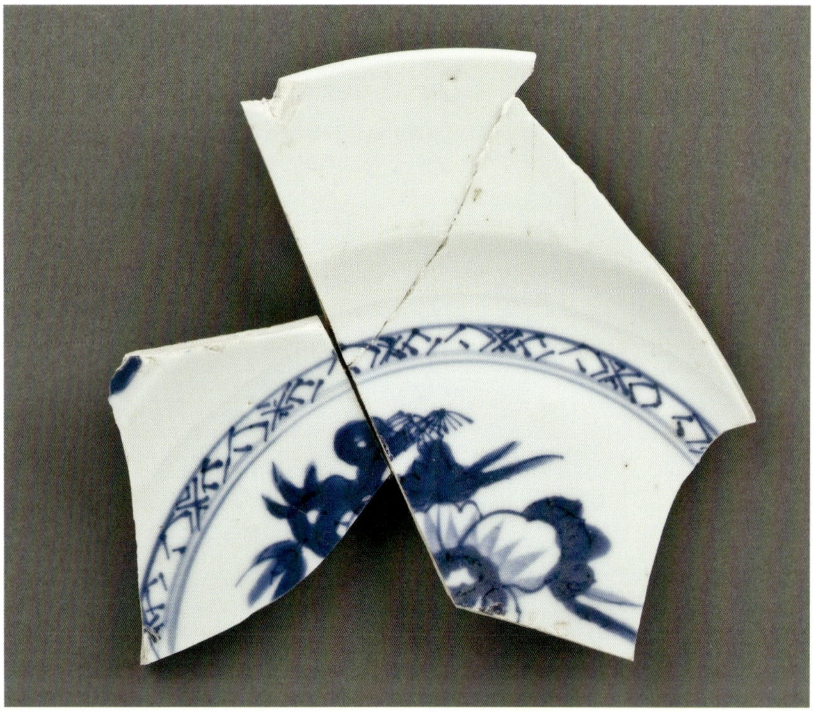

teapots were found as well as portions of a mid-eighteenth-century export plate (fig. 13). In addition, fragments from a Chinese porcelain dish decorated in the Japanese style with an overglaze application technique were recovered and date to about 1750–1765 (figs. 14, 15).[20] However, even

Figure 14 Plate fragment, China, 1750–1765, recovered from the Phillips trash pit. Hard-paste porcelain. This fragment, from an overglazed Japanese-style porcelain plate, shows the leg of a bird (see arrow) similar to one on the soup plate illustrated in fig. 15.

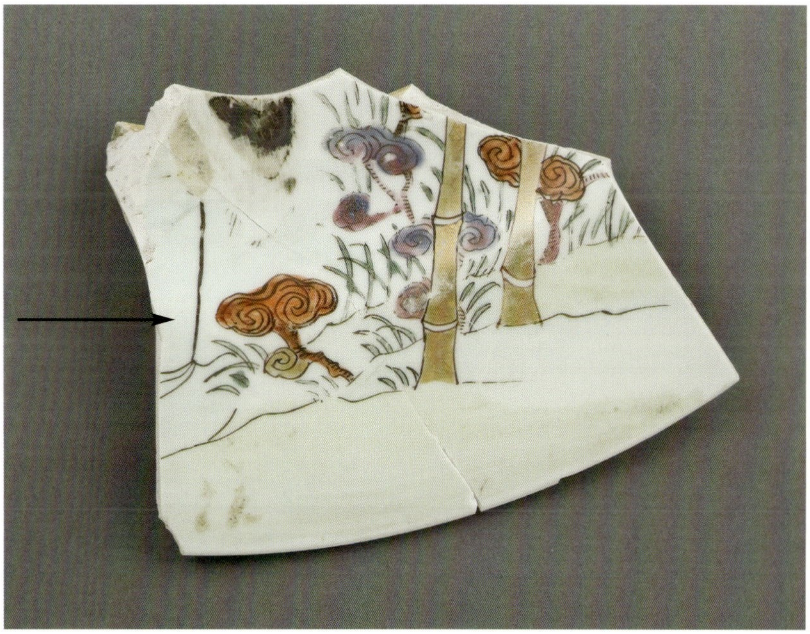

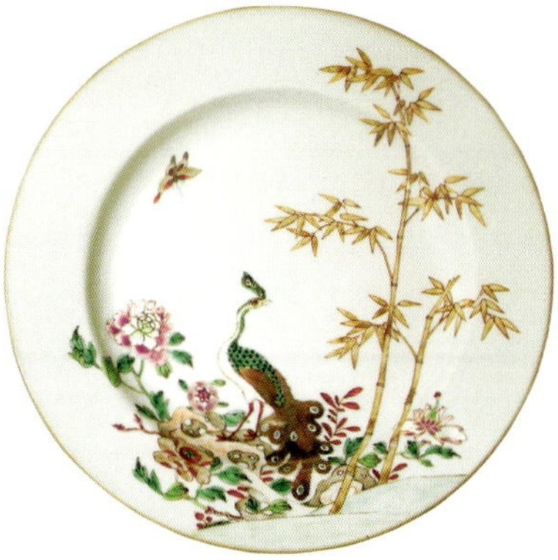

Figure 15 Soup plate, China, 1750–1765, illustrated in Daniel Nadler, *China to Order* (Paris: Vilo International, 2001), p. 83, pl. 74.

though blue-and-white porcelain (whether Canton or Nanking) graced the tables of the most prominent citizens of America—George Washington, for example—the mere presence of porcelain in the assemblage should not be viewed as a definite indicator of high social status or extreme wealth.

The trash pit contained a wide variety of porcelain and many low-quality pieces, and even though Captain Phillips was a successful ship captain and businessman, he was not in the tier of the wealthiest residents of Salem's golden age—Elias Hasket Derby, for example, who was the first millionaire in America, or William Gray, worth $3 million at the time of the 1807 Embargo, and George Crowninshield Sr., worth an estimated $16 million.[21]

Pearlware

Pearlware from the Phillips trash pit comprises 22 percent of the ceramic assemblage, or 339 sherds (72 vessels). This rarely marked type is the kind most commonly found in early nineteenth-century archaeological sites in the eastern United States.[22] The majority of the pearlware found in the trash pit had been made in the early nineteenth century. Approximately one-third of it has molded shell-edged patterns with blue or green underglaze painting (fig. 16). Shell-edged motifs appear as early as 1775 on creamware and were probably one of the first used to decorate pearlware.[23] Pearlware continued to be manufactured in great quantities during the nineteenth century, although the occurrence of green painting diminishes toward the end of the pearlware period (figs. 17, 18).

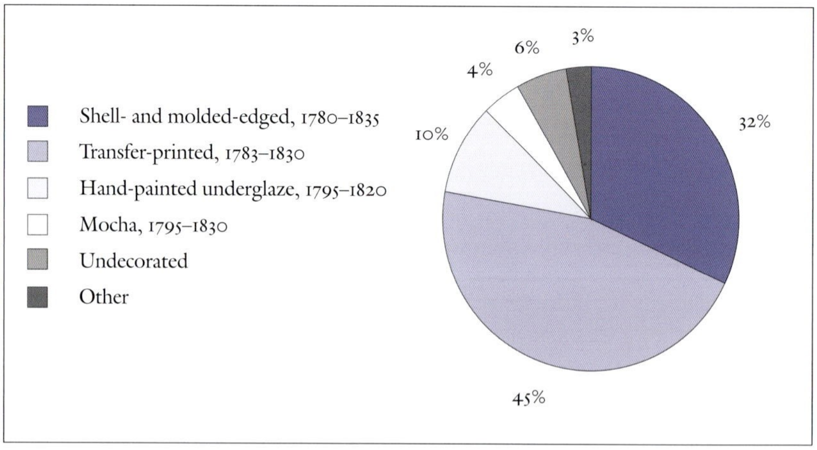

Figure 16 Types of pearlware recovered from the Phillips trash pit.

Molded shell-edged decoration was used primarily on plates, as well as on larger serving pieces such as platters and tureens. When pearlware was first introduced, the acquisition of a full molded shell-edged dinner service was considerably more expensive than the acquisition of a set of plates. One serving piece cost as much as a dozen plates, and therefore a large dinner service required some expense to maintain. By the end of the pearlware period, however, shell-edged and other molded patterns were used only on platters and plates.

The largest proportion of pearlware found within the trash pit (45 percent) is transfer-printed. The development of transfer-printing designs under the glaze revolutionized the Staffordshire ceramic industry by creating uniform market vessels. The technique allowed potters to apply complex decorations quickly and relatively inexpensively, thus creating thousands of

Figure 17 Plate, probably Staffordshire, England, 1780–1810, recovered from the Phillips trash pit. Pearlware. D. 7".

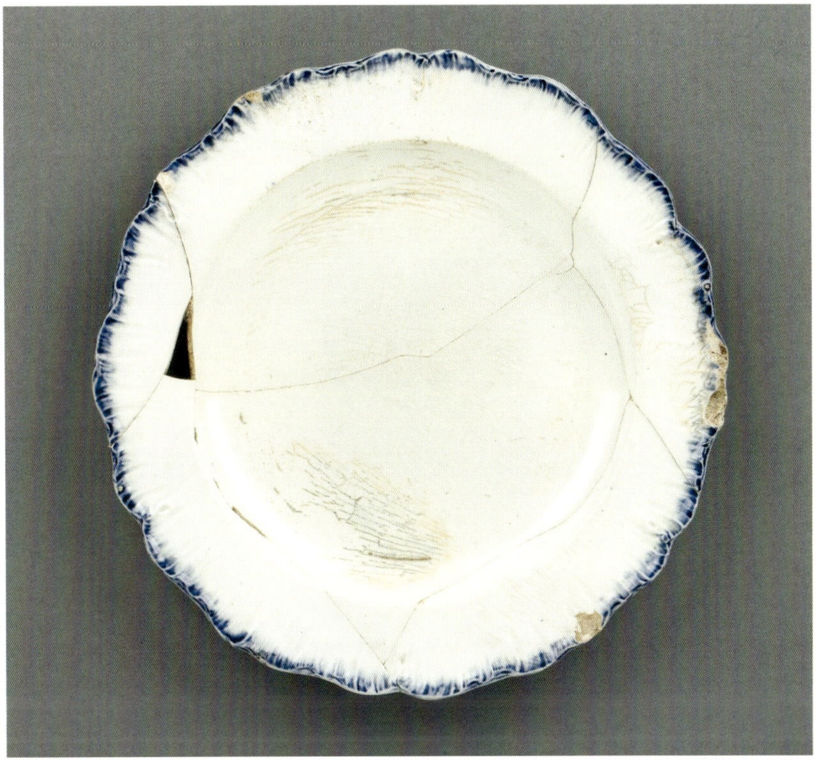

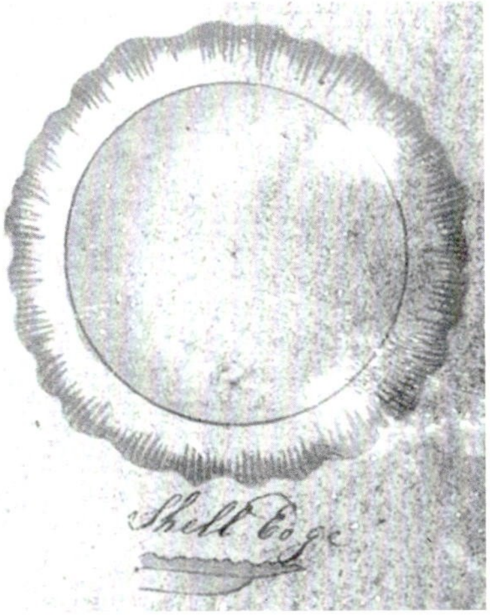

Figure 18 Shell-edged plate illustrated in Josiah Wedgwood's 1802 trade catalog. (Courtesy, Birmingham Museum of Art.)

designs for ceramics in a variety of colors and styles that were in demand by consumers both in England and abroad.[24] Some of the most popular and well-known patterns of the late eighteenth and early nineteenth centuries were Blue Willow, Asiatic Pheasants, and Canova.

In the years following the Revolutionary War, Staffordshire potters found an enthusiastic market in America, and pottery in hundreds of patterns made the journey across the Atlantic Ocean to grace the tables of the new republic. George Miller and his colleagues have shown that in the 1790s printed wares cost three to five times more than undecorated cream-colored earthenwares but by the first quarter of the nineteenth century had dropped to one or two times. Printed wares began to appear in great numbers in America after the War of 1812, probably the result of a major fall in ceramic prices.[25] Nearly 43 percent of the plates and soup plates ordered by New York merchants between 1838 and 1840 were printed.[26]

Almost all of the transfer-printed pearlware uncovered from the trash pit has chinoiserie-style patterns. With the advent of underglaze transfer printing, production of the complex landscapes and geometric borders common on Chinese porcelains became more cost efficient.[27] In the mid-eighteenth century, consumers who desired expensive Chinese porcelains but were unable to afford them could instead turn to painted renditions of Chinese-style designs on less costly ceramics, in particular tin-glazed earthenwares.[28] The whiteness of the newly developed pearlware body and glaze was well suited to the traditionally blue Chinese motifs, and the combination of Chinese-style designs and vessel forms was designed to appeal to those who would otherwise purchase Chinese porcelain.[29]

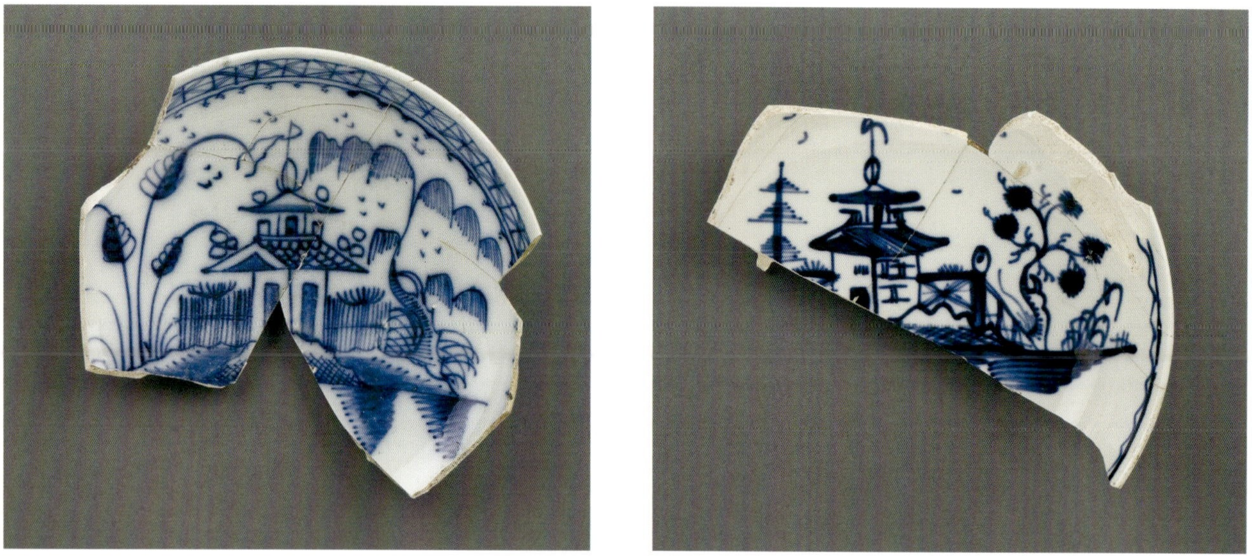

Figure 19 Saucers, England, 1785–1810, recovered from the Phillips trash pit. Pearlware or China glaze. D. 6".

There were also a few hand-painted examples of China glaze ware (fig. 19), a bluish tinted glazed earthenware with designs that imitated Chinese porcelain. It appears to have been first produced in 1775 in Staffordshire and was in use until the 1830s, although its peak production was between 1797 and 1814. China glaze ware resembles Chinese porcelain in its patterns, vessel shapes, blue-tinted glaze, and, of course, its name, "China glaze."[30]

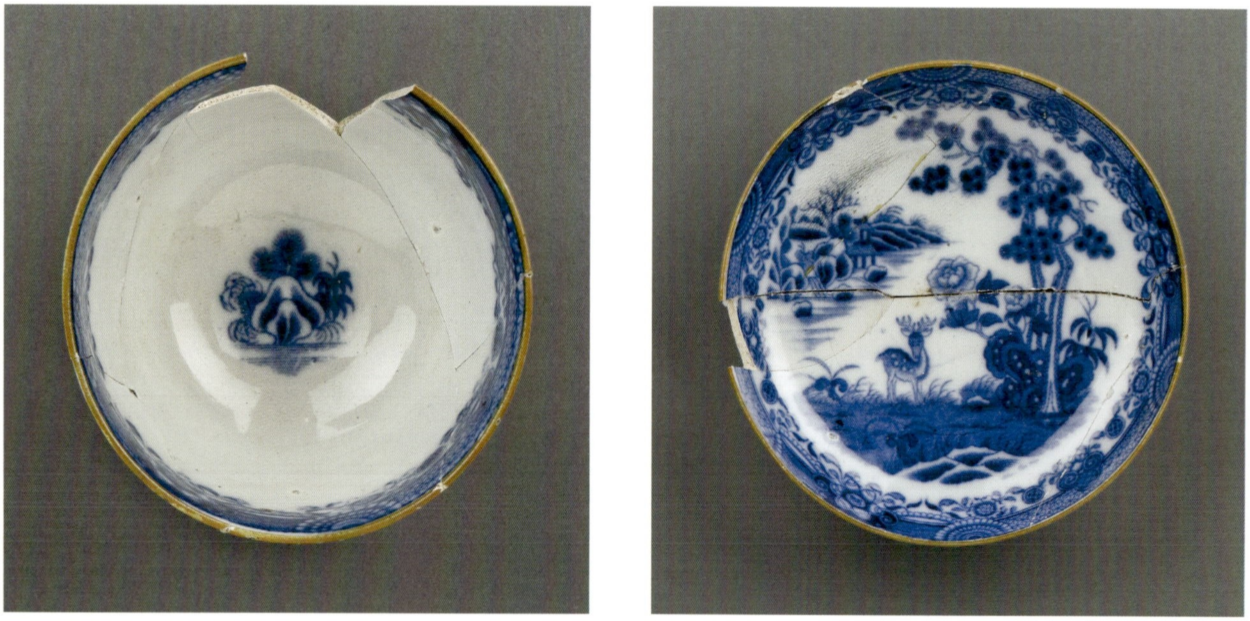

Figure 20 Teabowl and saucer, Staffordshire, England, 1810–?, recovered from the Phillips trash pit. Pearlware or China glaze. D. of teabowl 3½", D. of saucer 5¼".

The Chinese-style printed wares recovered from the trash pit all relate to tea service—bowls, saucers (fig. 20), and one teapot—and in fact represent the only pearlware tea service found in the trash pit.

Chinoiserie designs were most commonly produced between 1816 and 1836, whereupon figures in Western dress and Western architectural features began to appear.[31] Common decorative motifs found on printed

Chinese-style and chinoiserie earthenwares include pagodas, temples, weeping willow trees, cherry blossoms, orange trees, figures in Eastern dress, and junks and sampans.

One of the most interesting artifacts uncovered from the trash pit is a pearlware chamber pot with a medallion bearing a GR cipher for George III (fig. 21). English white salt-glaze washbasins and chamber pots were produced in large quantities until the 1760s, when imported German blue-and-gray stoneware began to infiltrate the American market.[32] In response to these new Westerwald wares, English salt-glaze potters created "scratch blue" chamber pots, mugs, and pitchers, which mimicked German-style wares.[33] The success of these English stoneware chamber pots led to the creation of pearlware copies; GR chamber pots, identical to the one from the Phillips trash pit, were made circa 1785–1810. Considering the intended contents of the vessel, the royal medallion was probably an expression of American patriotism, especially in light of the Revolutionary influence in Captain Phillips's life.[34]

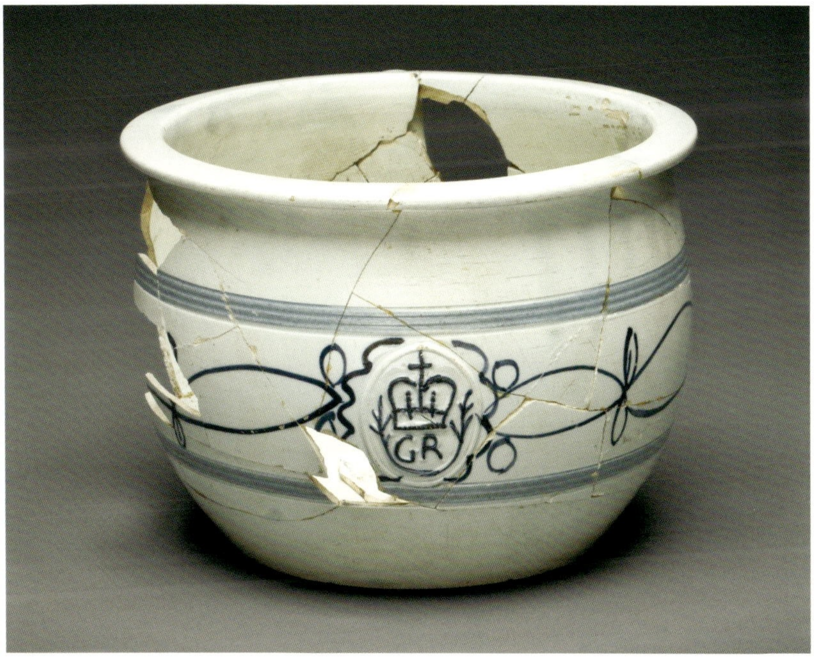

Figure 21 Chamber pot, possibly Staffordshire, England, 1785–1810, recovered from the Phillips trash pit. Pearlware in the scratch-blue style. H. 5½". Mark: "GR" on side.

The large deposits of pearlware found in the trash pit suggest that even though Chinese porcelain was sought after for its fine qualities, physical appearance, and representation of an international perspective (or an expression of the exotic), English earthenware (with forms that mimic Chinese wares) actually fulfilled a significant portion of that demand. Americans continued to look to England as well as France for guidance in fashion and refined taste even after the Revolutionary War and the War of 1812.[35] The presence of fashionable items in one's home, particularly those displaying scenes of faraway lands, conveyed messages about one's place in the world and one's knowledge of culture, history, and travel.[36]

Creamware

Ivor Noël Hume considers creamware to be "the most important development of the eighteenth century," because it was "a gradual perfection of a thin, hard-firing, pale-yellow or cream-colored earthenware, which after a preliminary firing, could be dipped in a clear glaze."[37] Seventeen percent of the ceramics recovered from the trash pit are creamware—209 sherds representing 55 vessels. Most of the uncovered forms (46 percent) are dinner plates, but also found were chamber pots, bowls, pitchers, basins, and a serving platter (fig. 22).

Figure 22 Types of creamware vessels recovered from the Phillips trash pit.

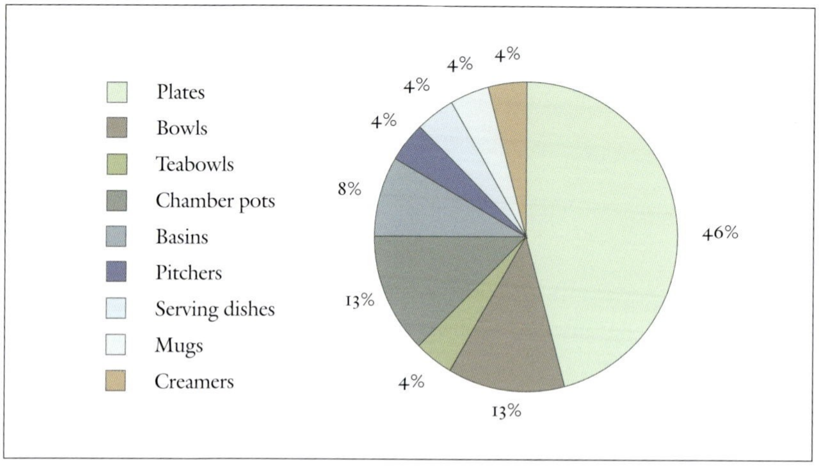

Figure 23 Plate, probably Staffordshire, England, 1790–1810, recovered from the Phillips trash pit. Creamware in the Flat Rim style. D. 8½".

Figure 24 Flat-rim creamware plate illustrated in Josiah Wedgwood's 1802 sale catalog. (Courtesy, Birmingham Museum of Art.)

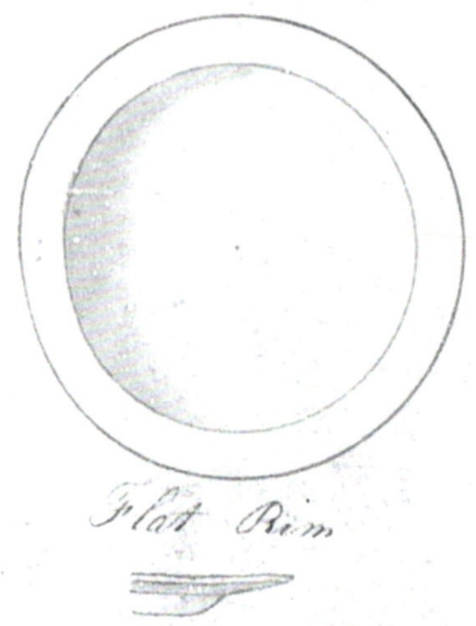

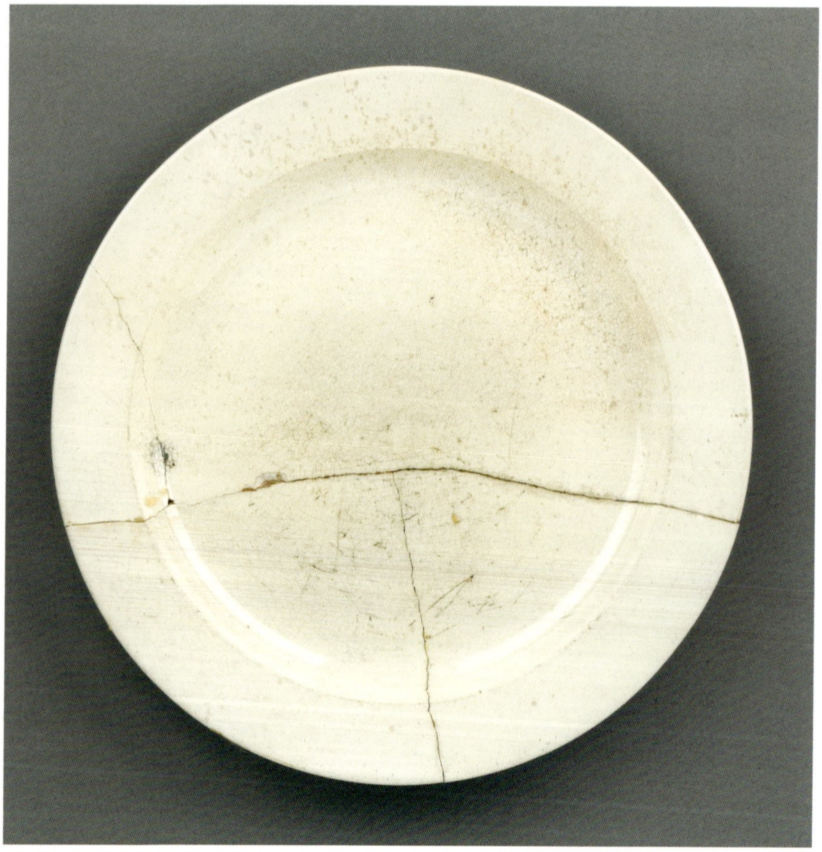

While various pattern books suggest that creamware plates were produced before 1783 with plain rims, they are generally found in American contexts of the 1790s and the early nineteenth century (figs. 23, 24). Although early creamware often has a deeper yellow color than later pieces (after 1785), this cannot be relied on for dating purposes: the trash pit yielded two plates decorated with the Royal Pattern (used as early as 1775) that had a light yellow glaze instead of the deeper yellow glaze commonly associated with the style (figs. 25, 26).

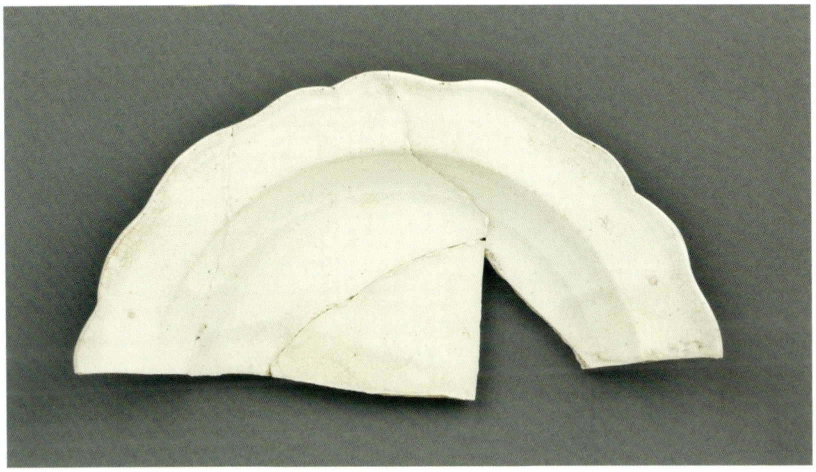

Figure 25 Plate, probably Staffordshire, England, 1790–1810, recovered from the Phillips trash pit. Creamware in the Royal style. D. 8".

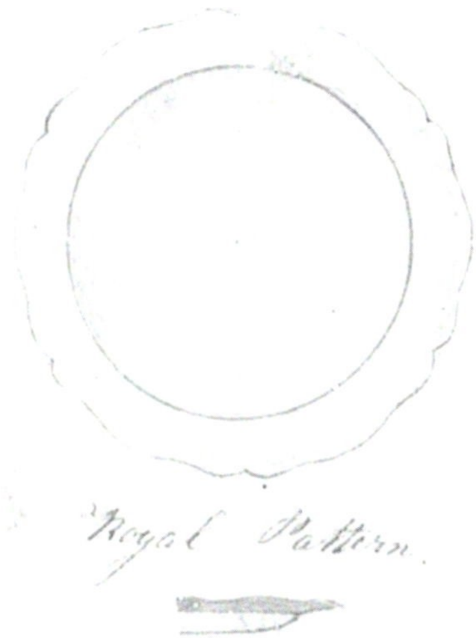

Figure 26 Royal Pattern creamware illustrated in Josiah Wedgwood's 1802 trade catalog. (Courtesy, Birmingham Museum of Art.)

The majority of the recovered creamware hollow ware, apart from one annular punch or slop bowl, is utilitarian and undecorated. Among these items are three chamber pots, which were constructed with simple, plain bulbous bodies, flat angular rims, and handles anchored at the top of the body just under the rim.[38] These pots are commonly associated with the last quarter of the eighteenth century, after which they were replaced by pearlware forms. The purely functional nature of the creamware vessel forms found in the trash pit reflects the decline of the ware's status in nineteenth-century America.

Glass

Perhaps the most intriguing find on the site is the fancy decorative glass found among the 133 sherds (71 vessels) of glass.[39] This glassware accounts for 20 percent of the glass discovered (fig. 27). There are seven identical chip-cut stemmed wineglasses with swagged and tasseled borders, three of which are almost complete vessels with bases. In the same style are a few straight-sided and barrel-shaped tumblers and one punch cup with a handle (fig. 28). This assemblage of decorative glassware is similar to patterns from a 1794 illustrated trade catalog of Bohemian glass distributor Johannes Schiefer (fig. 29), who operated an export and commission agency for trade to America, Russia, Spain, and other countries.[40] The catalog and price list were found among the possessions of John Gardiner of Gardiners Island, New York, but the catalog did not give manufacturer names.

Figure 27 Types of glass vessels recovered from the Phillips trash pit.

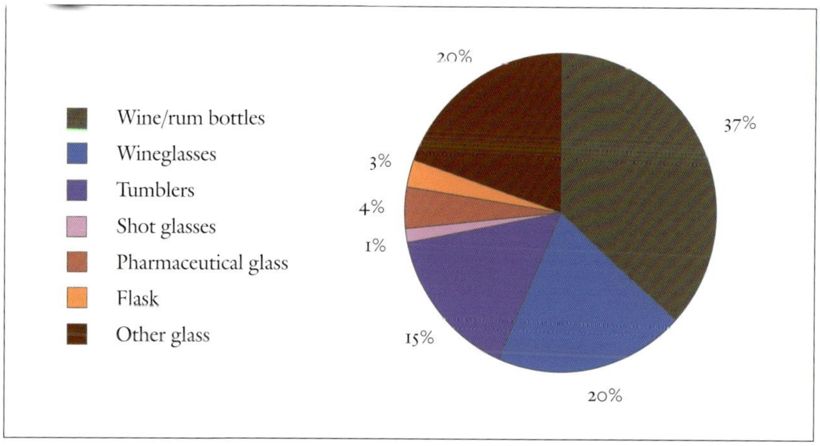

Figure 28 Wineglass, shot glass, and punch cup. Possibly Germany, 1790–?, recovered from the Phillips trash pit. Leaded glass. H. of wineglass 5½".

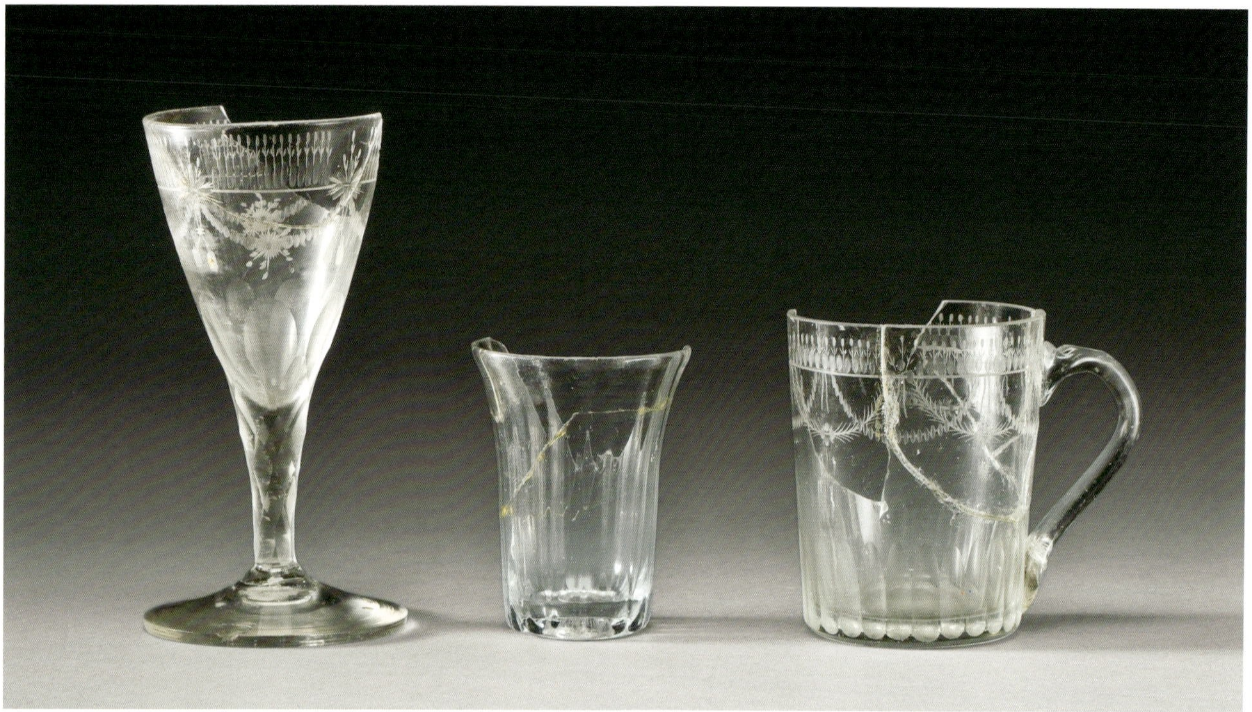

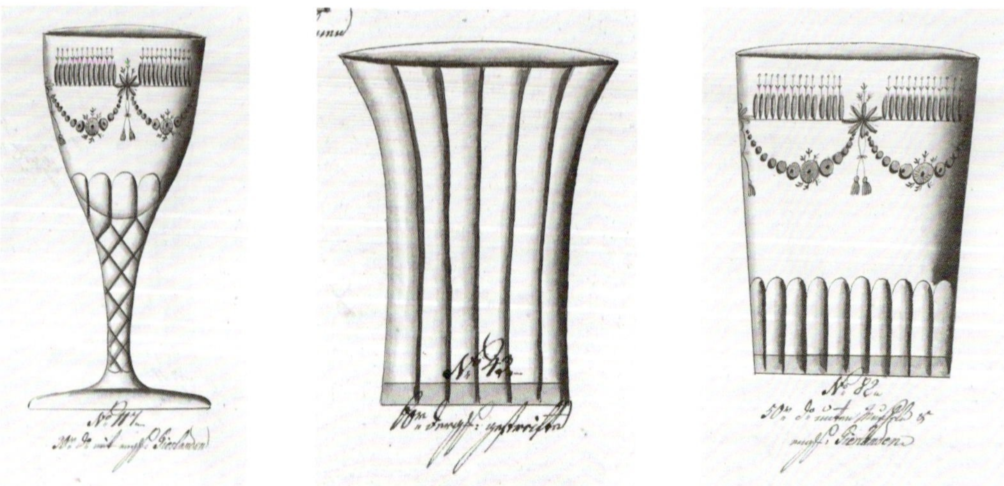

Figure 29 Glass forms illustrated in Johannes Schiefer, *Pattern Book for Glass,* ca. 1790–1825. (Courtesy, The Winterthur Library: Joseph Downs Collection of Manuscripts and Printed Ephemera.)

Vast amounts of glassware decorated with delicate engraving and facet cutting were imported from Europe to the United States after the Revolution. Most of it is recorded as "German," since it was made in glasshouses in Bohemia and other German states.[41] One might assume that this fine-quality glassware was quite costly, but in fact Bohemian glassware—made using a lightweight, nonleaded glass—was less expensive than English and Irish wares done in the same style. In 1813 a Philadelphia glass-and-china merchant shop advertised cut and engraved English wineglasses at $3.50 per dozen, whereas identical German models were only $2.00 per dozen.[42] The presence of such large numbers of Bohemian glass in the trash pit, as compared with English examples, suggests that Americans bypassed British wares in favor of the equally stylish yet significantly cheaper German glassware.

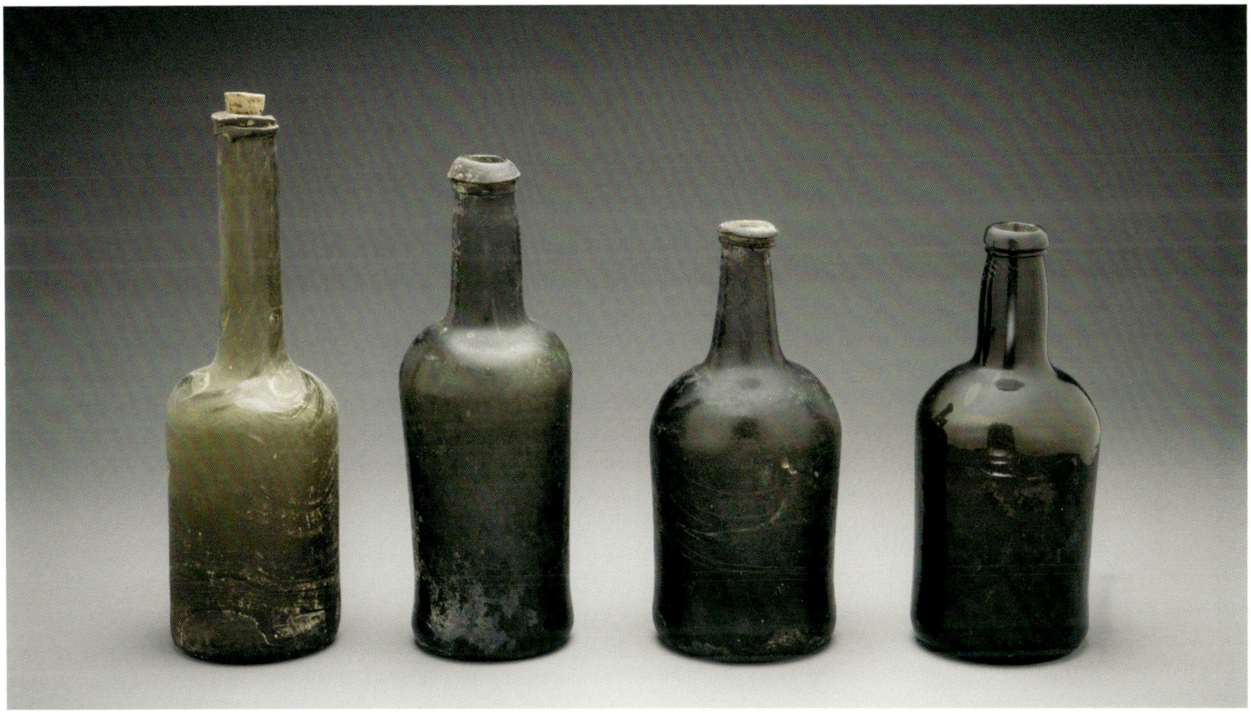

Figure 30 Wine bottles, France (*far left*) and England, 1780–1810, recovered from the Phillips trash pit. Blown glass. H. of tallest 13".

Most of the bottles found on New England historical sites are British-made dark green glass "wine" bottles, as were, but for one French bottle, those found in the Phillips trash pit. These vessels, similar to dated examples from the last quarter of the eighteenth century to the early nineteenth century, were used for shipping, storing, maturing, and serving a variety of liquids, such as ale, cider, distilled liquor, and wines (fig. 30).[43]

The large quantity of wine/rum bottles discovered in the Phillips trash pit (37 percent of the glass found) is in line with what is known about the popularity of liquor in Salem during its golden age. Small beer had been manufactured locally as early as 1637, and rum distilleries were established beginning with Emanuel Downing in 1648. By 1773 there were fifteen retailers in Salem selling spirits—St. Croix rum, Madeira wine, cherry brandy, and many other alcoholic beverages. The archaeological record and newspaper documentation reflect the role strong drinks played in daily food consumption. According to an address to the Essex County Prohibition Club in 1890 by historian Sidney Perley, "drunkenness was as common in the eighteenth as it had been in the seventeenth century. . . . After the revolution the production of liquor had greatly increased, and intemperance was more prevalent than ever."[44] Although it is known that Perley favored prohibition and might have been prone to exaggeration, the presence of twenty-six wine/rum bottles in the Phillips trash pit is certainly good evidence for his claims.

Coarse Earthenware

Thirteen percent of the ceramics found in the Phillips trash pit were coarse earthenware vessels—106 sherds (40 vessels). These locally made vessels are abundant on many seventeenth-and eighteenth-century sites in New England, often accounting for more than 85 percent of the total ceramic assemblage.[45] Approximately 175 potters lived and worked in Massachusetts between 1650 and 1798, since "the excellence and abundance of clay, the proximity of home markets, and the ready means for shipping the wares encouraged many craftsmen to enter the trade."[46] At the Salem Village Parsonage site (1681–1784) in Danvers, wheel-thrown plain earthenwares make up 89.6 percent of the total ceramic assemblage.[47] The majority of domestically produced redware found in colonial assemblages represents a relatively inexpensive commodity due to the ease in producing these wares and their frequent and hard use for dairy, kitchen, and food preparation.[48]

The potting industry reached its peak between 1775 and 1800. In 1775 seventy-five potteries operated in Danvers and Peabody, but by 1798 there were only twenty.[49] This decline in redware production, illustrated in the ceramic assemblage of the Phillips trash pit as well, is a direct result of several factors. The flood of European and Chinese ceramics into Salem forced the remaining redware potters to streamline their industry and abandon the manufacture of tablewares. Moreover, potters in eighteenth-century Massachusetts did not have large quantities of the raw materials—fine kaolin clays, for example—needed to produce the buff-bodied wares that had become popular, nor did they have "the sociopolitical organization, the economic resources, nor the authority for . . . importation."[50] After the

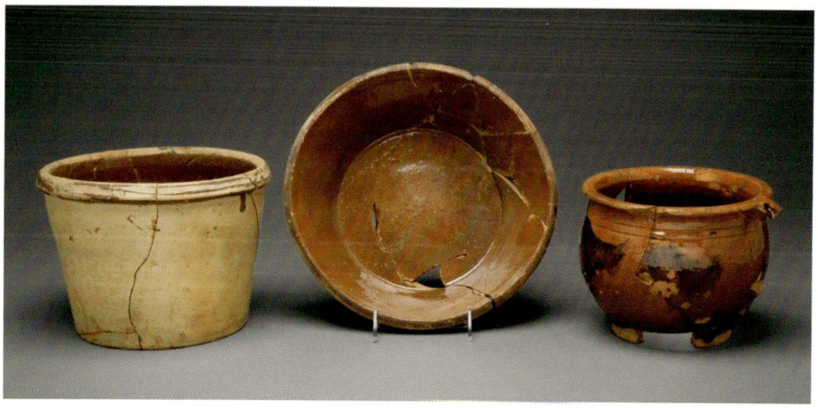

Figure 31 Storage jar, milk pan, and chamber pot, New England, 1800–1820, recovered from the Phillips trash pit. Lead-glazed earthenware. D. of milk pan 6".

War of 1812, approximately thirty earthenware potters remained in South Danvers, attesting to the lessening demand for these wares in comparison with refined English earthenwares.[51]

In 1810 Tench Coxe, a proponent of American-based potteries, noted, "the manufactory of ordinary ware of common potters' clay is very much extended in the United States. It is of great use in dairies, kitchens, larders, store-rooms, sale stores and manufactories."[52] Coxe's statement is supported by the archaeological assemblage from the trash pit, since the redware forms (except for three Buckley ware pitchers) are strictly utilitarian, used for storage, dairy production, and chamber pots (fig. 31).

Utilitarian Stoneware

For much of the eighteenth century British brown stonewares and German Westerwald blue-and-gray jugs, cylindrical tankards, and chamber pots were exported in very large quantities to America. Compared with other types recovered from the trash pit, the number of utilitarian stoneware vessels found (56 sherds, 26 vessels) is quite small, accounting for only 6 percent of the ceramics (fig. 32), and most is in the form of gin or mineral water bottles (fig. 32, left).

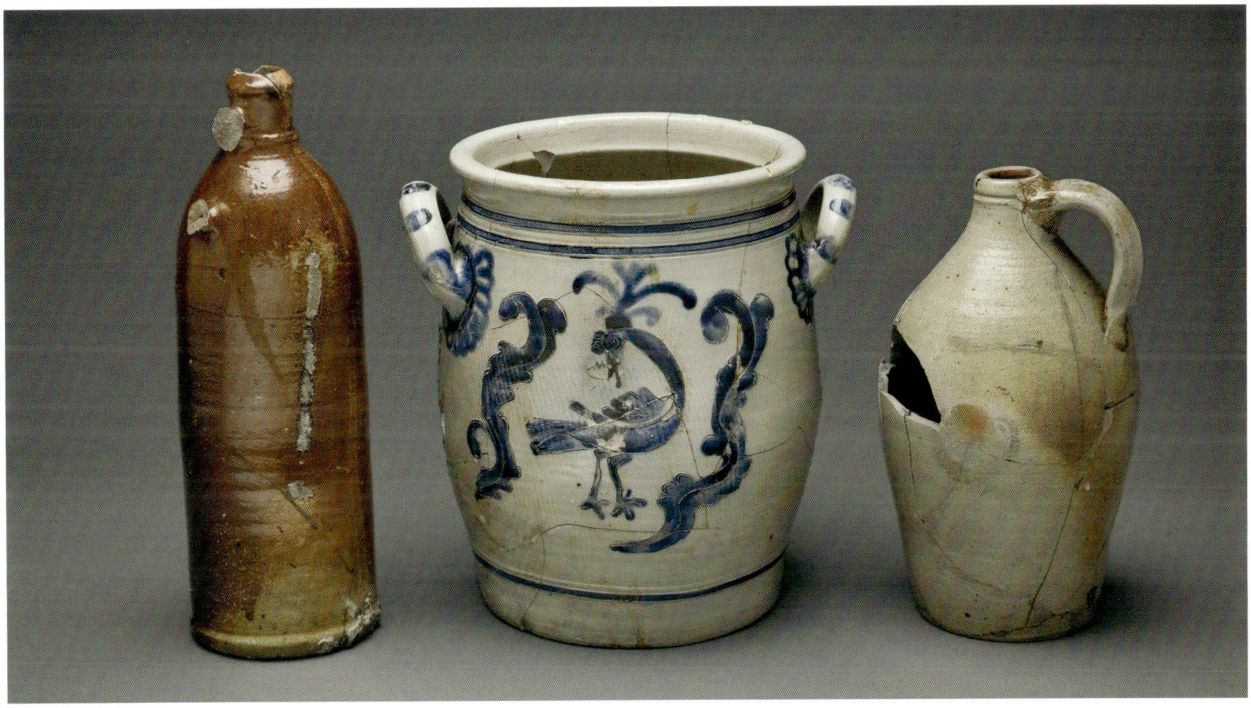

Figure 32 Gin/mineral water bottle (*left*), Germany, 1750–1820; Westerwald-style storage jar (*center*), Germany, 1780–1820; Jug (*right*), Connecticut, 1800–1820, all recovered from the Phillips trash pit. Salt-glazed stoneware. H. of storage jar 14¼".

Double-handled storage jars, made with horizontal rolled handles that were decorated with cobalt and stamped floral devices, were exported in small quantities to the American colonies in the second and third quarters of the eighteenth century. The most intact piece of stoneware recovered from the trash pit is an example, with handles, incised blue decoration, and a salt glaze (figs. 32 [center], 33). At first glance the somewhat crude execution of the jar made it a candidate for an American example made in kilns

Figure 33 Detail of the inscribed mark on the body of the storage jar illustrated in fig. 32.

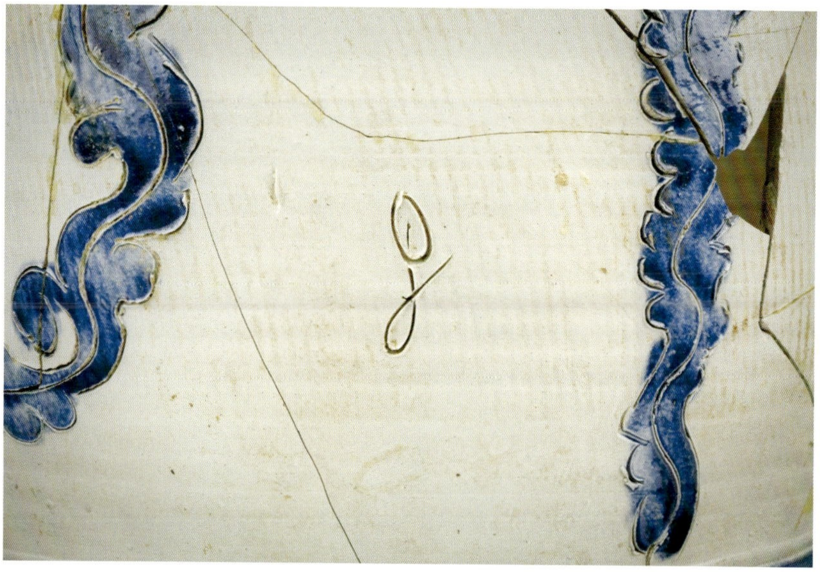

in Manhattan or New Jersey.[53] While the vessel shares many characteristics with early American examples, the presence of shallow chatter marks—regularly spaced cordations in the surface of the clay resulting from the use of a template blade during the forming of the pot—has been seen only on Westerwald products.[54] Regardless of its German origin, this type of Westerwald is rarely seen in American archaeological contexts.

Discussion

The plethora and variety of ceramics and glass found in the Phillips trash pit are not surprising in the context of Salem during its golden age, a period in which increased global mercantile activity brought many worldly goods to the city. From 1785 to 1807, 1,500 crates of Chinese porcelain were imported into Salem (with one-half then transported to other ports); 1,364 crates of European earthenware (84 percent of it creamware and pearlware) were imported into both Salem and Marblehead.[55] This ratio, very similar to that of the porcelain as compared with the creamware and pearlware in the Phillips trash pit, is a good example of archaeology reflecting history (fig. 34). It shows how vast the ceramic market of Salem was, a direct effect

Figure 34 Ceramic types recovered from the Phillips trash pit.

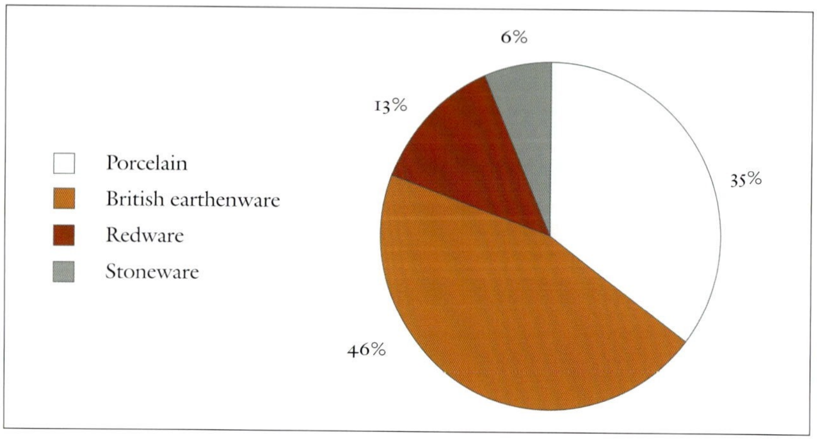

of Salem's establishment as a cosmopolitan city after the Revolutionary War. For this reason we find a multitude of porcelain, pearlware, creamware, other ceramic types, and glassware in the inventories of homes, the probate records of merchants, and now, for the first time, in the archaeological assemblage of a trash pit found on the property of a successful Salem merchant.

This picture becomes clearer in an examination of the numerous local advertisements published in Salem papers from the 1780s to the 1830s (fig. 35), which shows that Salem residents had access to a wide array of

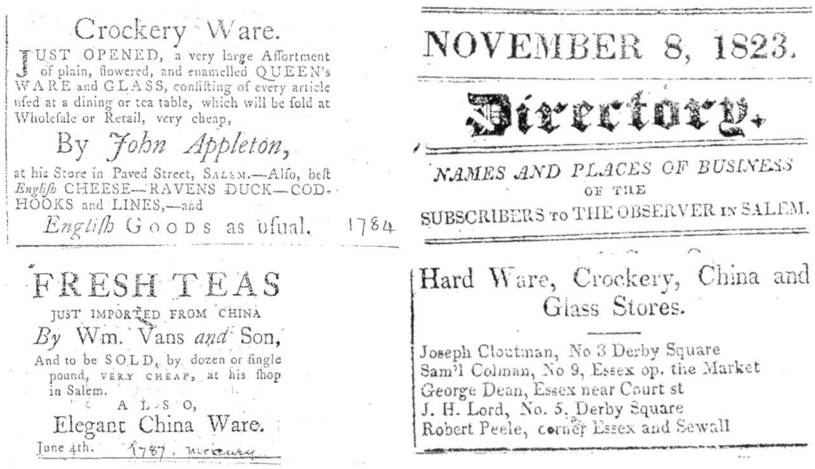

Figure 35 Advertisements and listings published in late eighteenth- and early nineteenth-century Salem newspapers.

ceramics and glassware. In 1784 one merchant advertised in the *Salem Register* "a very large Assortment of plain, flowered, and enamelled QUEEN's WARE and GLASS, consisting of every article used at a dining or tea table, which will be sold at Wholesale or Retail, very cheap, By *John Appleton*, at his Store. . . ." On June 4, 1787, about one year after the return of the *Grand Turk*, the first Salem ship to sail to China, William Vans and Son advertised "fresh" China teas for sale and "Elegant China Ware." Even in 1823, by which time the port of Salem had started to decline as a center of overseas trade, the *Observer*'s published directory of the names and businesses of its subscribers included numerous "Hard Ware, Crockery, China and Glass Stores."

Tea drinking became normative behavior in America after the 1750s, by which time one-third of the population "at a moderate consumption, [drank] tea twice a day."[56] A by-product of increased mercantilism and the controlled manners of the merchant elite, tea consumption was closely tied to the ceramic trade. With this in mind, it is not surprising to find large numbers of teabowls, saucers, and teapots in the Phillips trash pit. According to archaeologist Lorinda Goodwin, tea services became a required necessity in connection with the social act of drinking, and the presence of multiple services in the Phillips trash pit suggests that the family entertained a larger social circle of friends and business partners.[57] Although most of the teaware from the trash pit is Chinese porcelain, there is a fair amount of pearlware, indicating that although Chinese porcelain was considered the most elegant—and therefore preferred—service for tea, a family of some significant social status relied on earthenware equivalents as acceptable substitutes.

Conclusion

Jonathan Goldstein, an economic historian who wrote about the Philadelphia China trade, argued that porcelain was "the single largest traditional Chinese decorative item which the average American could afford."[58] From the data already discussed—its prevalence in the Salem market, the relatively inexpensive price for blue-and-white ware, and its presence in the trash pit of a merchant family—the statement appears to be accurate. After the Revolution, Salem merchants imported more porcelain than they re-exported, pointing to an increase in consumption by local residents.[59] While the elite Salemites owned many porcelain items, it is not unreasonable to think working- or lower-middle-class families might have owned a set or two of tea service. Archaeological evidence of their homes during Salem's golden age is required to confirm this claim.

The Phillips family's story represents a micro-historical study of Salem during its golden age. What began as one man's curiosity has produced a significant contribution to Salem's archaeological record, filling a major hole in the archaeological excavations of households that have been undertaken to date. The collection of artifacts from the Captain Stephen Phillips house is important for reassessing Salem's past and offers a baseline for future archaeological and historical endeavors. It also provides an indication of the taste preferences of Salem residents, specifically the merchant class. Like Captain Phillips, Salem was a "go straight ahead"[60] city after the Revolutionary War. The city's merchant class created global trading networks on their own, establishing direct routes between Salem and the Far East. These lines of exchange brought significant wealth to the city, making it the sixth largest port in America by 1790, and allowed its citizens access to worldly goods. As a result, a larger percentage of Salem residents were able to live a cosmopolitan lifestyle, having access to spices for meals, being able to have tea in Chinese porcelain and English services, and drinking spirits from glassware made in distant lands. The archaeological record tells this story through a Salem sea captain who rose through the ranks to become a successful merchant, who participated in the importation of foreign goods by captaining some of the first ships to venture into Far Eastern ports, and who was a central participant in the city's flourishing economy.

ACKNOWLEDGMENTS This project would not have a life had local artist Racket Shreve not discovered the artifacts at his home on Chestnut Street in the mid-1970s. He provided immense enthusiasm, support, and interest in my work. I would like to thank the many staff members at the Peabody Essex Museum for their help along the way: Dr. Daniel Finamore, Russell W. Knight Curator of Maritime Art and History, for directing me to the collection of artifacts from 17 Chestnut Street and for his editorial comments; William Sargent and Karina Corrigan, H. A. Crosby Forbes Curator of Asian Export Art, for their help in the identification of

Chinese export porcelain; and Dr. Jeffrey P. Brain, for his help in analyzing stoneware and glass material from the trash pit. I am grateful for the background on several members of the Phillips family provided by Margherita Desy, historian at the Naval History & Heritage Command Detachment in Boston, and for the archaeological insight of Professors Stephen Mrozowski, Stephen Silliman, and David Landon of the University of Massachusetts, Boston. I thank Rob Hunter for his continued interest in this piece.

1. Geoffrey P. Moran, Edward F. Zimmer, and Anne E. Yentsch, *Archaeological Investigations at the Narbonne House: Salem Maritime National Historic Site, Massachusetts*, Cultural Resources Management Study No. 6 (Boston: National Park Service, 1982).

2. Roger D. Howlett, "In Praise of Trash," in *Hamilton Hall Antiques Show* (Salem, Mass.: Peabody Museum of Salem, 1975).

3. Salem's golden age is considered to be a period following the Revolutionary War to about the 1820s, when Salem ships were trading directly with the Far East.

4. Since a trained archaeologist did not oversee the excavation, stratigraphy was not recorded, nor were soil changes and levels noted. Without these forms of data, no depositional matrix could be constructed, and artifacts from different depositional layers could not be discerned as homogeneous or not. The loss of these methods to place the artifacts in specific contexts is disappointing but did not significantly detract from the overall analysis of the site. The artifacts were still contextualized within a structure that was used specifically for dumping refuse items, which to some degree offsets the lack of stratigraphic data.

5. George Schwartz, "'Go Straight Ahead': Salem's Rise to Global Entrepôt as Reflected in the Archaeological Assemblage from the Captain Stephen Phillips House" (master's thesis, University of Massachusetts, Boston, 2005).

6. His father, Deacon Stephen Phillips, was an integral part of the growing patriotic movement throughout Marblehead. In 1770 he was voted moderator of the town meeting, one of the most powerful and influential political positions in the town and one he held throughout most of the Revolutionary period. Leading meetings at which many patriotic measures foreshadowing the coming revolution were passed or discussed, he was able to steer local legislation and appoint committees. It is against this backdrop of revolution that Captain Stephen Phillips was educated. For more on the life of Captain Phillips, see ibid., pp. 21–33.

7. Richard Hall Wiswall, "Notes on the Building of Chestnut Street," *Essex Institute Historical Collections* 75 (1939): 206.

8. The construction of the barn sealed off the trash pit.

9. Stephen A. Mrozowski, "Prospects and Perspectives on an Archaeology of the Household," *Man in the Northeast* 27 (1984): 41.

10. Minimal amounts of earlier tin-glaze wares and later whitewares were found in the trash pit.

11. Ivor Noël Hume, *A Guide to Artifacts of Colonial America* (New York: Alfred A. Knopf, 1969), p. 256.

12. H. A. Crosby Forbes, *Hills and Streams: Landscape Decoration on Chinese Export Blue and White Porcelain*, exh. cat. (Washington, D.C.: International Exhibitions Foundation, 1982), p. 6.

13. Jean McClure Mudge, *Chinese Export Porcelain for the American Trade, 1785–1835* (Newark: University of Delaware Press, 1962), p. 70.

14. Ibid.

15. Ibid.

16. Five Fitzhugh pattern plates were also in the trash pit.

17. Ibid., p. 71.

18. Variation in the quality of Canton ware is not a direct correlation to the time in which it was produced. It is often said, however, that after the first quarter of the nineteenth century the quality of Canton deteriorated.

19. Quoted in Mudge, *Chinese Export Porcelain for the American Trade*, p. 75.

20. Personal communication with William Sargent.

21. *Salem: Maritime Salem in the Age of Sail*, National Park Service Handbook 126 (Washington, D.C.: National Park Service, 1987).

22. Lynne Sussman, "Changes in Pearlware Dinnerware, 1780–1830," *Historical Archaeology* 11 (1977): 111.

23. Ibid., p. 107. George L. Miller and Robert R. Hunter Jr., "English Shell-Edged Earthenware: Alias Leeds Ware, Alias Feather Edge," in *The Consumer Revolution in 18th-Century English Pottery*, Proceedings of the Wedgwood International Seminar, no. 35 (n.p.: Wedgwood International Seminar, 1990), pp. 107–36; and Robert H. Hunter Jr. and George L. Miller, "English Shell-Edged Earthenware," *The Magazine Antiques* 145 (March 1994): 432–43.

24. Patricia M. Samford, "Response to a Market: Dating English Underglaze Transfer-Printed Wares," *Historical Archaeology* 31, no. 2 (1997): 1.

25. George L. Miller, Ann Smart Martin, and Nancy S. Dickinson, "Changing Consumption Patterns: English Ceramics and the American Market from 1770 to 1840," in *Everyday Life in the Early Republic*, edited by Catherine E. Hutchins (Winterthur, Del.: Henry Francis du Pont Winterthur Museum, 1994), pp. 234–38.

26. Ibid., p. 234.

27. Samford, "Response to a Market," p. 3.

28. Miller, Martin, and Dickinson, "Changing Consumption Patterns," p. 234.

29. Samford, "Response to a Market," p. 8.

30. George L. Miller and Robert R. Hunter, "How Creamware Got the Blues: The Origins of China Glaze and Pearlware," in *Ceramics in America*, edited by Hunter (Hanover, N.H.: University Press of New England for the Chipstone Foundation, 2001), pp. 135–61.

31. Only a few of the sherds recovered had these types of designs. Samford, "Response to a Market," p. 1.

32. Hume, *Guide to Artifacts of Colonial America*, p. 118.

33. Ibid.

34. James Duncan Phillips, "Captain Stephen Phillips, 1764–1838," *Essex Institute Historical Collections* 76 (1940): 97–135. It is against this backdrop of revolution that Captain Stephen Phillips was educated. His adolescence was heavily influenced by the continual warfare he witnessed in the North Shore beginning at the age of eleven. At this point in his life, Colonel Leslie's regiment had landed in Marblehead for a raid on Salem. Family legend has it that young Stephen, at the age of thirteen, offered his services to John Paul Jones for the *Ranger*'s voyage to France in October 1777. Phillips was turned away at the last moment because he was considered too young, an experience that he recounted later in his life to his grandson Stephen Henry Phillips as "a great disappointment in my life."

35. Samford, "Response to a Market," p. 7.

36. Ibid.

37. Hume, *Guide to Artifacts of Colonial America*, p. 123.

38. Ibid., p. 148.

39. Examples of less expensive glasses of about the same period, blown but uncut and undecorated, include two straight-sided tumblers and the shot or cordial glass with blown paneled sides.

40. Arlene M. Palmer, *Glass in Early America: Selections from the Henry Francis du Pont Winterthur Museum* (Winterthur, Del.: Henry Francis du Pont Winterthur Museum, 1993), p. 72.

41. Ibid.

42. Ibid.

43. Olive R. Jones, *Cylindrical English Wine and Beer Bottles, 1735–1850* (Ottawa, Ont.: National Historic Parks and Sites Branch, 1986), p. 17.

44. An address delivered by Sidney Perley at the quarterly meeting of the Essex County Prohibition Club at Topsfield, April 16, 1890, published in the *Salem Gazette*, April 22, 1890.

45. Sarah Peabody Turnbaugh, "17th and 18th Century Lead-Glazed Redwares in the Massachusetts Bay Colony," *Historical Archaeology* 17, no. 1 (1983): 3.

46. Ibid., pp. 4, 13.

47. Ibid., p. 3.

48. Ibid., p. 4.

49. Jessica Lanier, "The Post-Revolutionary Ceramics Trade in Salem, Massachusetts, 1783–1812" (master's thesis, Bard Graduate Center, New York, 2004), p. 191.

50. Turnbaugh, "17th and 18th Century Lead-Glazed Redwares in the Massachusetts Bay Colony," p. 14.

51. Lanier, "Post-Revolutionary Ceramics Trade in Salem," p. 203.

52. Quoted in ibid., p. 202.

53. Arthur F. Goldberg, Peter Warwick, and Leslie Warwick, "The Eighteenth-Century New Jersey Stoneware of Captain James Morgan and the Kemple Family," in *Ceramics in America*, edited by Robert Hunter (Easthampton, Mass.: Antique Collectors' Club, 2008), pp. 2–40; and Meta F. Janowitz, "New York City Stonewares from the African Burial Ground," in ibid., pp. 41–66.

54. Ivor Noël Hume, *If These Pots Could Talk: Collecting 2,000 Years of British Household Pottery* (Milwaukee, Wis.: Chipstone Foundation, 2001), pp. 106–7.

55. Lanier, "Post-Revolutionary Ceramics Trade in Salem," p. 177.

56. Lorinda B. R. Goodwin, *An Archaeology of Manners: The Polite World of the Merchant Elite of Colonial Massachusetts* (New York: Kluwer Academic/Plenum Publishers, 1999), p. 31.

57. Ibid., p. 123.

58. Jonathan Goldstein, *Philadelphia and the China Trade, 1682–1846: Commercial, Cultural, and Attitudinal Effects* (University Park: Pennsylvania State University Press, 1978), pp. 38–39.

59. Lanier, "Post-Revolutionary Ceramics Trade in Salem," p. 212.

60. The expression "go straight ahead" was used by Captain Phillips to describe a progressive person who is continually moving forward, striving for better things for himself and the society in which he lives while not being held down by the mentalities and biases of the past. The captain's son, Stephen Clarendon, refers to his father's maxim in a letter to his own son Stephen Henry a few days after the captain's death in 1838; Phillips House collection, Salem, Massachusetts.

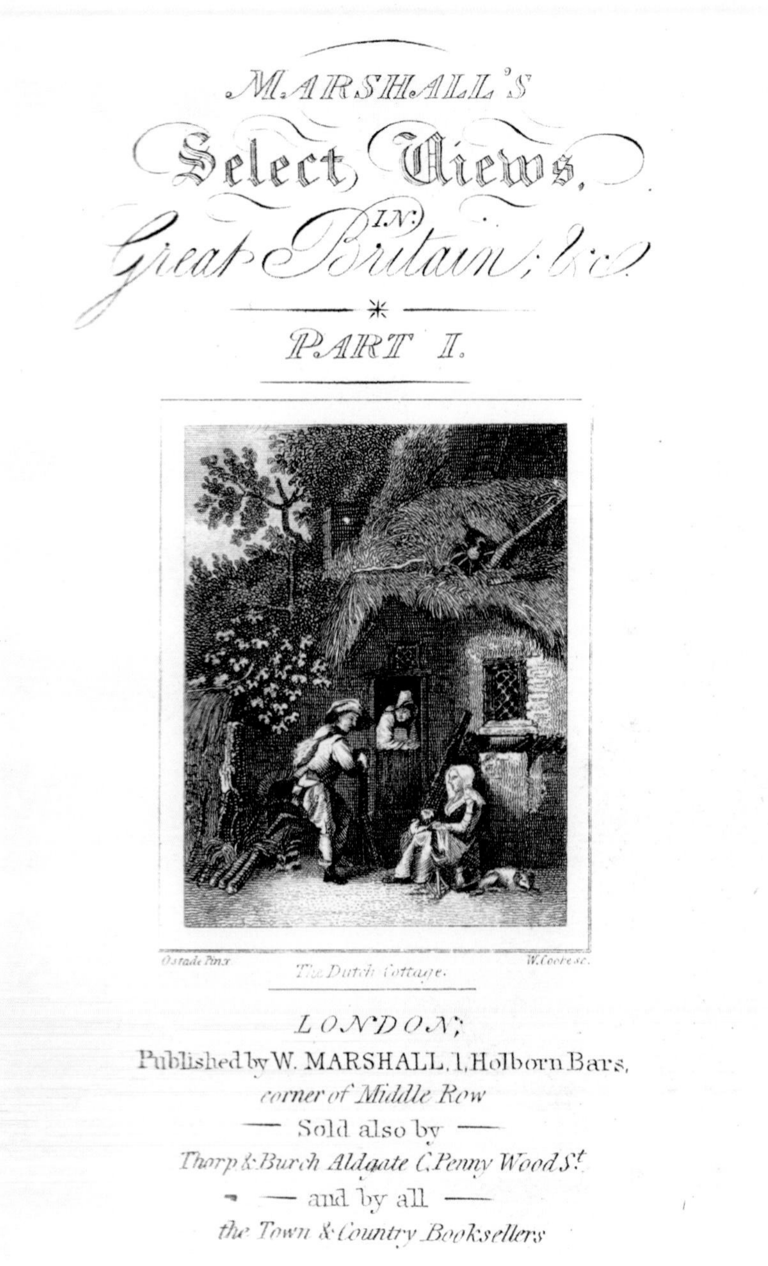

Figure 1 Title page, *Marshall's Select Views, in: Great Britain; &c., Part I* (London: W. Marshall, 1825).

Dick Henrywood

The States Border Series by Ralph and James Clews

▼ IN THE NINETEENTH century the Staffordshire ceramic industry developed a huge trade to North America. Some potters seem to have concentrated on the American market—the Jackson Brothers, for example, whose wares are rarely found in England—and tailored their products specifically for American tastes. Exports included pots of every description, but one staple was transfer-printed ware, often referred to as transferware. In the 1820s printed views were produced in dark blue (a shade that found little favor in England), and from about 1830 other colors predominated—red (or pink), purple, green, brown, black, and occasionally even yellow, all colors not common in England.

Several firms came to dominate the export market for transferware, and major names included William Adams of Stoke, Ralph Hall of Tunstall, Enoch Wood & Sons of Burslem, the Stevensons of Cobridge, and Ralph and James Clews, also of Cobridge. The Clews brothers produced a very wide range of patterns, but of specific interest here is a rather enigmatic series of untitled views known as the States Border series, sometimes alternatively called the America and Independence series. The central views, which are relatively small and placed within a scroll frame, are flanked by two female figures on plinths labeled, respectively, "AMERICA AND" and "INDEPENDENCE." The first figure, with a square and compasses on her apron, wears a blindfold on which is written "JUSTICE" and holds an oval portrait of George Washington. The second figure, who is kneeling, holds a staff topped with a cap on which is written "LIBERTY." Surrounding the central view and the figures of America and Independence is decorative floral work bordered by scalloped panels bearing the names of, in most examples, fifteen American states (New York, Delaware, Virginia, New Hampshire, Massachusetts, Rhode Island, Connecticut, North Carolina, South Carolina, Georgia, Vermont, Kentucky, New Jersey, Pennsylvania, and Maryland), much in the style of the patriotic Chain of States motif. Some jugs include Ohio and/or Tennessee. Most are marked, usually with an impressed "CLEWS WARRANTED STAFFORDSHIRE" circling a crown; less common is an impressed crown with "CLEWS / WARRANTED / STAFFORDSHIRE" curving beneath.

The States Border series is found on a wide variety of vessel forms, all printed in dark blue for export to America, where, by historical Staffordshire terms, they are relatively common. A single exception to printing in blue is a green shell-edged dish illustrated by David and Linda Arman in their *Anglo-American Ceramics, Part 1*.[1] The center of their dish, which is decorated in gray and black with just a vignette view and the two flanking figures, is hand-painted rather than transfer-printed. It is not known whether the dish is marked, and speculation as to its significance is not appropriate here.

The source for all except one of the central views can now be revealed as *Marshall's Select Views, in: Great Britain; &c.*, published by William Marshall in London from 1825 to 1828 (fig. 1). Small sections of this serial work were issued monthly for four years, then assembled into a bound volume at the end of each year. Each section contained two full-page engravings and four smaller engravings on two pages, all accompanied by descriptive text. Although the page size was very small—about 4½ by 2¾ inches—Marshall, who was not averse to promotional hyperbole, advertised in 1826:

> This Work (the Plates alone of the Two Volumes published having cost One Thousand Guineas) may be deemed the most complete and cheapest Topographical History of remarkable Places and beautiful Seats in the Kingdom. . . . Collectors of Fine Engravings will find Marshall's Prices for Prints considerably cheaper than any other House in the Kingdom.[2]

Despite these claims, the engravings are not of the highest quality and, because of their small size, do not compare well against competitors' products.

Most of the engravings in these volumes are anonymous, but a few of the larger ones bear the name of the artist and/or the engraver. It must be pointed out that Marshall's prints are often surprisingly similar to views found in other, earlier contemporary publications, notably John Preston Neale's *Views of the Seats* (1818–1829), William Angus's *The Seats of the Nobility and Gentry in Great Britain and Wales* (1787, 1815), and F. W. L. Stockdale's *Excursions in the County of Cornwall* (1824). For their States Border series, the Clews brothers used half-page engravings with one exception, a full-size example titled *Town Hall and Quay, Yarmouth* (see fig. 10), which was drawn by John Preston Neale and engraved by Samuel Lacey.

An issue date in mid-1826 is suggested by the fact that all of the views are copied from the first fifteen issues of Marshall's publication, which were released from 1825 to the first three months of 1826. As 1826 was the fiftieth anniversary of the signing of the Declaration of Independence in 1776, it is reasonable to surmise that the series was created to commemorate that occasion—hence the prominence of the emblematic figures America and Independence. One has to wonder, though, why the Clews brothers used British views for the series; possibly they were under pressure to manufacture and ship the wares and had no suitable American views at hand. This would likely explain the small size and lack of titles, since they probably would not have wanted to draw attention to the fact that not only were these not American views, they were, in fact, English prints.

As of this writing, the States Border series comprises fourteen different views, all but one of which are copied from Marshall's prints. The scenes were comprehensively cataloged (but not identified) by David and Linda Arman in 1977; now their information can be paired with the identified views, as shown in Table 1.[3]

As with all the best puzzles, we are left with one enigma. It is hoped that the source print for the sole unidentified view will eventually come to light.

Table 1. Connecting the Armans' descriptive title to the source print of the States Border series.

ARMANS' DESCRIPTIVE TITLE	SOURCE PRINT
"Two Storey Building with Curved Drive"	*Bear Forest, Ireland*

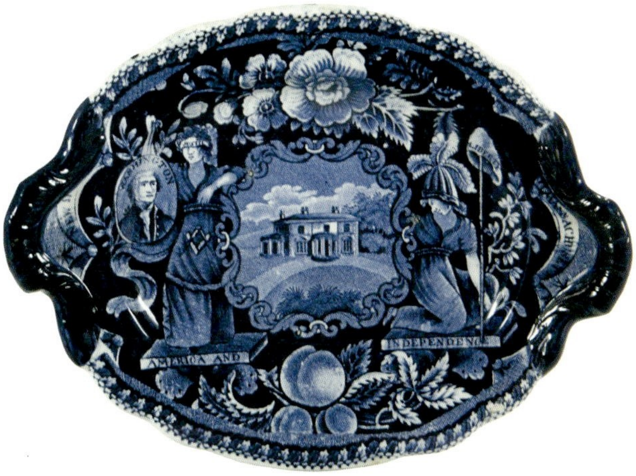

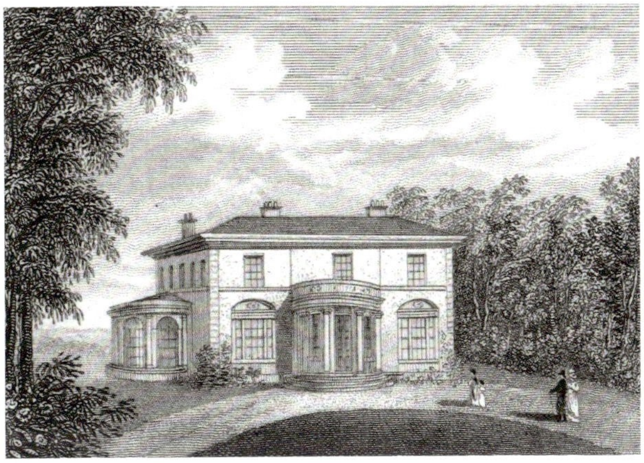

Figure 2 *Left*: Sauce tureen stand, Ralph and James Clews, Cobridge, England, 1826–1834. Lead-glazed earthenware L. 8⅛". *Right*: *Bear Forest, Ireland*, copperplate engraving.

This scene is also found on plates, a mug, and a cup plate.

"Mansion, Winding Drive"	*Belton House, Lincolnshire*

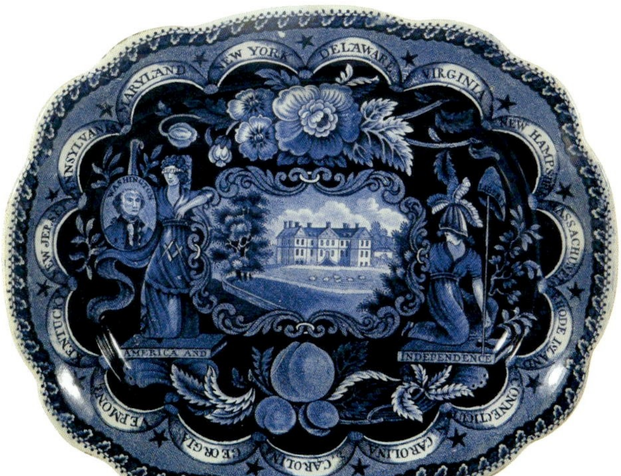

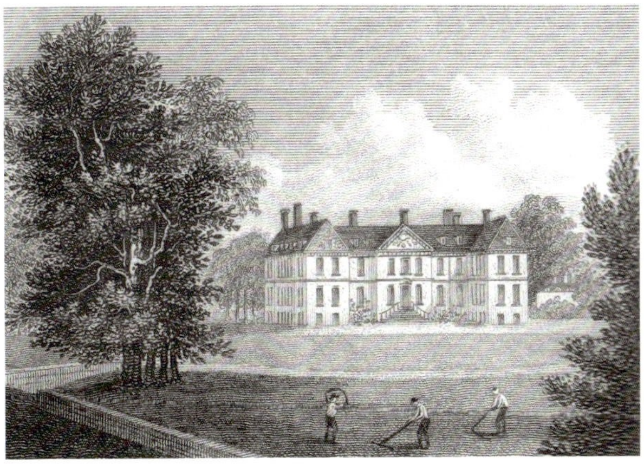

Figure 3 *Left*: Dish, Ralph and James Clews, Cobridge, England, 1826–1834. Lead-glazed earthenware L. 9⅝". *Right*: *Belton House, Lincolnshire*, copperplate engraving.

This scene has also been found on deep dishes, a jug, and a mug.

ARMANS' DESCRIPTIVE TITLE	SOURCE PRINT
"Building, Fishermen with Net"	*Bishton Hall, Staffordshire*

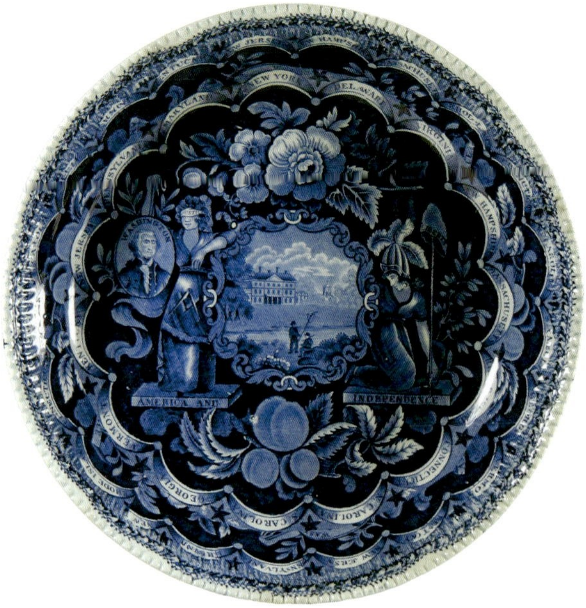

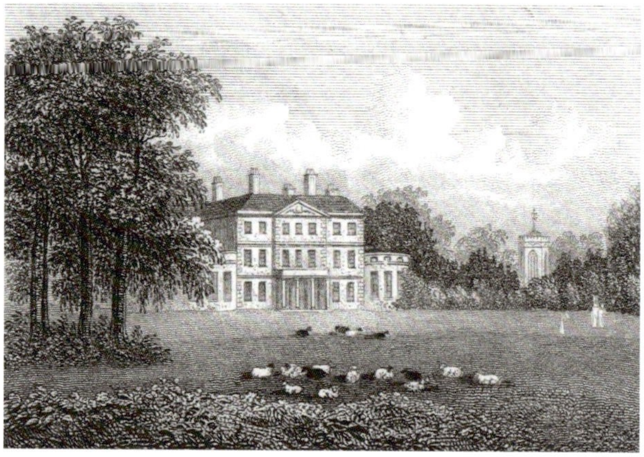

Figure 4 *Left*: Bowl, Ralph and James Clews, Cobridge, England, 1826–1834. Lead-glazed earthenware. D. 12½". *Right*: *Bishton Hall, Staffordshire*, copperplate engraving.

This scene is also found on plates.

"Building in Distance, Women in Foreground"	*Burley, Rutlandshire*

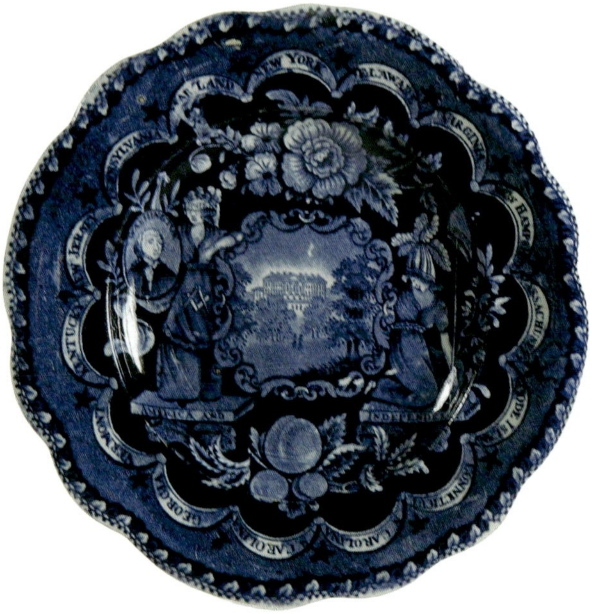

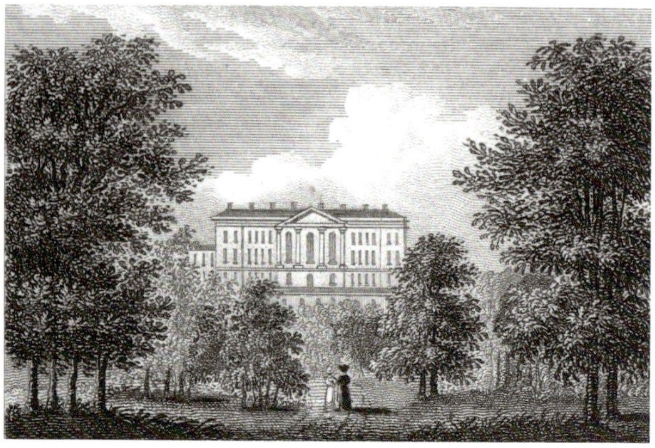

Figure 5 *Left*: Plate, Ralph and James Clews, Cobridge, England, 1826–1834. Lead-glazed earthenware. D. 5⅝". *Right*: *Burley, Rutlandshire*, copperplate engraving.

ARMANS' DESCRIPTIVE TITLE	SOURCE PRINT
"Building, Deer on Lawn"	*Bushy Park, Middlesex*

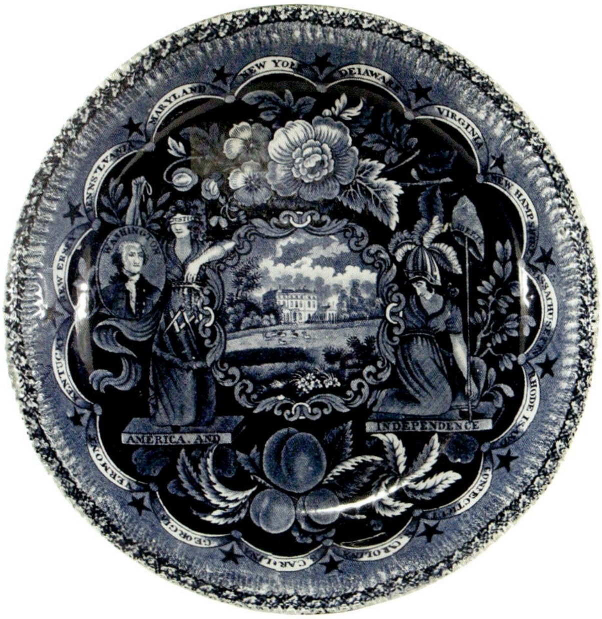

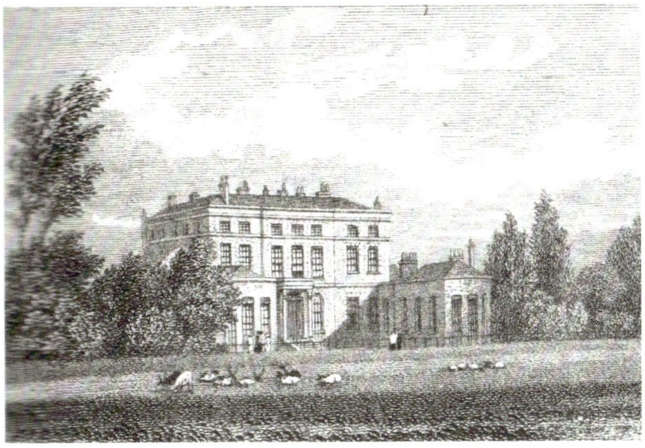

Figure 6 *Left*: Bowl, Ralph and James Clews, Cobridge, England, 1826–1834. Lead-glazed earthenware. D. 9⅝". *Right*: *Bushy Park, Middlesex*, copperplate engraving.

This scene is also found on plates as well as on the green shell-edged dish illustrated by David and Linda Arman, p. 14, fig. 2.13.

"Three Storey Building, Two Wings and Center Section"	*Carlton Hall, Northamptonshire*

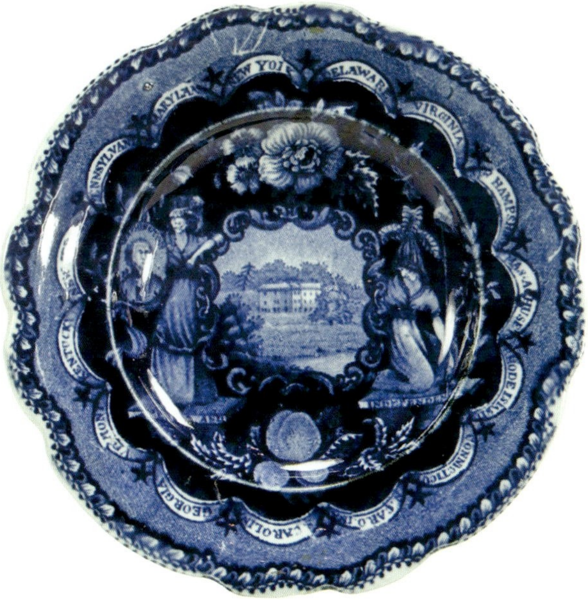

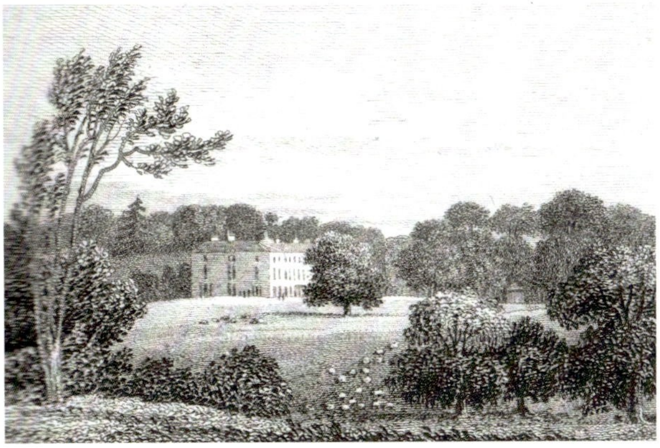

Figure 7 *Left*: Plate, Ralph and James Clews, Cobridge, England, 1826–1834. Lead-glazed earthenware. D. 5⅜". *Right*: *Carlton Hall, Northamptonshire*, copperplate engraving.

This scene is also found on cup plates.

ARMANS' DESCRIPTIVE TITLE	SOURCE PRINT
"Mansion, Circular Drive"	*Coghill Hall, Yorkshire*

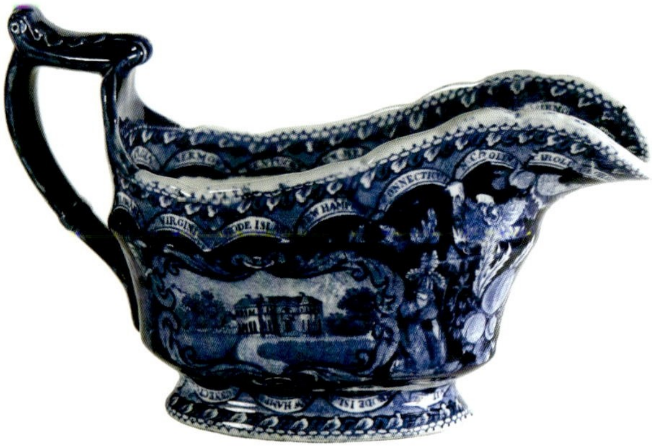

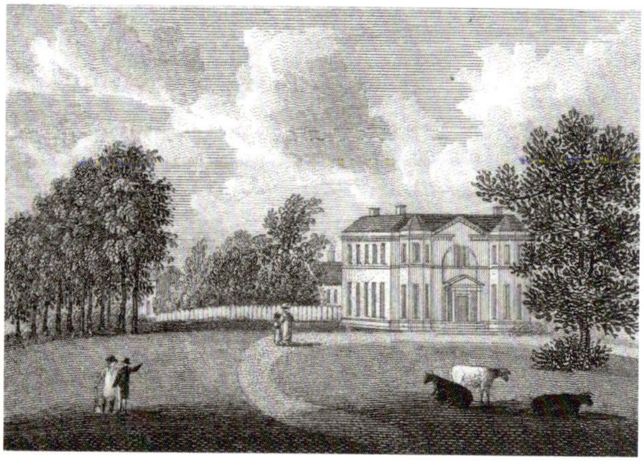

Figure 8 *Left*: Sauce boat, Ralph and James Clews, Cobridge, England, 1826–1834. Lead-glazed earthenware. L. $7\frac{5}{8}$". *Right*: *Coghill Hall, Yorkshire*, copperplate engraving.

This scene was used quite widely on other items.

"Building, Two Wings, Water in Foreground"	*Doveridge Hall, Derbyshire*

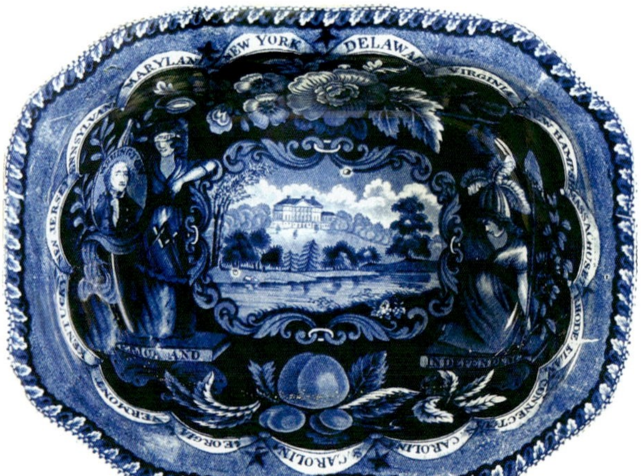

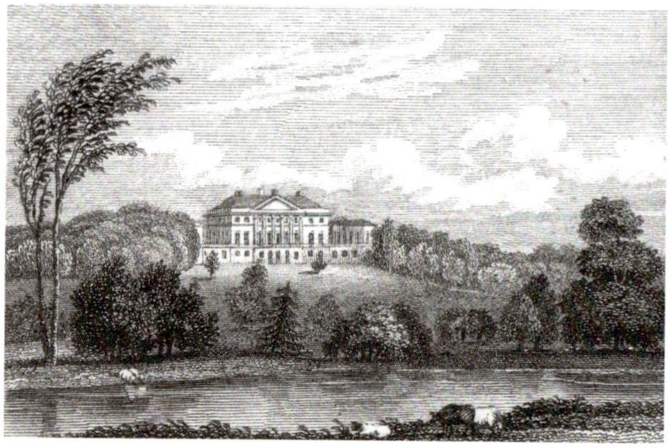

Figure 9 *Left*: Baker, Ralph and James Clews, Cobridge, England, 1826–1834. Lead-glazed earthenware. L. $9\frac{3}{8}$". *Right*: *Doveridge Hall, Derbyshire*, copperplate engraving.

This scene has also been found on a small dish, a conical bowl, and various jugs. Some of the jugs include Ohio and/or Tennessee in the border of state names.

ARMANS' DESCRIPTIVE TITLE	SOURCE PRINT
"Dock, with Large Building and Ships"	*Town Hall and Quay, Yarmouth*

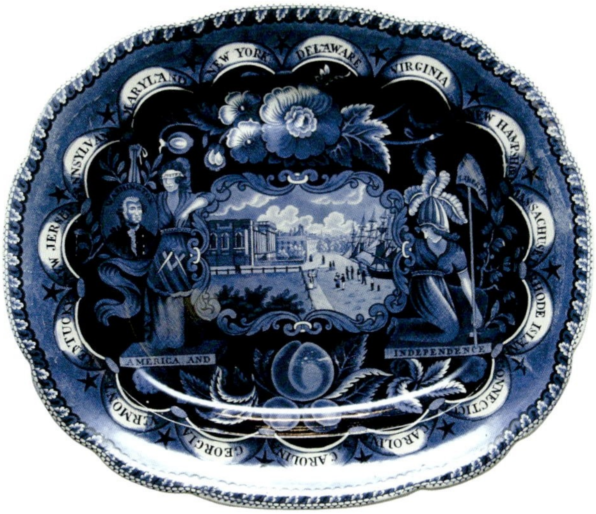

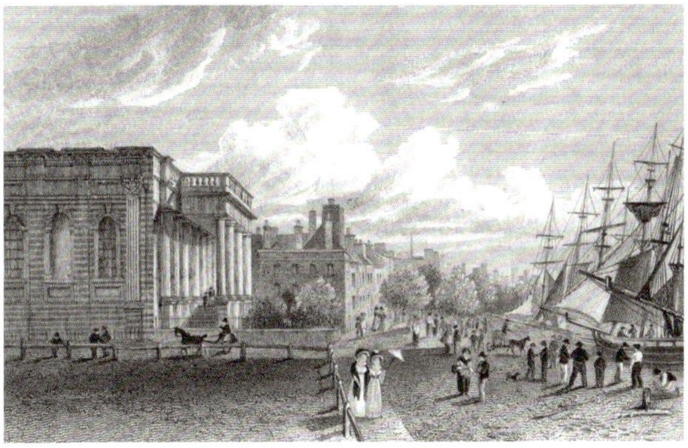

Figure 10 *Left*: Dish, Ralph and James Clews, Cobridge, England, 1826–1834. Lead-glazed earthenware. L. 18⅝". *Right*: *Town Hall and Quay, Yarmouth*, signed by "J[ohn] P[reston] Neale"; copperplate engraving signed "S[amuel] Lacey."

Previous tentative and optimistic suggestions that the scene depicted is the New York Customs House are clearly incorrect.

"Mansion, Small Boat with Flag in Foreground"	*Gunnersbury House, Middlesex*

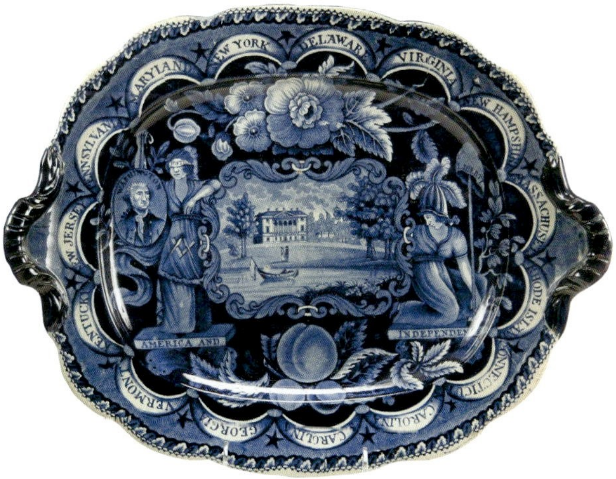

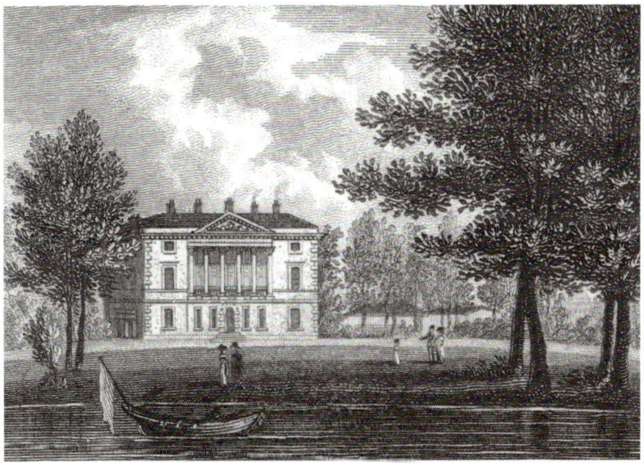

Figure 11 *Left*: Soup tureen stand, Ralph and James Clews, Cobridge, England, 1826–1834. Lead-glazed earthenware. L. 16". *Right*: *Gunnersbury House, Middlesex*, copperplate engraving.

This scene has also been found on dishes and a range of other wares.

ARMANS' DESCRIPTIVE TITLE	SOURCE PRINT
"Building, Sheep on Lawn"	*Lartington, Yorkshire*

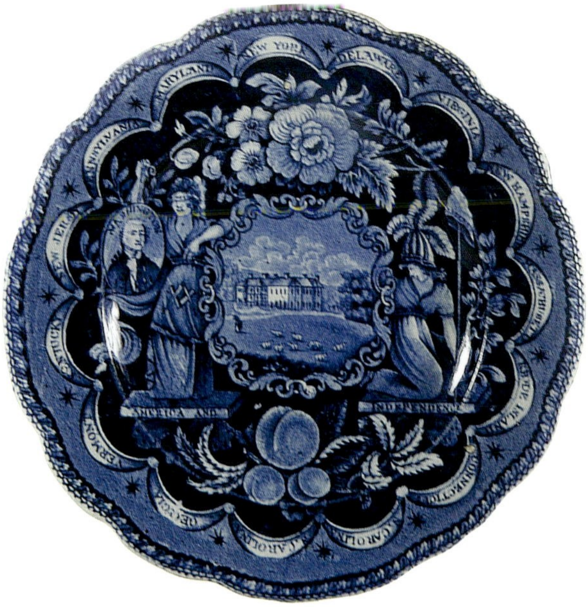

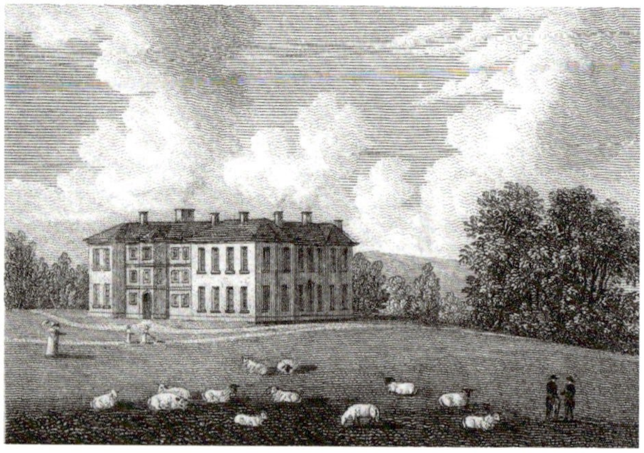

Figure 12 *Left*: Plate, Ralph and James Clews, Cobridge, England, 1826–1834. Lead-glazed earthenware. D. 8¾". *Right*: *Lartington, Yorkshire*, copperplate engraving.

This scene is also known on a small dish and a bowl.

"Three Storey Mansion, Small Extension to Left"	*St. Woolston's, Kildare*

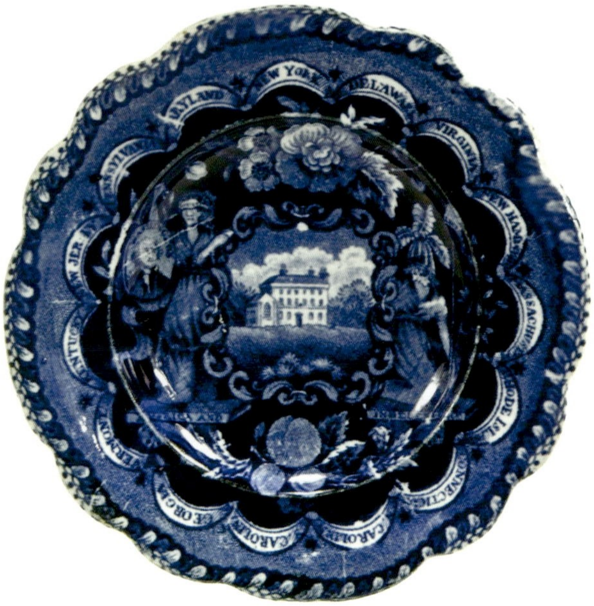

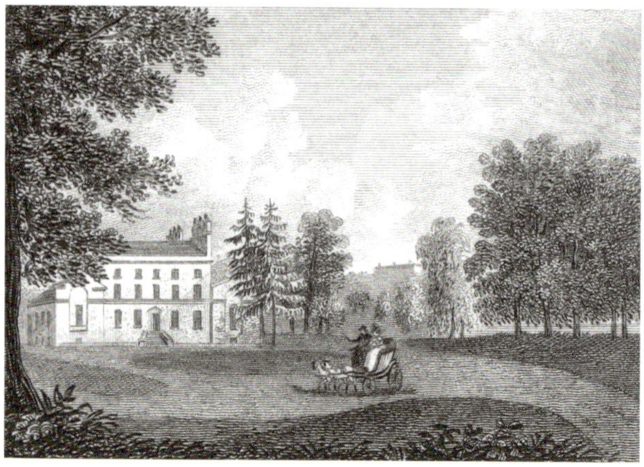

Figure 13 *Left*: Cup plate, Ralph and James Clews, Cobridge, England, 1826–1834. Lead-glazed earthenware. D. 4½". *Right*: *St. Woolston's, Kildare*, copperplate engraving.

ARMANS' DESCRIPTIVE TITLE | SOURCE PRINT

"Castle with Flag, Boats in Foreground" — *Trematon Castle, Cornwall*

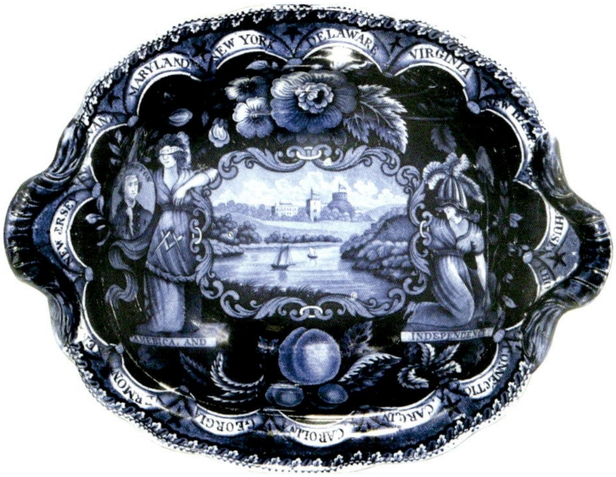

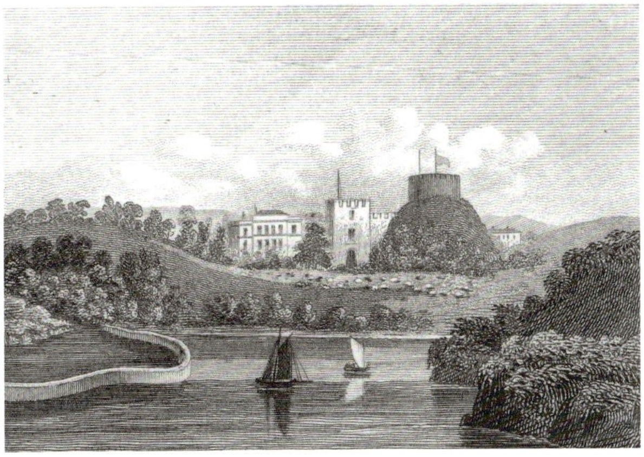

Figure 14 *Left*: Serving dish, Ralph and James Clews, Cobridge, England, 1826–1834. Lead-glazed earthenware. L. 7⅝". *Right*: *Trematon Castle, Cornwall*, copperplate engraving.

This scene was used widely on other items.

"Mansion, Foreground a Lake with Swans" — Unidentified

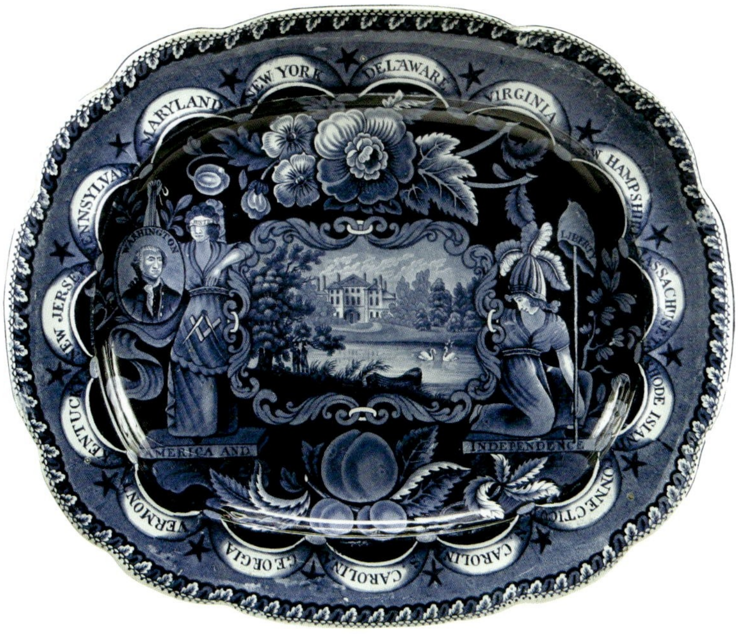

Figure 15 Dish, Ralph and James Clews, Cobridge, England, 1826–1834. Lead-glazed earthenware. L. 16⅛".

This scene is not among the views engraved by Marshall, nor has it been located in other illustrated books of the period.

ACKNOWLEDGMENTS I thank Terry Neale-Sheppard for providing access to his copies of *Marshall's Select Views, in: Great Britain; &c.*, and Chet Creutzburg and Kurt O'Hare for images of the wares themselves. Without their unstinting help, this article would not have been possible.

1. David Arman and Linda Arman, *Anglo-American Ceramics, Part 1: Transfer Printed Creamware and Pearlware for the American Market, 1760–1860* (Portsmouth, R.I.: Oakland Press, 1998), p. 14.

2. An advertisement in *The Examiner, A Sunday Paper, on Politics, Domestic Economy, and Theatricals, for the Year 1826*, no. 964 (July 23, 1826): 480, reads:

> Just Published, No. 20 of Marshall's Select Views in Great Britain, containing Six fine Engravings, with Letter-Press Descriptions, price only 6d. Two Volumes of this Work are completed, each containing Seventy Two Prints, price 6s 6d each, in boards. Two numbers will be regularly published monthly, until the Work is finished. This Work (the Plates alone of the Two Volumes published having cost One Thousand Guineas) may be deemed the most complete and cheapest Topographical History of remarkable Places and beautiful Seats in the Kingdom. Sold by all booksellers, and Published by W. Marshall, No. 1 Holborn Bars; where are on Sale upwards of Three Hundred Views of Towns, Cities, and Gentlemen's Seats, Dutch Subjects, splendid Prints by Stothard, engraved by Heath, &c. calculated for Illustration or Scrap Books. Collectors of Fine Engravings will find Marshall's Prices for Prints considerably cheaper than any other House in the Kingdom.

3. David Arman and Linda Arman, *Historical Staffordshire: An Illustrated Check List* (Danville, Va.: Arman Enterprises, 1974); and idem, *Historical Staffordshire: An Illustrated Check List, First Supplement* (Danville, Va.: Arman Enterprises, 1977).

Richard Duez and Don Horvath

with Brenda Hornsby Heindl

The Stoneware Years of the Thompson Potters of Morgantown, West Virginia, 1854–1890

▼ OUR 2004 *Ceramics in America* article, "The Potters and Pottery of Morgan's Town, Virginia: The Earthenware Years, Circa 1796–1854," introduced readers to the early earthenware pottery that was manufactured by the Thompson family of potters. Through archaeological discoveries and studies of their glazing techniques and the forms they made, the potters of Morgantown became better known and their products recognizable. Readers also learned of the succession of potters and other associates, apprentices, and family members who attempted to operate a pottery manufacture along the Monongahela River.[1] The Thompson family, who gained a stronghold in earthenware manufacture in Morgantown though few of their earthenware pieces survive, would be the potters to carry the industry through to stoneware.

Since the publication of that article, research into the Morgantown potters' stoneware has continued. The variations of form, the decorative styles, and the tools they used to create their wares provide important information about the American pottery industry in the mid- to late nineteenth century and, more specifically, about the operations of this particular one.

Many of the decorative techniques used on the stoneware are attributed to David Greenland Thompson, a son of John Wood Thompson, and were introduced as the Morgantown pottery transitioned to stoneware. Given that this stoneware production occurred over a period of approximately three decades, it is not possible to identify the thrower and decorator of every object. It is reasonably certain that two of them are David and his father, however, and this article focuses on their skill and creative development.

"We are gratified to state the experiment has proven entirely successful"

By 1850 the Morgantown Pottery, whose original location on Lot 88 had burned down in 1830, was firmly established on Lot 1, next to the wharf on the Monongahela River (fig. 1).[2] The 1850 census shows four Thompson men working at the pottery: John W., his brother Francis M., and two of his sons—Francis (Frank W.) and David Greenland. In 1853 the Morgantown Pottery passed from John W. to his son James Jackson Thompson, then later to James's brother Francis (Frank W.). During this period, the ware production moved from the "porous terra cotta covered with a yellow lead glaze" to salt-glazed stoneware.[3]

A rare insight into this transition is documented on a jar bearing the inscription "First 2 jar for Francis Madiera June the 2nd, 1854" (figs. 2, 3).[4] The body is primarily a salt-glazed gray, with ruddy streams of red running

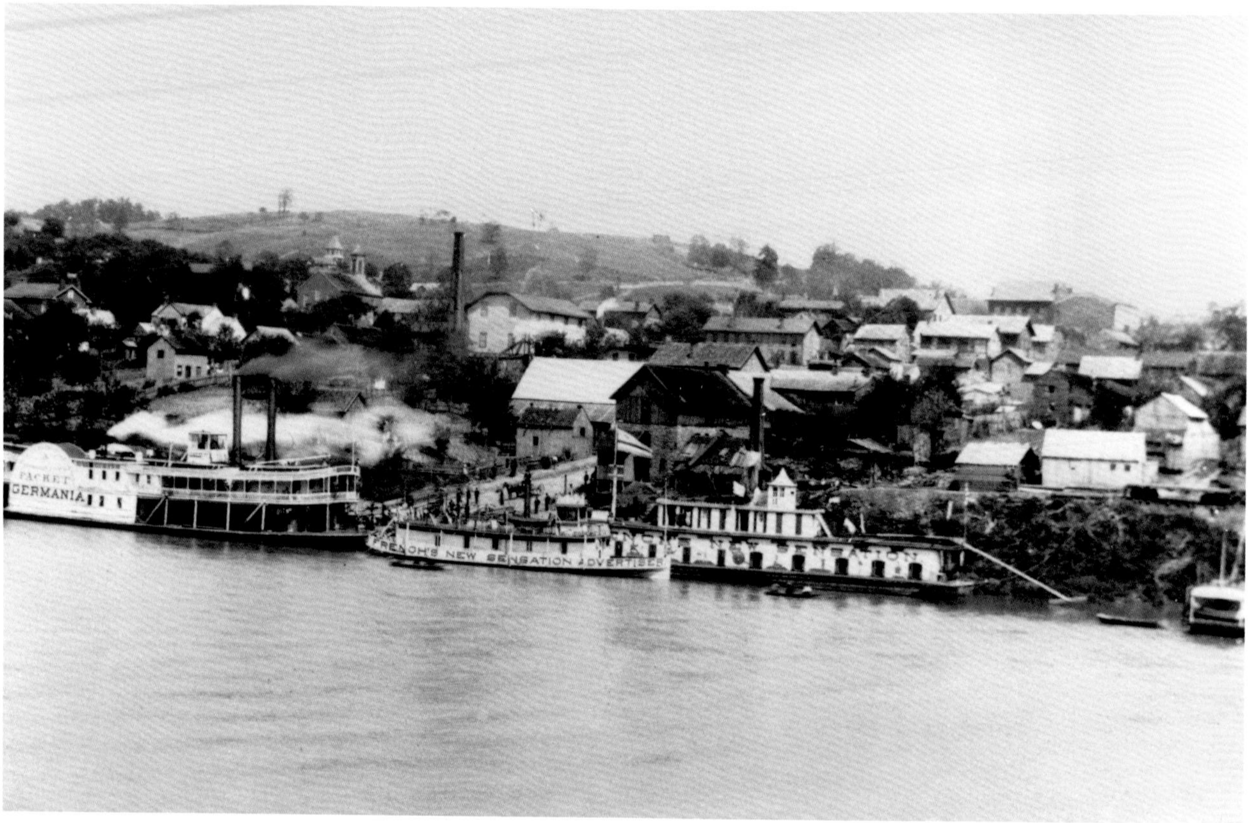

Figure 1 Photograph, Morgantown, West Virginia, ca. 1870. View of Lot 1 at the wharf on the Monongahela River. (Courtesy, West Virginia University.)

Figure 2 Jar, attributed to David Greenland Thompson, Morgantown, West Virginia, 1854. Salt-glazed stoneware. H. 10". Inscribed on shoulder: "First 2 jar for Francis Madiera June the 2nd, 1854." (Duez Collection; unless otherwise noted, photos by Gavin Ashworth.) Francis Madiera was the postmaster in Morgantown at that time. The same cobalt script "M N" (for Morgantown) occurs on two pieces made by David Greenland Thompson.

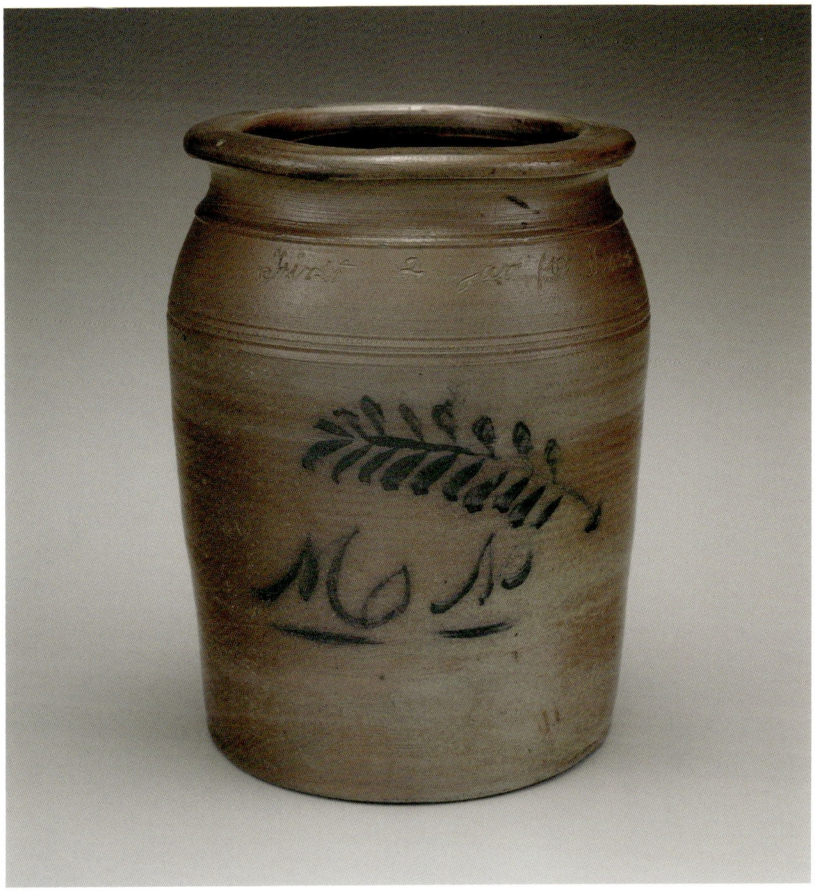

Figure 3 Details of the inscription on the jar illustrated in fig. 2.

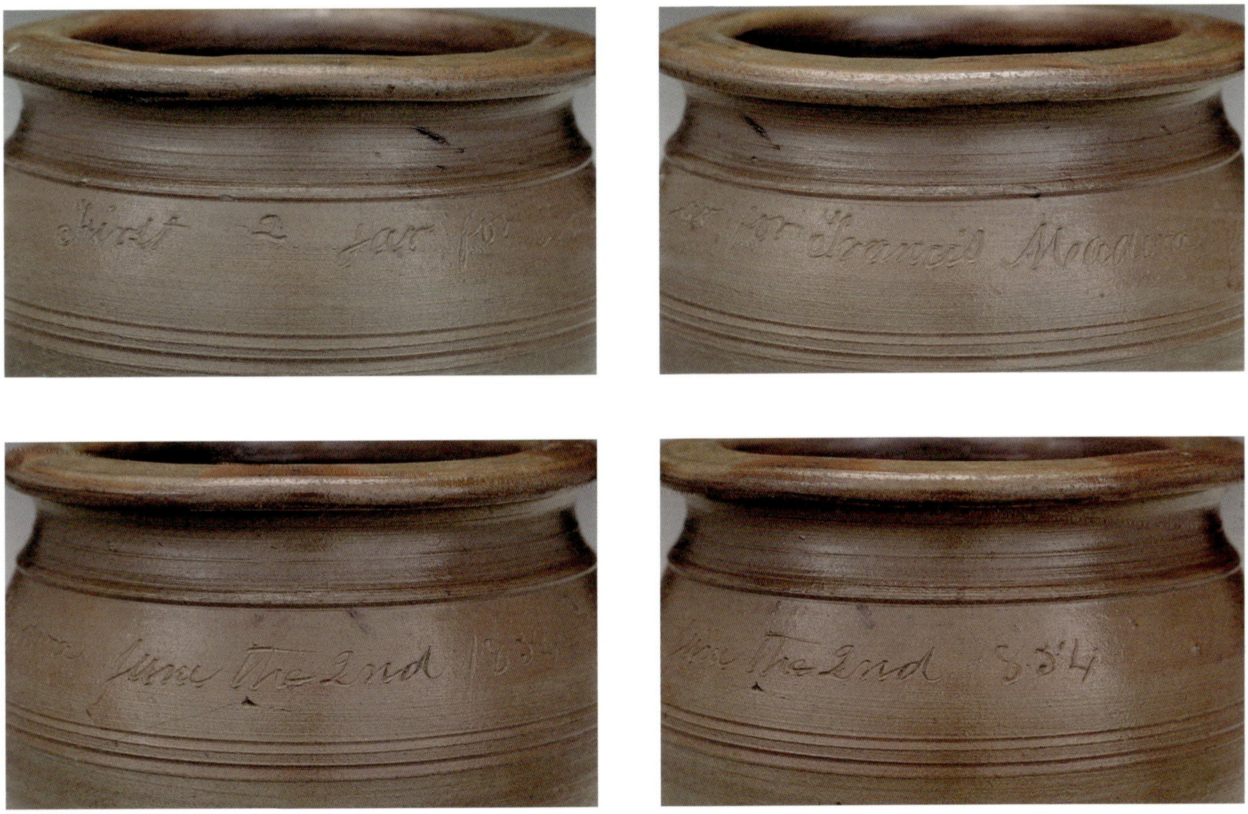

STONE-WARE—The first kiln of Stone-ware ever manufactured in Morgantown was burned this week at the Pottery of the Messrs. Thompson and from the specimens taken from the kiln, we are gratified to state the experiment was proven entirely successful. The enterprising proprietors of this well known establishment have put up the appliances necessary for the manufacture of every variety of Stone-ware, so that hereafter that article will be an article of export instead of import, as formerly, at our wharf.

Figure 4 Newspaper announcement, *The Monongalia Mirror* (Morgantown, Virginia), July 22, 1854, p. 2 (reproduced). (Courtesy, West Virginia and Regional History Collection, West Virginia University Libraries.)

through it. On the side in cobalt blue are a fernlike decoration and an inscribed "M" and "N" (for Morgantown). The identical initials occur on two pieces signed by David Greenland Thompson, which might indicate that he was the maker or decorator of this magnificent piece.

In the *Monongalia Mirror* of July 22, 1854, the Thompsons announced production of their first stoneware: "The first kiln of Stone-ware ever manufactured in Morgantown was burned this week at the Pottery of the Messrs. Thompson and from the specimens taken from the kiln, we are gratified to state the experiment has proven entirely successful. . . . [H]ereafter that article will be an article of export instead of import, as formerly, at our wharf" (fig. 4).[5]

By the time of the 1860 census, James J. had joined the other four family members at the pottery (fig. 5). Three years later John W. died, and at the outbreak of the Civil War his sons James J. and Francis W. enlisted as officers in the Union army, leaving David Greenland to run the pottery. The Civil War caused a shift not only in the American economy and indus-

1668	1617	John W Thompson	76	M	Potter	3,000	600	Md
		Deborah "	71	F				Va
		David G	39	M	Do			"
		James J	33	"	Do			"
		Francis W	30	"	Do			"
		Francis M	30	"	Do			"
		Aleinda Robertson	25	F	Domestic			"
		Jacob W Windows	19	M	Day Laborer			"
		Daniel Knox	30	"	Wegoner			"

Figure 5 Population census, Monongalia County, Virginia, 1860 (reproduced). The original document, which is accessible online through www.ancestry.com, lists the owner of the pottery shop, John W. Thompson, and his four sons as "potter."

try but in political boundaries as well, one of which resulted in the creation of West Virginia. It is thought that many of the more complex decorative techniques, such as the use of roulettes, stamps, and painted images, were implemented in the workshop during John W. Thompson's lifetime and ended with his death (fig. 6).

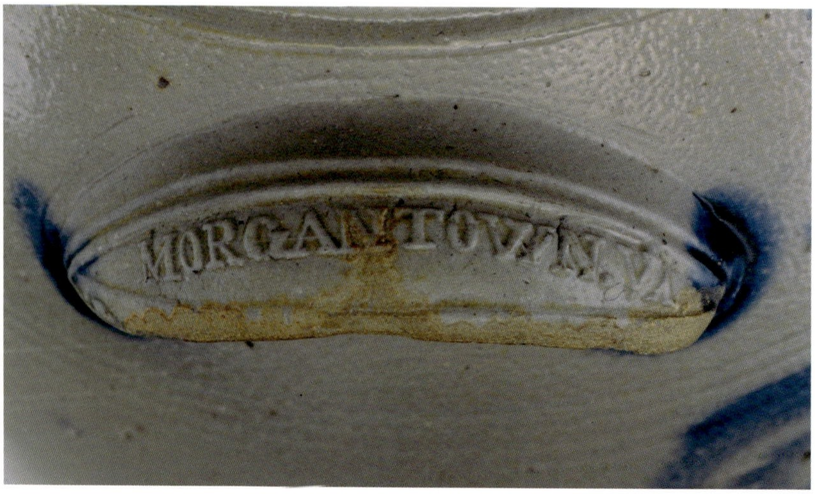

Figure 6 Detail of the handle on a 15", 4-gallon stoneware jar. Mark, embossed on handle: MORGANTOWN, VA. (Duez Collection.)

The Thompsons' transition to stoneware manufacture was made possible by an earlier discovery of a suitable stoneware clay seam in the Monongahela Valley of Pennsylvania, just across the river from Morgantown in the second quarter of the nineteenth century. As a result, a multitude of Pennsylvania potteries shifted from earthenware to stoneware.[6] As early as the late 1840s and early 1850s area potters began to produce salt-glazed stoneware from the "very fine bluish white clay."[7] Three miles north of Morgantown, in Collins Ferry on the Monongahela River, William Crichfield was making stoneware as early as 1844.[8] Alexander Boughner, who began stoneware production in the late 1840s, was one of the earliest to do so in the Greensboro area of Pennsylvania.[9]

In addition to their accessibility to suitable stoneware clay, the Thompsons benefited from their proximity to the region's water-based transportation system. Use of the river greatly increased transportability and availability of the materials needed for stoneware production (fig. 7). A canal connecting

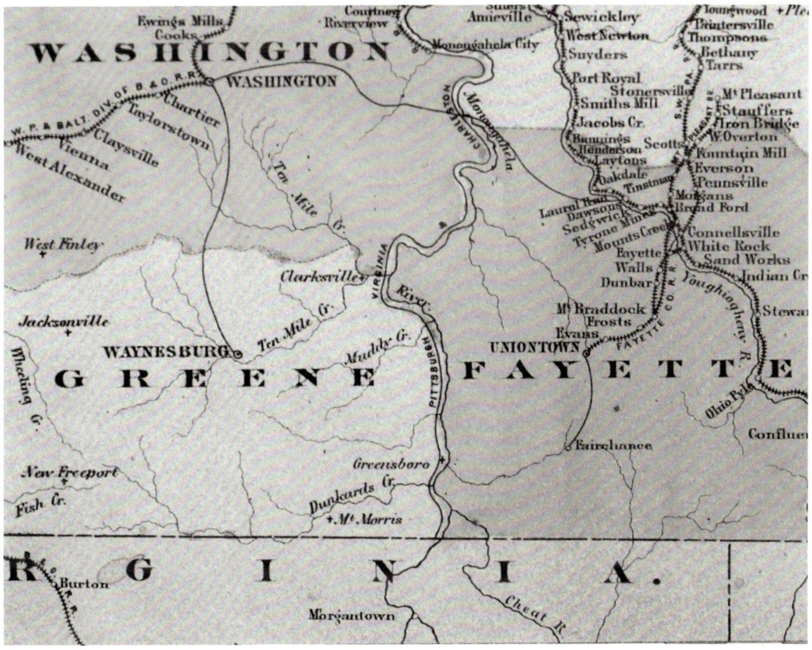

Figure 7 Map of Greene and Fayette Counties, Pennsylvania, illustrated in *Caldwell's Illustrated Combination Centennial Atlas of Greene Co. Pennsylvania* (J. A. Caldwell, 1876). (Duez Collection; photo, Dick Duez.) Uniontown, Greensboro, and Morgantown share regional proximity.

the Monongahela River to other waterways in Pittsburgh opened in the late 1820s. This became useful for more rapid transportation and access to other cities along the rivers. The canal also expanded the market of many potters, allowing them to transport their pots to Ohio, Indiana, as far south as New Orleans, and wider markets in the East. Shipping materials on the river also made salt glazing more cost effective. Readily accessible salt came in large quantities and could replace the use of lead for glazes, known as a hazard to both the consumer and the producer.[10]

Family Connections: Potters of Uniontown and Connellsville, Pennsylvania

For this article it is important to recognize family ties. Many characteristic Morgantown pieces are attributed to potters who worked in Uniontown and Connellsville, towns in bordering Fayette County, Pennsylvania (fig. 8).

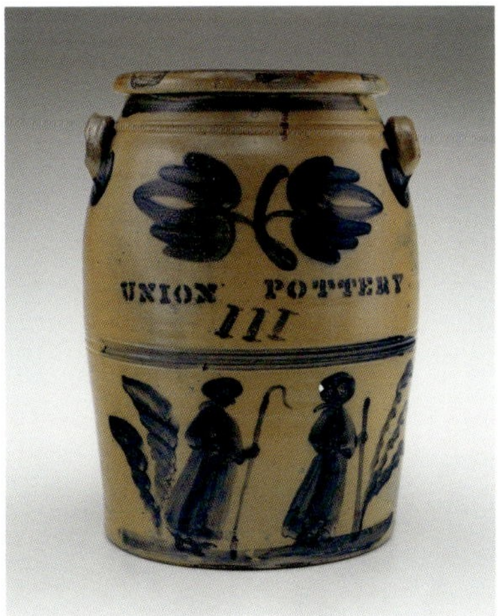

Figure 8 Jar, Union Pottery, Uniontown, Pennsylvania, ca. 1870. Salt-glazed stoneware. H. not recorded. Marks: UNION POTTERY / III. (Private collection.)

Abner Greenland was a well-known, early earthenware potter of Uniontown. In 1806 he married Jane Thompson, sister of John W. Thompson,[11] and they had two sons, John and Norval. John Greenland was listed as working with his father in the late 1820s shortly before his father's death in 1830.[12] In 1859 he was listed as running a store in Connellsville, Pennsylvania, and in 1872 as operating a firm called Greenland & Son, Manufacturers of Stoneware.[13] Norval Greenland remained in Uniontown, but in an 1859 directory was listed as a wagonmaker.[14] In 1864 Norval formed a partnership with one Cromwell Hall and developed a stoneware production business under the name Hall & Greenland. The partnership ended by the 1870s, but Norval Greenland did not end production until 1882.[15]

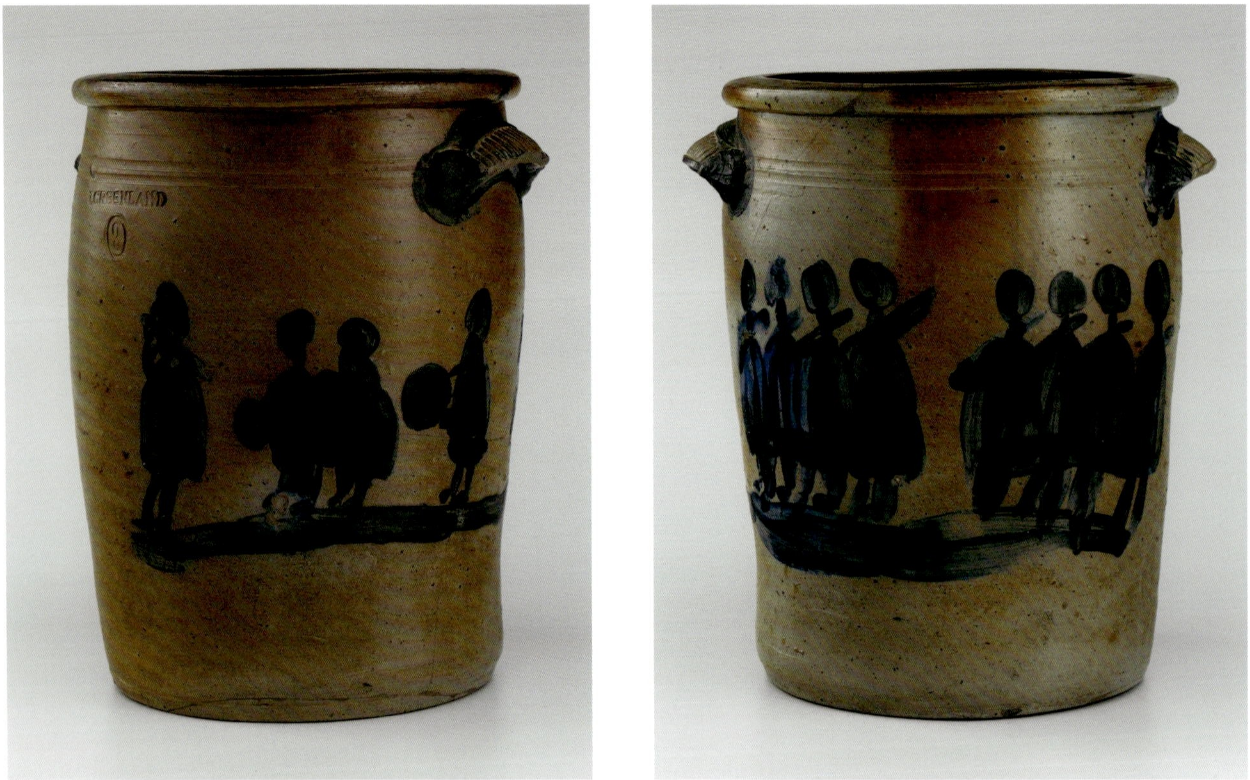

Figure 9 Jar, Union Pottery, Connellsville, Pennsylvania, ca. 1865. Salt-glazed stoneware. H. 12¾". Marks: impressed on shoulder, J. GREENLAND; 3 [gallon capacity stamp]. (Private collection.) The cobalt decoration depicts a marching band of soldiers wearing great coats and kepis.

John Greenland made one of the best-known Connellsville pottery pieces (figs. 9, 10). The estimated date of manufacture is circa 1865, and the decoration likely commemorates the end of the Civil War. The cobalt-blue decoration on the jar displays human figures marching with guns, led by a small instrument band complete with a leader. The images on the jar as well

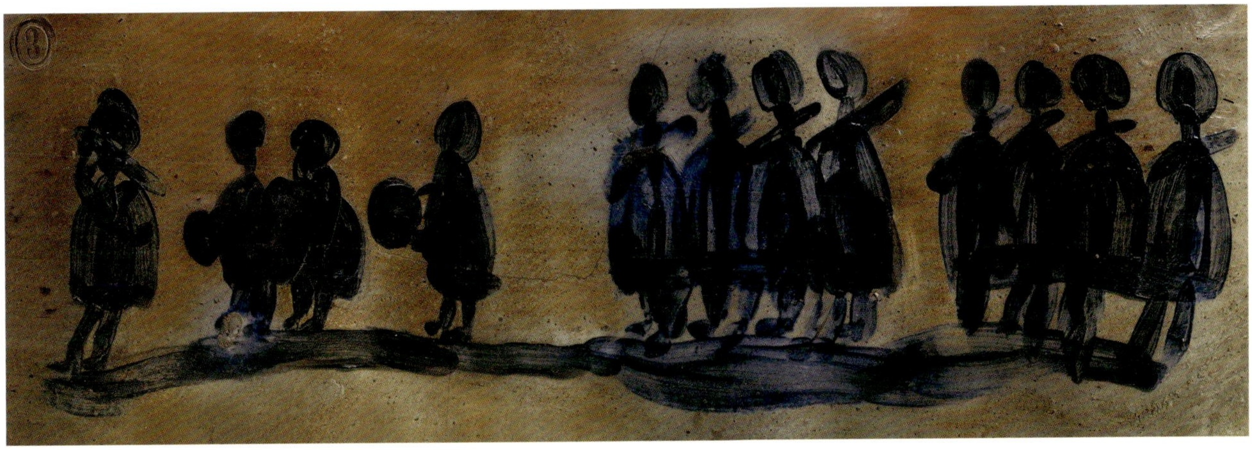

Figure 10 Detail of the decoration on the jar illustrated in fig. 9; composite image of cobalt designs on the jar.

as the form, which bears a capacity stamp (fig. 11), attest to the similarities of Connellsville, Uniontown, and Morgantown pieces. Marking the capacity with signature stamps and designs became a trademark of the Morgantown pottery and can be seen on almost every piece produced from 1855 to 1870. The transition of the Morgantown potters to stoneware production was accompanied by a remarkable expansion of the use of roulettes and the acquisition of unique molds for handles, or lugs.

Figure 11 Detail of the impressed marks on the jar illustrated in fig. 9.

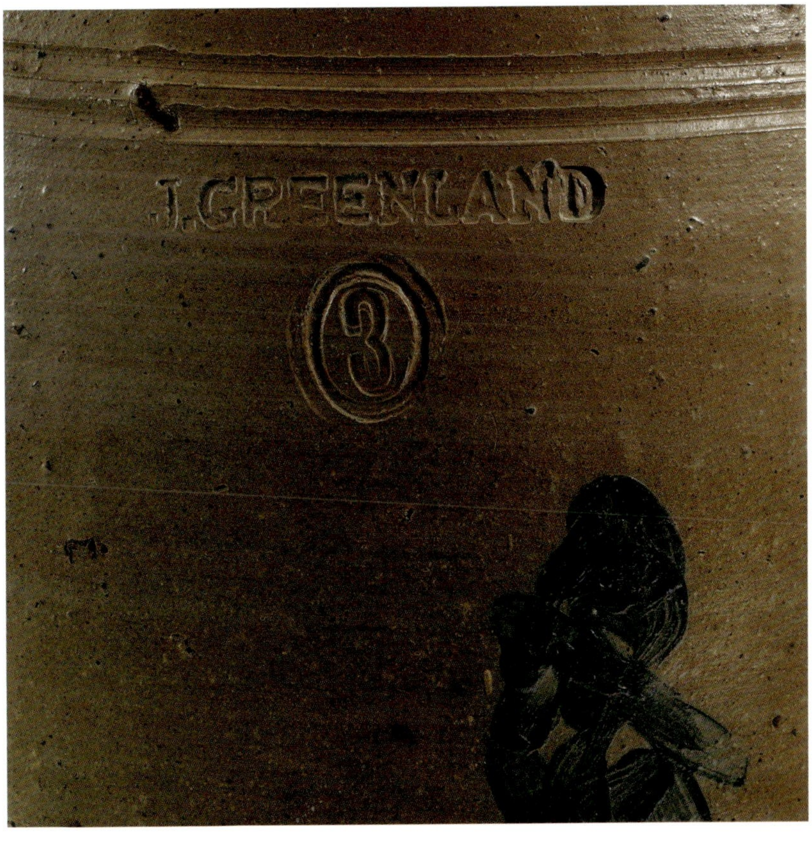

"Wares of the Second Period"

In 1901 Walter Hough, assistant curator in the Smithsonian Institution's ethnology division, published a report entitled "An Early West Virginia Pottery" in which he labeled David Greenland Thompson's stoneware products as "wares of the second period."[16] Ceramics historians are in great debt to Hough, a West Virginia native who worked to collect the tools of the Morgantown pottery shortly after it closed. These extremely rare survivals went to the Smithsonian, as did some examples of stoneware and earthenware—several of which are stamped "D. G. THOMPSON" (figs. 12, 13).

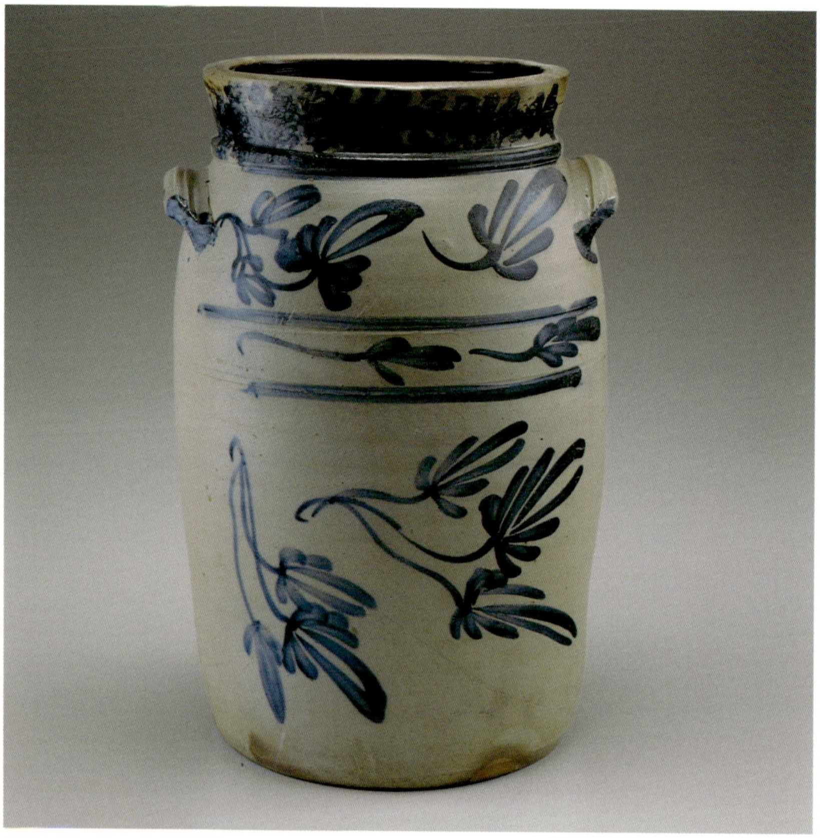

Figure 12 Jar, David Greenland Thompson, Morgantown, West Virginia, 1870. Salt-glazed stoneware. H. 15". Marks: impressed on shoulder, D. G. THOMPSON; 4 [gallon capacity stamp within a beaded circle]; Morgantown. (Duez Collection.)

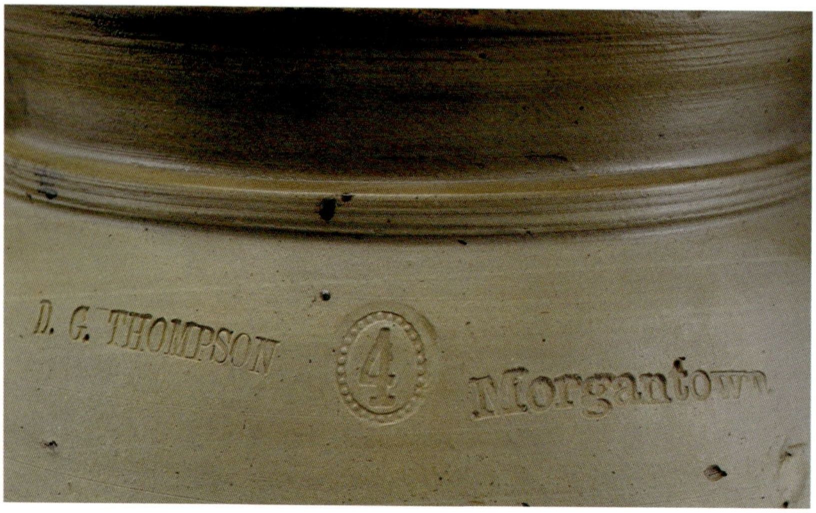

Figure 13 Detail of the marks on the jar illustrated in fig. 12.

The tools that Hough collected highlight the variety of techniques employed by the Morgantown potters. A number of shaping ribs (used to smooth the sides of pieces) are included; some are actual ribs of animals but most are of wood. There are throwing tools for incised lines, trimming

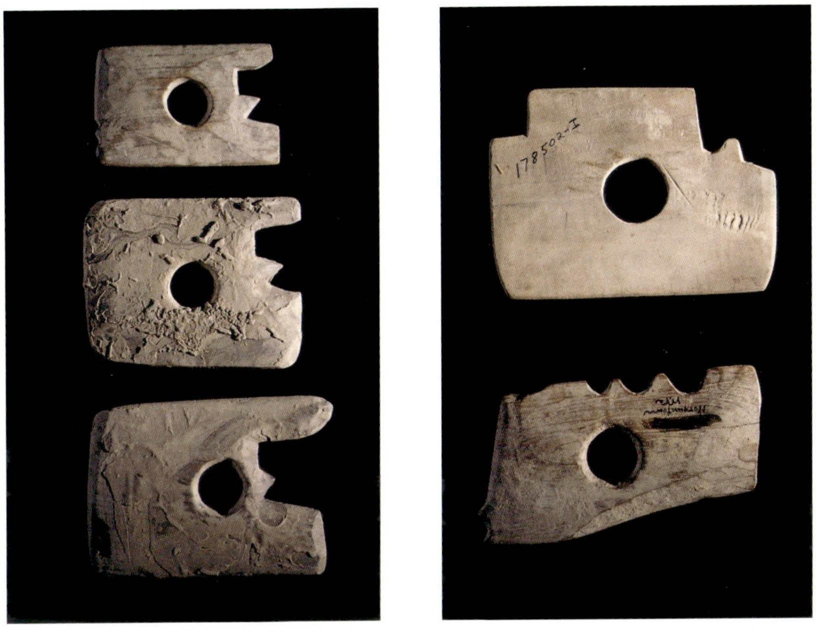

Figure 14 Ribs, Thompson Pottery, Morgantown, West Virginia, 1850–1870. Wood. (Courtesy, National Museum of American History.) These ribs would have been used to make the heavy rims of the crocks and other pottery pieces.

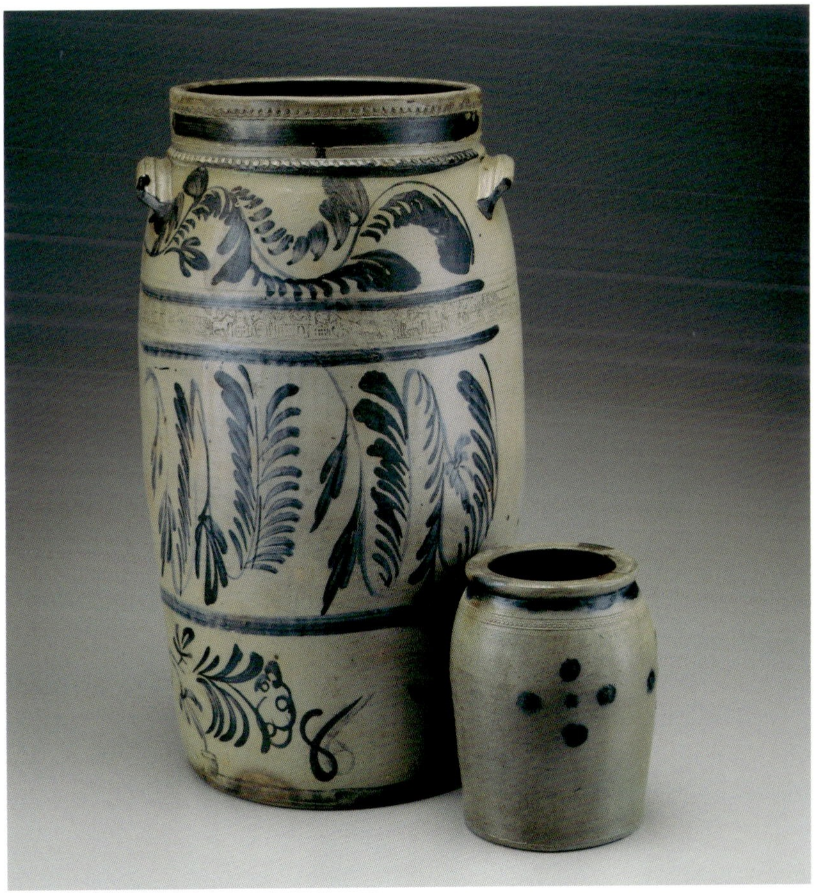

Figure 15 Jars, Morgantown, West Virginia, (*left*) ca. 1865, (*right*) ca. 1860. Salt-glazed stoneware. H. (*left*) 20½", (*right*) 7½". (Duez Collection.)

Figure 16 Crock, Morgantown, West Virginia, ca. 1870. Salt-glazed stoneware. H. 7½". (Private collection.)

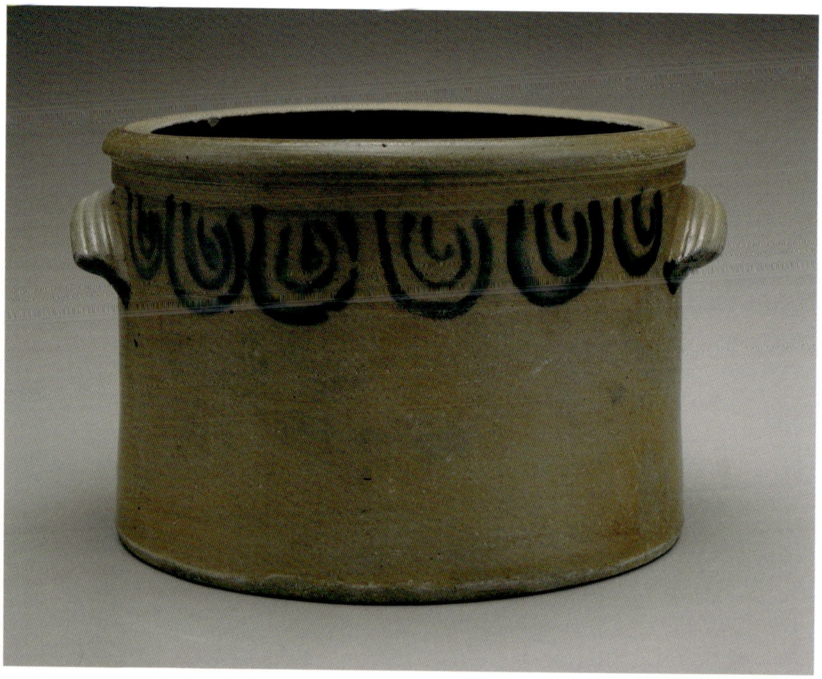

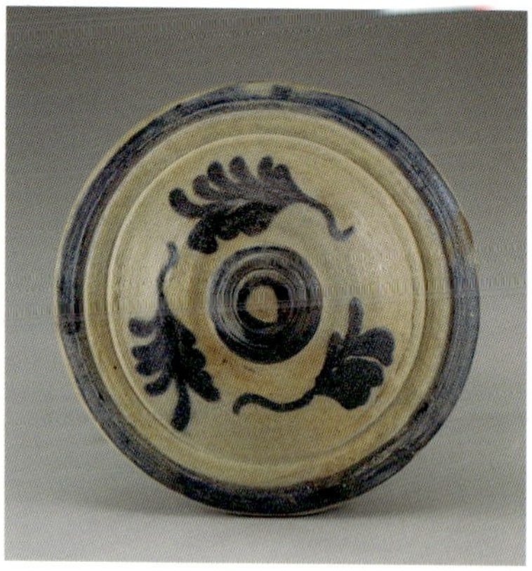

Figure 17 Crock lid, Morgantown, West Virginia, ca. 1870. Salt-glazed stoneware. D. 7". (Duez Collection.) This is the only known example of a stoneware lid made at Morgantown. It was found in three pieces at the Morgantown wharf.

Figure 18 Preserve jars, Morgantown, West Virginia, ca. 1870. Salt-glazed stoneware. H. (*left to right*) 8½", 8⅝", 8½". (Private collection.)

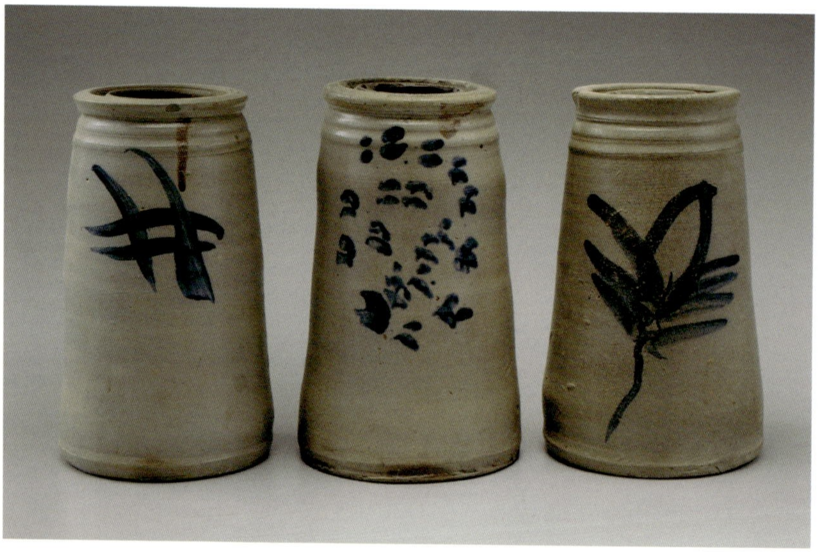

Figure 19 Preserve jar, attributed to David Greenland Thompson, Morgantown, West Virginia, 1869. Salt-glazed stoneware. H. 8½". Marks, inscribed on side in cobalt blue: Thompson / Morgantown / 1869. (Duez Collection.) This piece might have been made to commemorate David Greenland Thompson's fiftieth birthday.

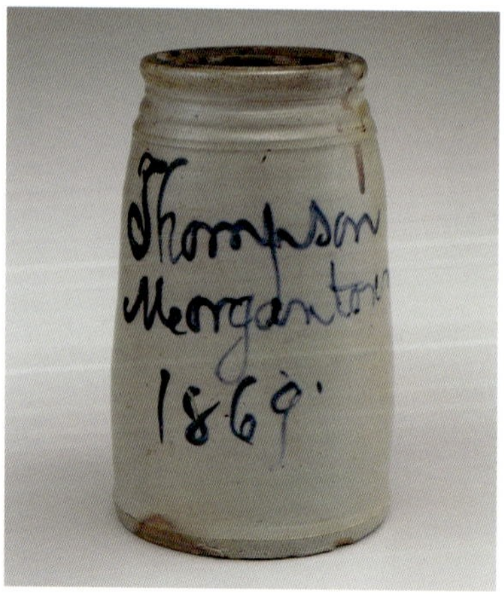

clay, and tools used to lift the pieces of pottery from the wheel. There are other wood implements used to shape the rims and molding during final finishing. The profile ribs illustrated in figure 14 assisted in forming the heavy banded and often angled rims seen on the Morgantown jars and crocks.

Crocks and jars dominate the Thompson factory's surviving forms, although the potters were skilled at making a variety of forms in a variety of sizes (fig. 15). Jar sizes from one-half gallon to ten gallons have been recorded. Surviving cake crocks are somewhat rare (fig. 16), and the only known lid comes from an example pieced together from fragments found on the site of the pottery (fig. 17). A particularly distinctive form in the Morgantown production is the tapered preserve jar, sometimes called a wax sealer (figs. 18, 19).

Bottles and jugs with high, rounded shoulders and gently flared feet are known (fig. 20). Many have cobalt-blue decoration, and some have a rouletted pattern following the shoulder (figs. 21, 22). Many of the pitchers have bellies similar to those of the tapered preserve jar (figs. 23, 24). An unusual example with cobalt-blue floral decoration has the stamp of David Greenland Thompson (fig. 25).

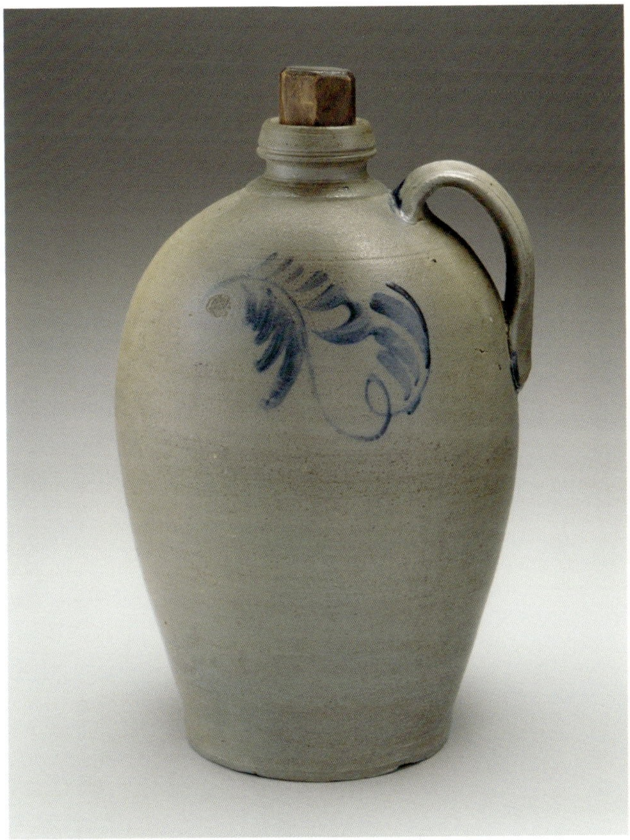

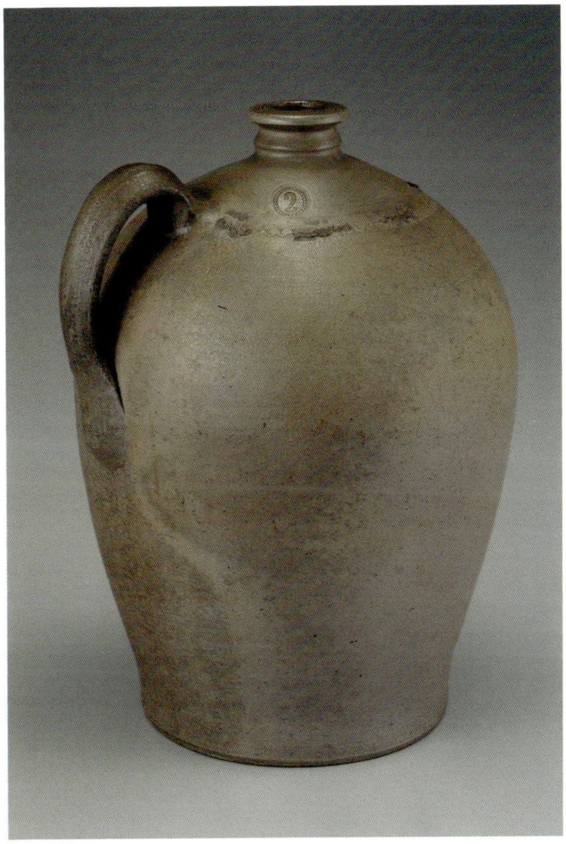

Figure 20 Bottle, Morgantown, West Virginia, ca. 1860. Salt-glazed stoneware. H. 13½". (Duez Collection.)

Figure 21 Bottle, Morgantown, West Virginia, 1854–1860. Salt-glazed stoneware. H. 12". Mark: 2 [gallon capacity stamp within a beaded circle]. (Duez Collection.) This bottle, which was found in Daybrook, Monongalia County, West Virginia, has a rouletted pattern on its shoulder (see fig. 22).

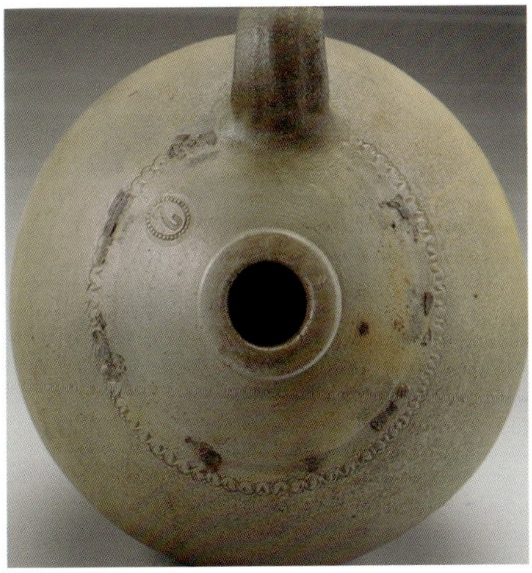

Figure 22 Detail of the top of the bottle illustrated in fig. 21, showing the rouletted decoration.

Figure 23 Pitcher, Morgantown, West Virginia, ca. 1870. Salt-glazed stoneware. H. 11". (Private collection.) This pitcher has a conical base, cobalt decoration, and a street-scene rouletted design. It was recovered in Grafton, West Virginia, in the early 1970s by a street crew.

Figure 24 Pitcher, Morgantown, West Virginia, ca. 1870. Salt-glazed stoneware. H. 11". (Private collection.)

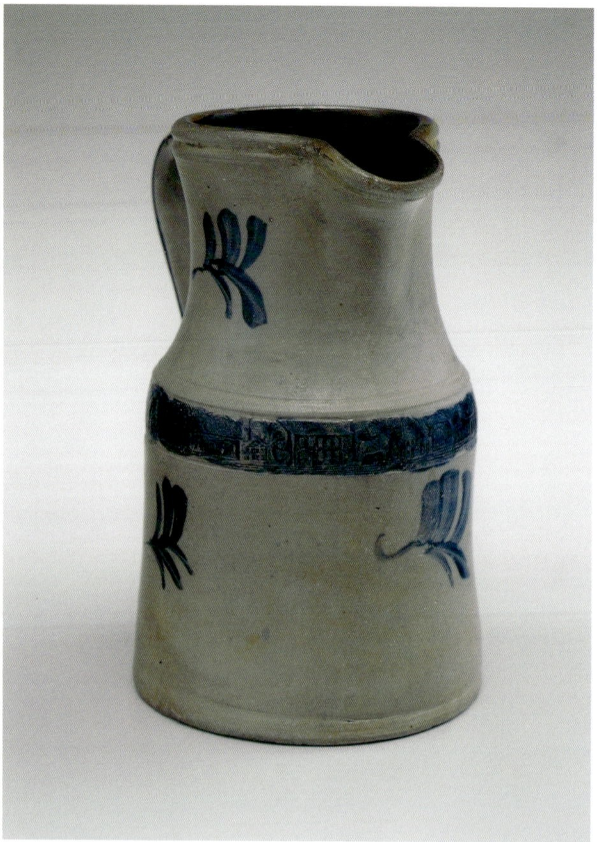

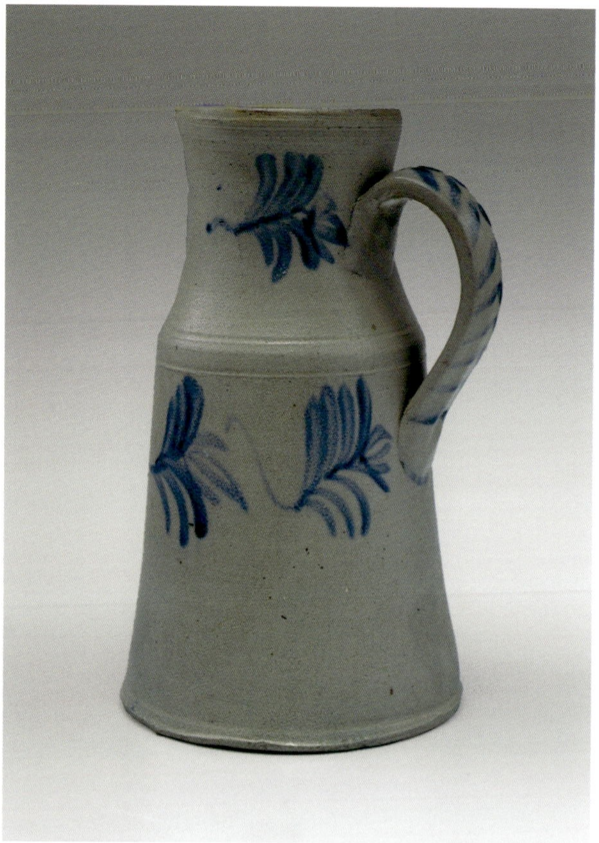

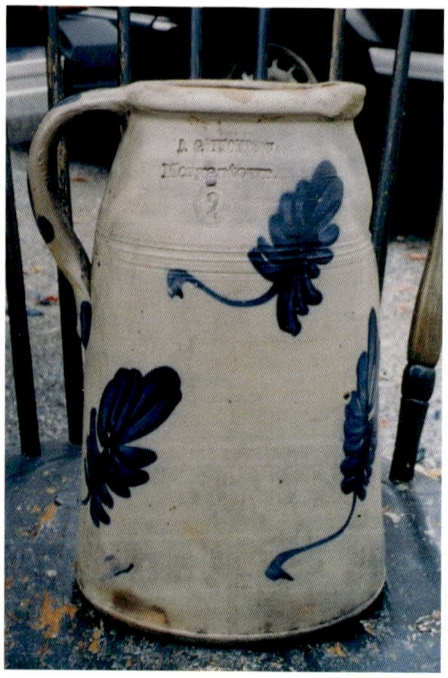

Figure 25 Pitcher, David Greenland Thompson, Morgantown, West Virginia, ca. 1870. Salt-glazed stoneware. H. 9". Marks, impressed on shoulder: D. G. THOMPSON / Morgantown; 2 [gallon capacity stamp within a beaded circle]. (Private collection; photo, Don Horvath.)

Figure 26 Detail of an extruded handle, Morgantown, West Virginia, ca. 1870. Salt-glazed stoneware. (Duez Collection.)

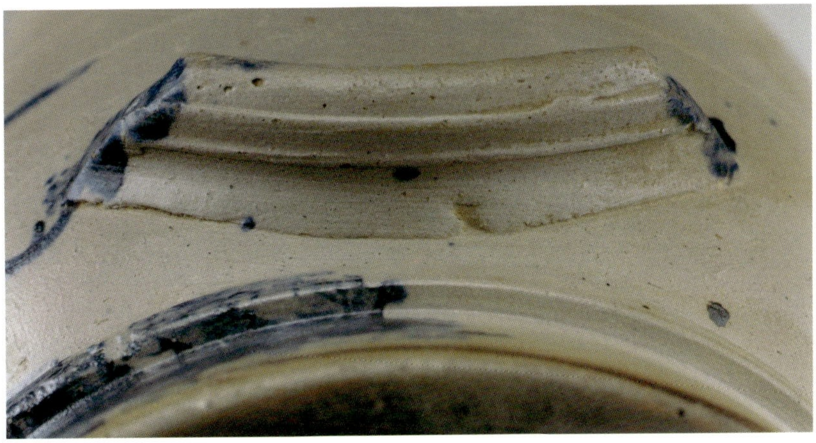

The Smithsonian collection of tools from Morgantown includes implements for making the handles for the stoneware jars and jugs.[17] Extruded handles made with precut dies created a uniform design and sizes; most of the jar handles were extruded with fairly simple profiles (fig. 26).[18] Some of the handles were embossed after having been extruded and applied to the jar (figs. 27, 28). The attention to detail across the top of each lug handle is exquisite and generally seen only in Morgantown examples (fig. 29).

Figure 27 Jar, Morgantown, West Virginia, 1860s. Salt-glazed stoneware. H. 15". Mark: 4 [gallon capacity stamp]. (Duez Collection.) The tulip decoration is unusual for Morgantown stoneware; it is more commonly seen on wares by A. V. Boughner of Greensboro, Pennsylvania.

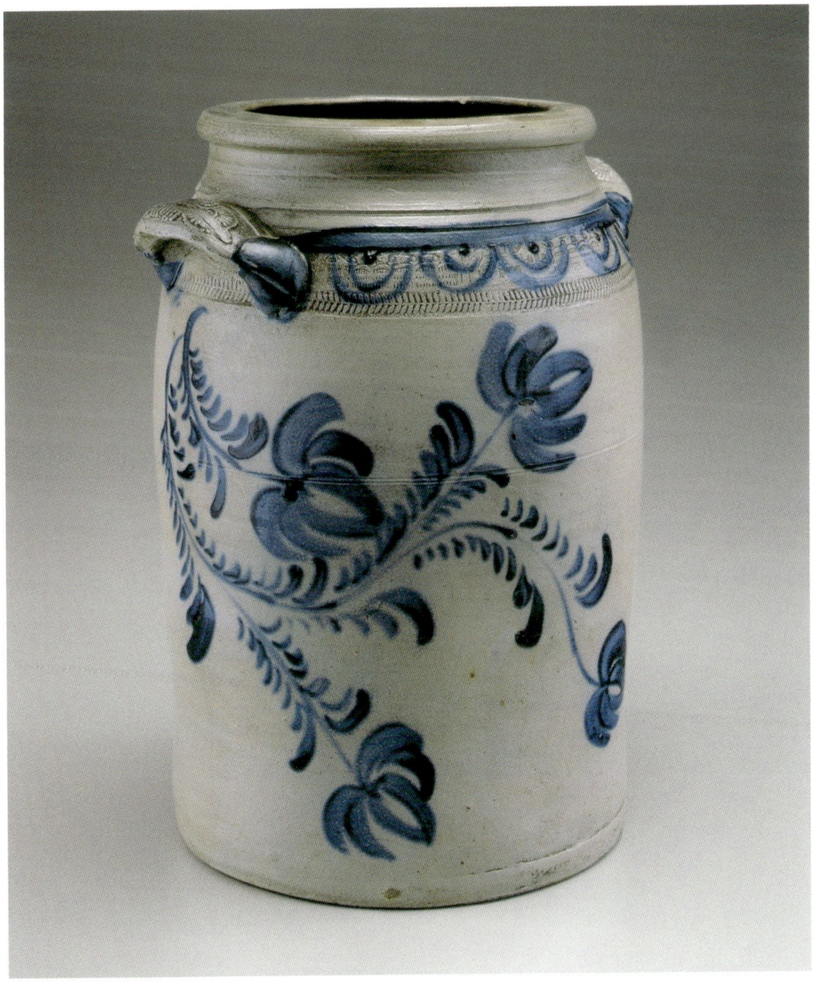

Figure 28 Detail of the embossed handles of the jar illustrated in fig. 27.

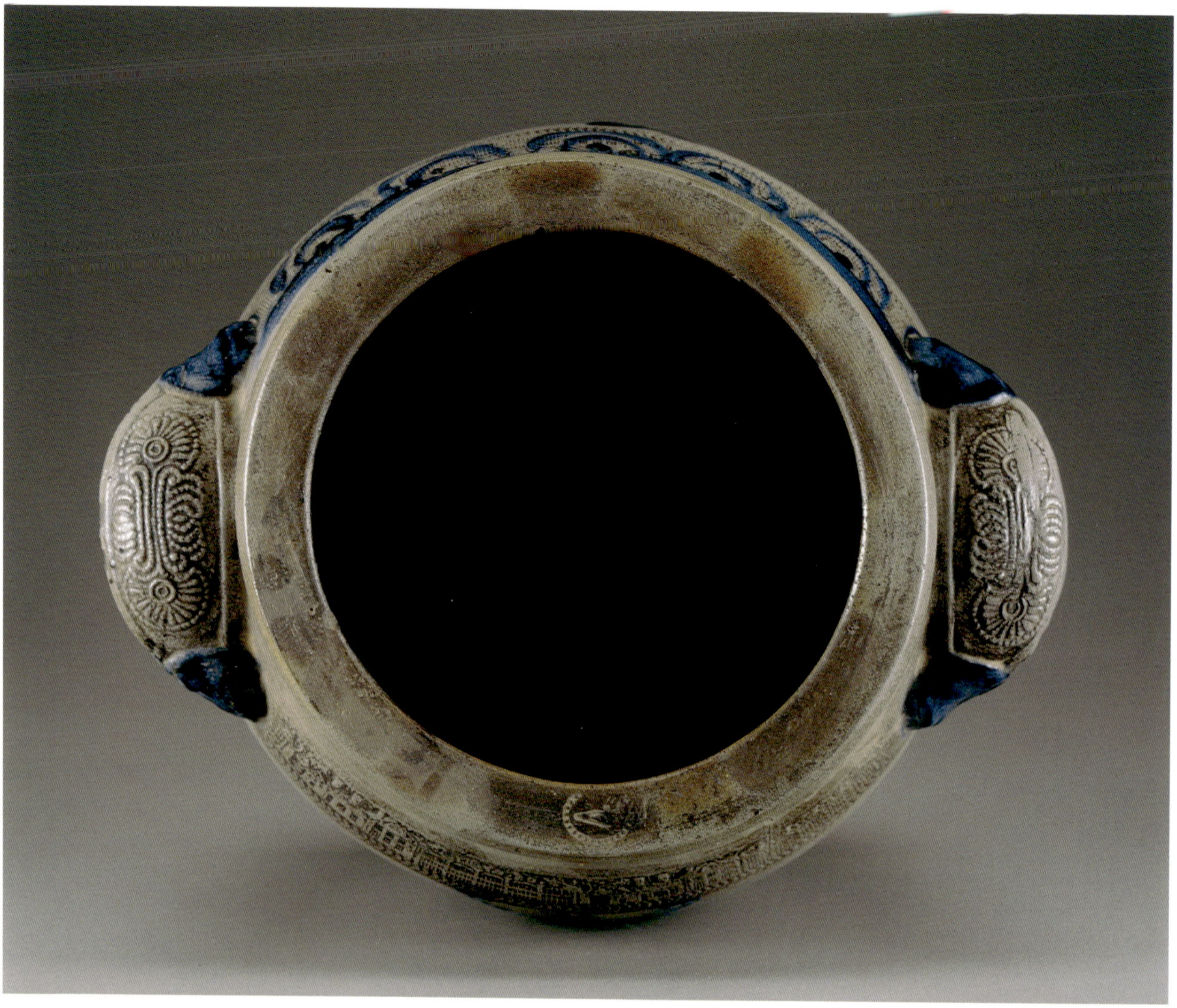

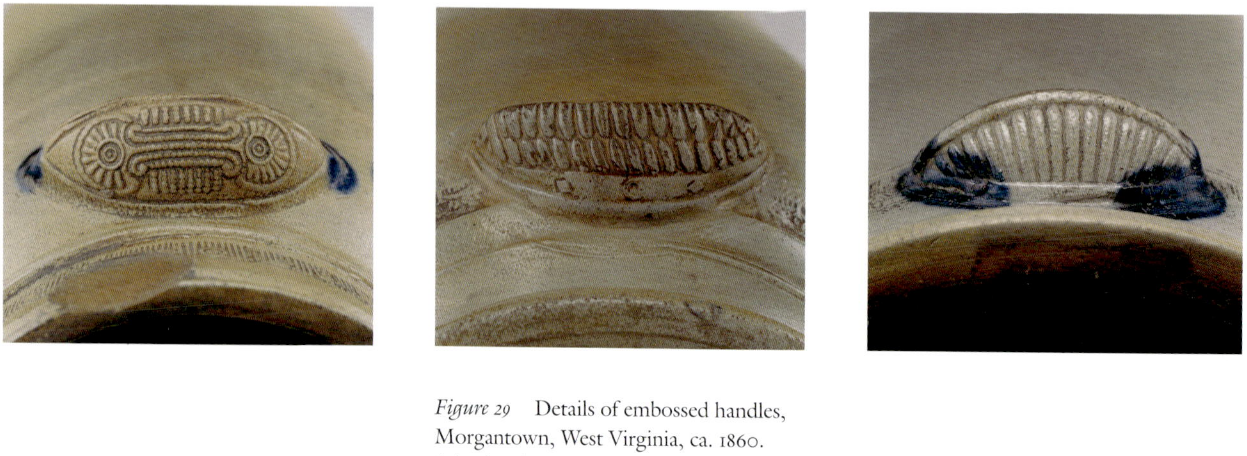

Figure 29 Details of embossed handles, Morgantown, West Virginia, ca. 1860. Salt-glazed stoneware. (Duez Collection.)

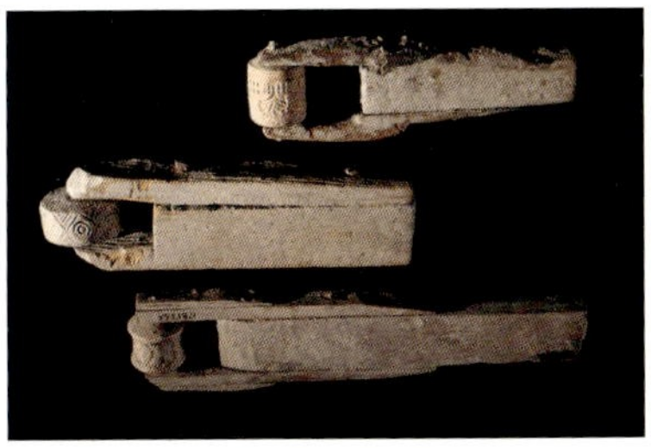

Figure 30 Roulettes, Thompson Pottery, Morgantown, West Virginia, 1850–1870. Wood and bisque clay. (Courtesy, National Museum of American History.)

The decorative technique of rouletting was used on Morgantown earthenware before it was applied to stoneware. Precisely when and under what circumstances David Greenland Thompson obtained the concept of roulette-wheel decoration is not known. William Ketchum has noted that potteries in Massachusetts, New Jersey, and New York were employing such decorations as early as 1830.[19] Perhaps Thompson had traveled to one of those manufacturers, or perhaps, given his proximity to the river systems, he had seen wares with this characteristic being transported or traded.

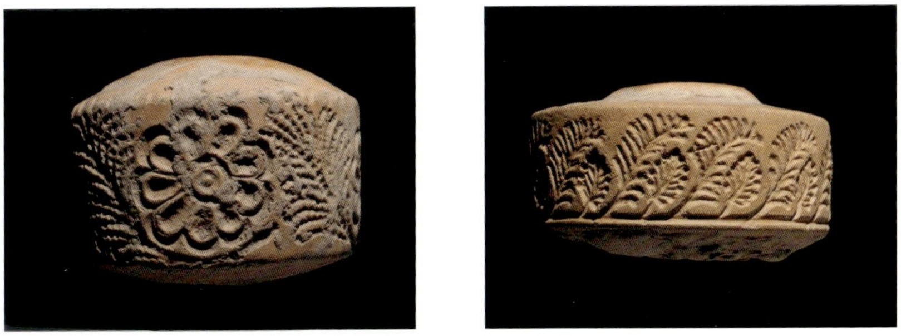

Figure 31 Roulettes, Thompson Pottery, Morgantown, West Virginia, 1850–1870. Bisque clay. (Courtesy, National Museum of American History.)

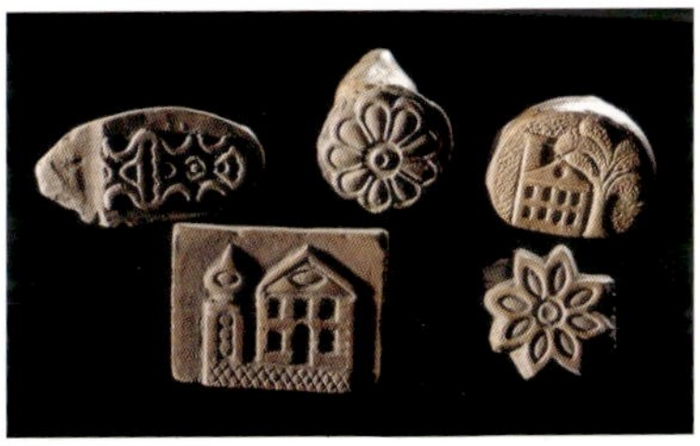

Figure 32 Pottery stamps, Thompson Pottery, Morgantown, West Virginia, 1850–1870. Bisque clay. (Courtesy, National Museum of American History.) These stamps were used to impress small decorative elements onto the wares.

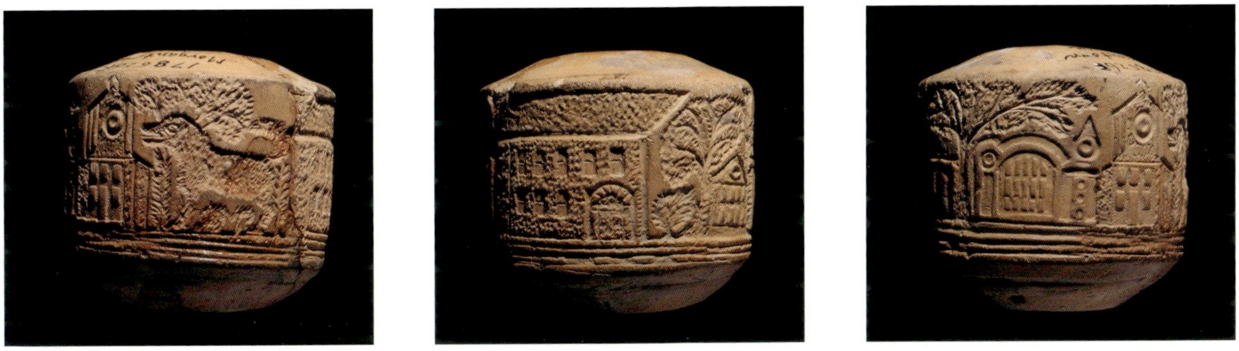

Figure 33 Roulette, Thompson Pottery, Morgantown, West Virginia, 1850–1870. Bisque clay. (Courtesy, National Museum of American History.) This stamp would have been held between two fingers in order to roll the design onto the ware.

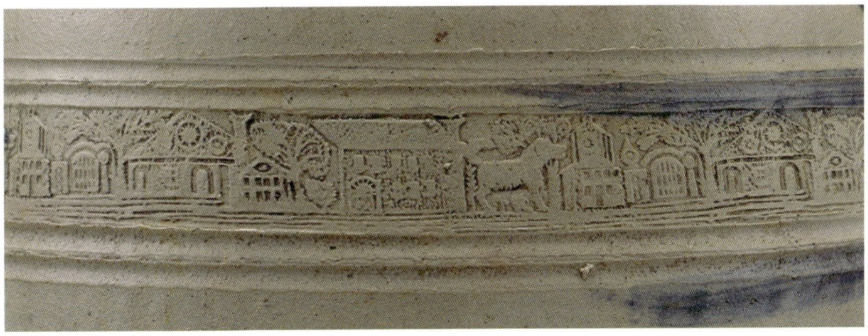

Figure 34 Detail of the impressed design made from the roulette illustrated in fig. 33.

These roulette wheels are made of a bisque clay material and fitted in wooden handles (figs. 30, 31). A number of small decorative stamps are part of the tool assemblage as well (fig. 32) and might have served as the positive masters to create the rouletted designs. Many of the designs carved onto the roulettes are detailed and intricate. There are several street scenes with well-defined small buildings and animals (figs. 33, 34). Some create a pastoral landscape complete with hay wagon, small structures, and trees (figs. 35, 36).

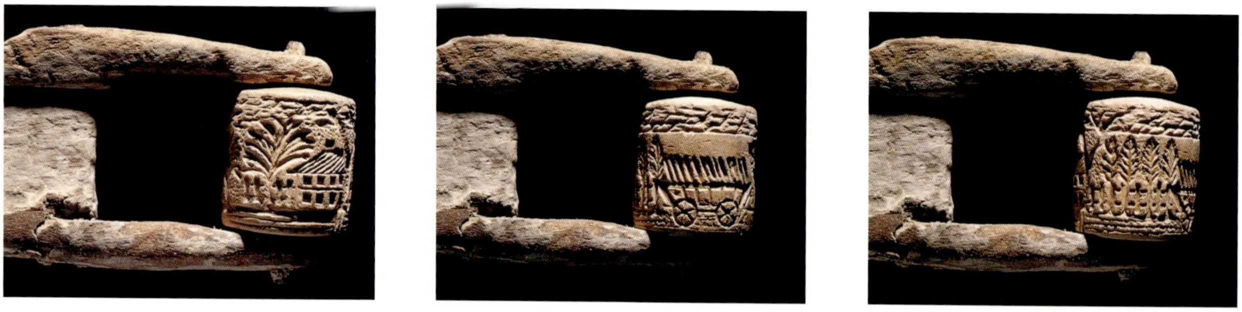

Figure 35 Roulette, Thompson Pottery, Morgantown, West Virginia, 1850–1870. Wood and bisque clay. (Courtesy, National Museum of American History.)

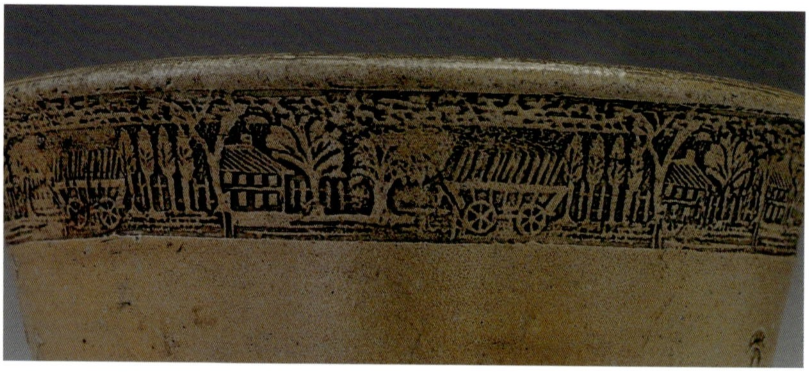

Figure 36 Detail of the impressed design made from the roulette illustrated in fig. 35.

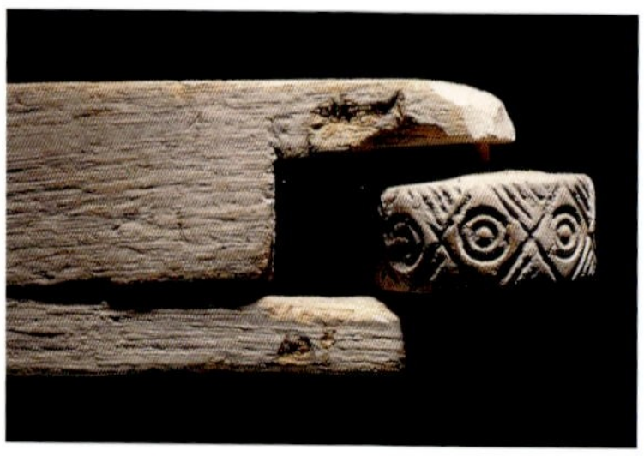

Figure 37 Roulette, Thompson Pottery, Morgantown, West Virginia, 1850–1870. Wood and bisque clay. (Courtesy, National Museum of American History.)

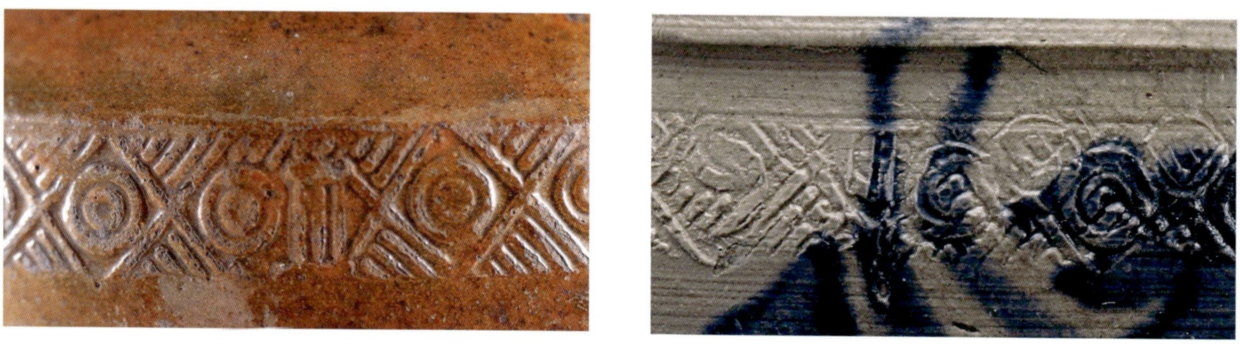

Figure 38 (*Left*) Detail of a rouletted design on an earthenware jar, Morgantown, West Virginia, ca. 1840–1854. (Private collection.) (*Right*) Detail of a rouletted design on a stoneware jar, Morgantown, West Virginia, 1860–1870. (Private collection.) The roulette tool used to make the design on both jars is illustrated in fig. 37.

The use of these elaborate roulettes on Morgantown seems to be restricted to the 1850s and 1860s. Some rouletted patterns carry over from the earthenware period (figs. 37, 38). After David Greenland Thompson acquired the shop in the 1850s, the designs became more varied and were widely used on multiple forms. After the mid-1860s the use of rouletted decoration on Morgantown pots waned, perhaps becoming too labor intensive to use in conjunction with these utilitarian objects.

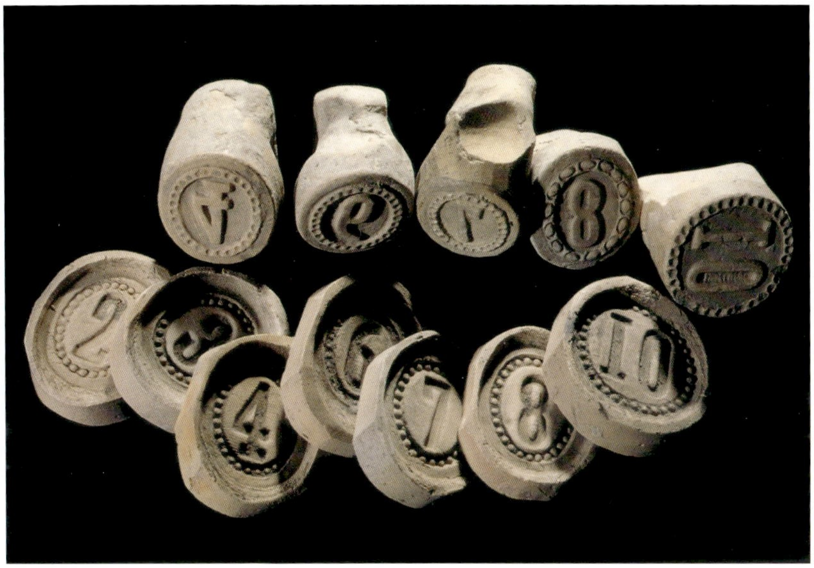

Figure 39 Capacity stamps and master stamp molds, Thompson Pottery, Morgantown, West Virginia, 1850–1890. Bisque clay. (Courtesy, National Museum of American History.)

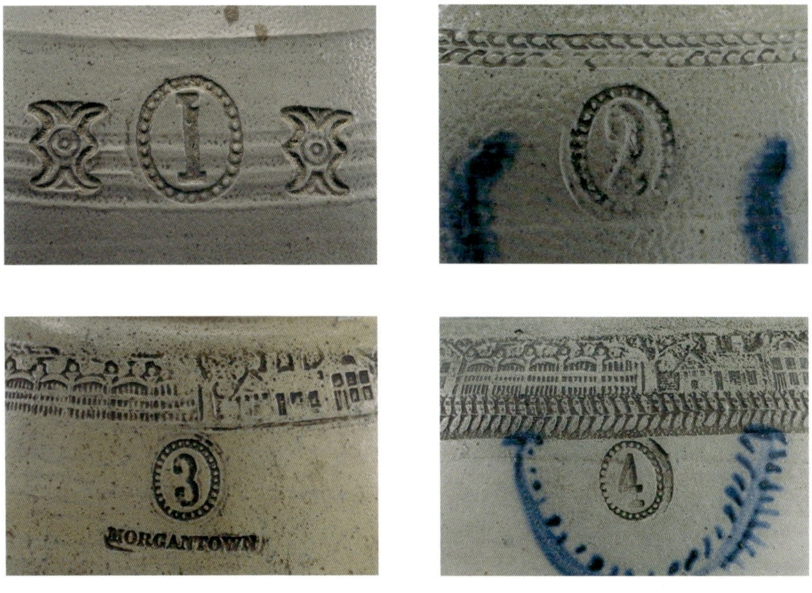

Figure 40 Details of gallon capacity stamps from various Morgantown jars.

In addition to the roulette wheels, the tool collection in the Smithsonian includes beaded stamps used to mark the capacity of the wares (figs. 39, 40). These stamps are associated primarily with the Morgantown pottery, although some of the Uniontown pots bear similar markings. The beaded capacity stamps continued to be used through the 1870s.

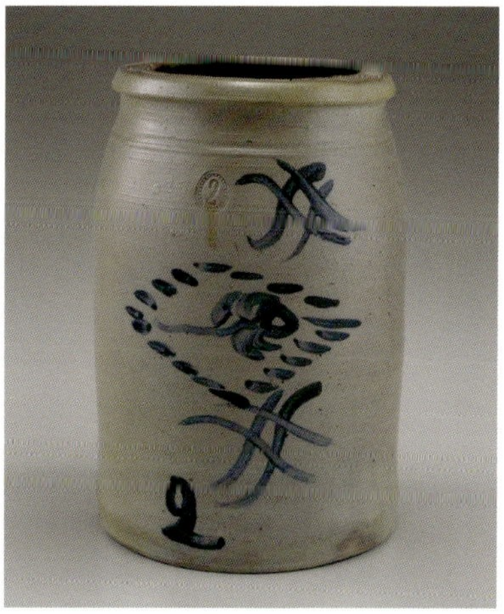

Figure 41 Jar, Morgantown, West Virginia, 1854–1860. Salt-glazed stoneware. H. 8½". Marks: 2 [impressed gallon capacity stamp]; 2 [inscribed in cobalt]; Steam / Bote [inscribed on bottom]. (Duez Collection.) This jar is decorated with a cobalt Tic-Tac-Toe motif of the type used during the earthenware period.

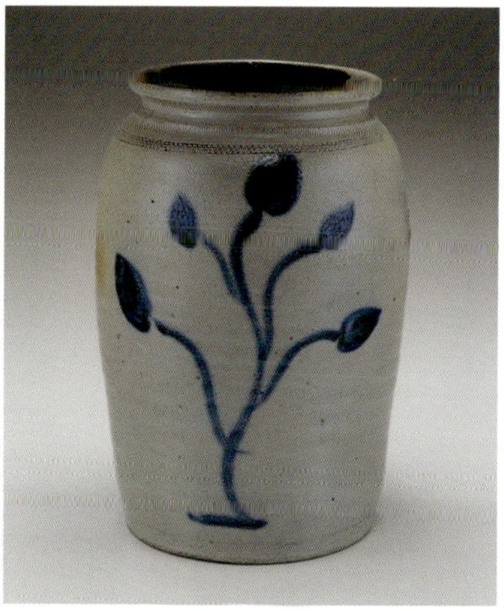

Figure 42 Jar, Morgantown, West Virginia, ca. 1860. Salt-glazed stoneware. H. 12¼". Mark: 2 [beaded gallon capacity stamp]. (Duez Collection.) A budding branch decorates this stoneware jar, which has a history in Marion County, West Virginia.

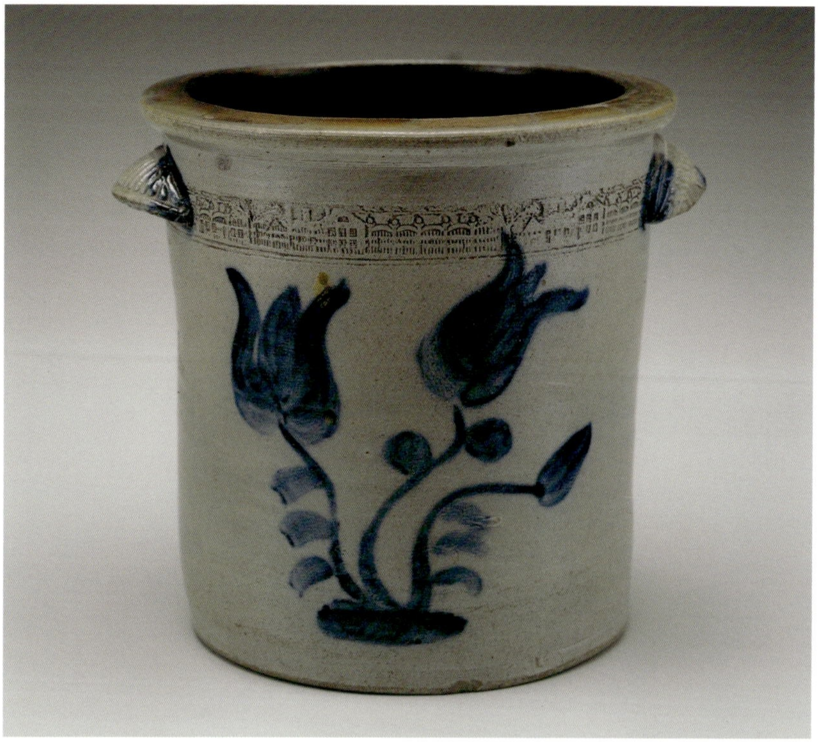

Figure 43 Crock, Morgantown, West Virginia, 1860–1870. Salt-glazed stoneware. H. 9½". Mark: 2 [impressed gallon capacity stamp within a beaded circle]. (Duez Collection.) The floral decoration on this handled crock is enhanced with a rouletted street-scene pattern. The crock was found in Kingwood, West Virginia, where it had long been used in the making of buckwheat cakes.

While beaded stamps continued to be used, floral cobalt decoration dominated through the 1870s on the Morgantown products. The decorative fern on the 1854 jar illustrated in figure 2 reflects the shift toward cobalt decoration within the budding stoneware industry of the Morgantown pottery. These fernlike decorations had not been widely used on the earthenware forms, although the so-called Tic-Tac-Toe design is found on both earthenware and stoneware (fig. 41). Cobalt was the most expensive commodity in a potter's studio in the nineteenth century, and potters would often mix it with a slip made from the clay body they were using in order to make it last longer.[20] Typically, cobalt was applied with a brush, though a sponge was used on several pots attributed to the Morgantown pottery.

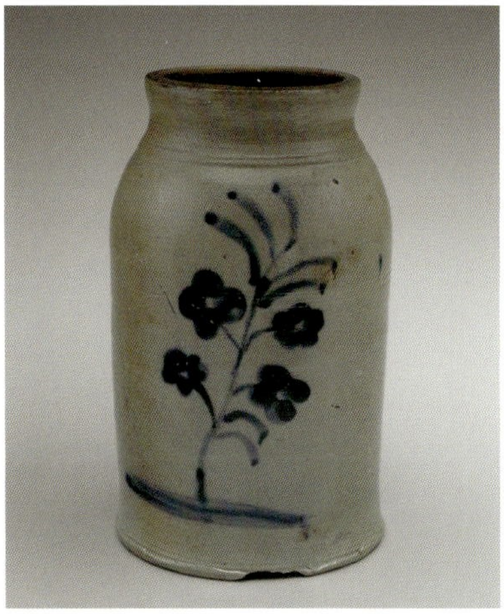

Figure 44 Churn, Morgantown, West Virginia, 1860–1870. Salt-glazed stoneware. H. 8¾". (Private collection; photo, Ken Landreth.)

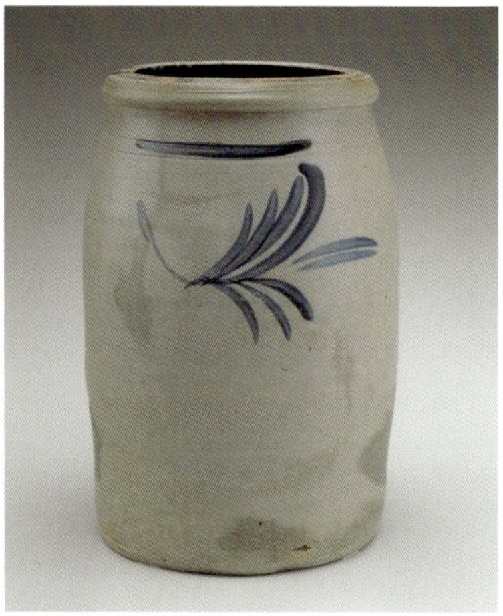

Figure 45 Jar, attributed to Frank W. Thompson, Morgantown, West Virginia, 1865–1875. Salt-glazed stoneware. H. 12¼". Mark, impressed on shoulder: F. W. Thompson / Morgantown. (Ken Landreth Collection.) This jar is the only known piece that bears what seems to be a reference to Frank W. Thompson.

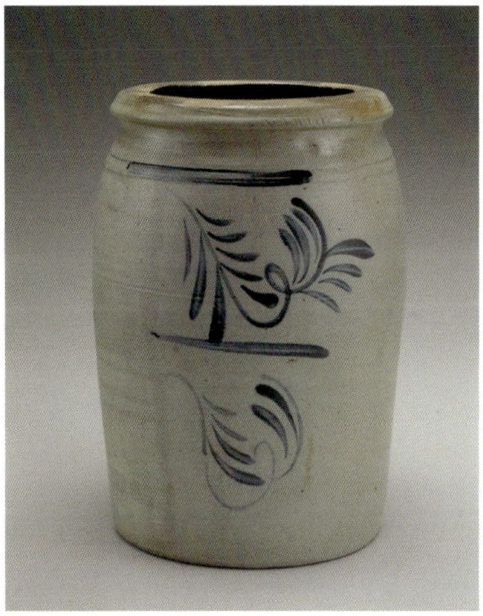

Figure 46 Jar, attributed to Frank W. Thompson, Morgantown, West Virginia, 1865–1875. Salt-glazed stoneware. H. 11¼". Mark: 2 [gallon capacity stamp within a beaded circle]. (Private collection.) Based on its similarities to the decorated jar illustrated in fig. 45, this example was probably decorated by Frank W. Thompson as well.

Wares from southwestern Pennsylvania and areas bordering West Virginia were known for their ample foliage and floral decorations—tulips, fuchsias, and, in Morgantown, ferns (figs. 42–44). The Uniontown pottery wares often display lavish trumpet flowers, a decoration not found among Morgantown's wares.[21] The more sparse decoration seen on the jars illustrated in figures 45 and 46 are attributed to Frank W. Thompson, who had returned to the business following the Civil War. Floral decoration on the stoneware continued well into the 1870s.

The transition to stoneware also included the introduction of a series of whimsical images of people, animals, and birds. A small group of pots by the Thompson, Uniontown, and Greenland potters depicts roughly painted images of people surrounded by foliage or trees (figs. 47, 48). Many of these pots, affectionately referred to as "people pots," also incorporate David Greenland Thompson's roulettes, stamps, and impressed designs as well as the whimsical characters and animals.[22]

While many believe that "people pots" originated in the Uniontown area, we feel strongly that John W. Thompson introduced these depictions on pots in Morgantown. The pots with definite Morgantown characteristics can be closely dated to between 1854 and 1863, when Thompson died.[23]

Figure 47 Jar, Morgantown, West Virginia, ca. 1860. Salt-glazed stoneware. (Private collection.) Rendered in cobalt slip on the side of this jar is a woman holding a flag, with trees on either side of her. The scene is framed at the top by a feathery arc, also in cobalt slip.

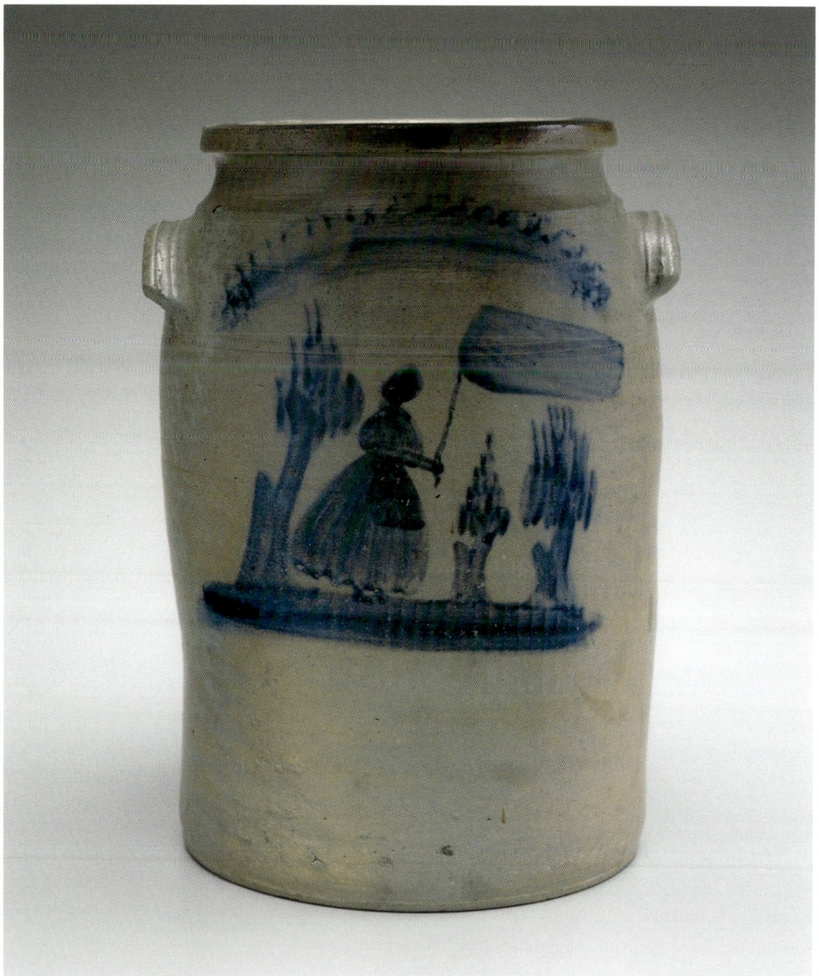

Figure 48 Jar, Morgantown, West Virginia, ca. 1860. Salt-glazed stoneware. H. 9¾". (Private collection.)

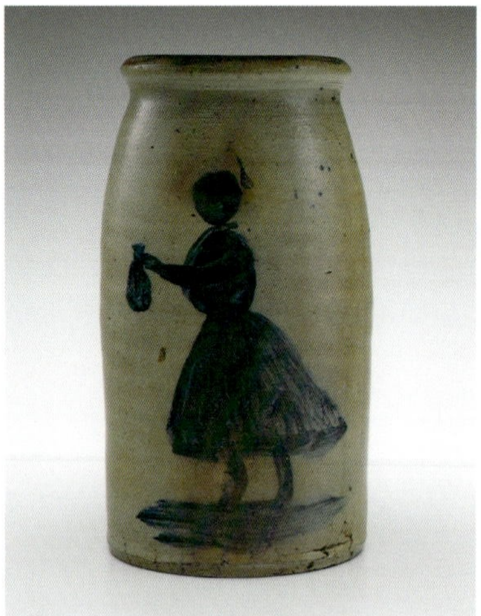

Figure 49 Cream pitcher, Morgantown, West Virginia, ca. 1860. Salt-glazed stoneware. H. 6¼". (Private collection.) This pitcher, purchased in Grafton, West Virginia, is decorated with an elaborate sunburst, four human figures, and a duck or goose.

Norval Greenland of Uniontown did not begin his pottery factory until the early 1860s, and so must have been influenced by John W. Thompson's designs. By 1865 the characteristic "people pots" of Uniontown seem to disappear also, and decorations gravitated toward floral designs. Similarities between the Uniontown, Greenland, and Morgantown pot decorations make it more difficult to distinguish where each pot was made.

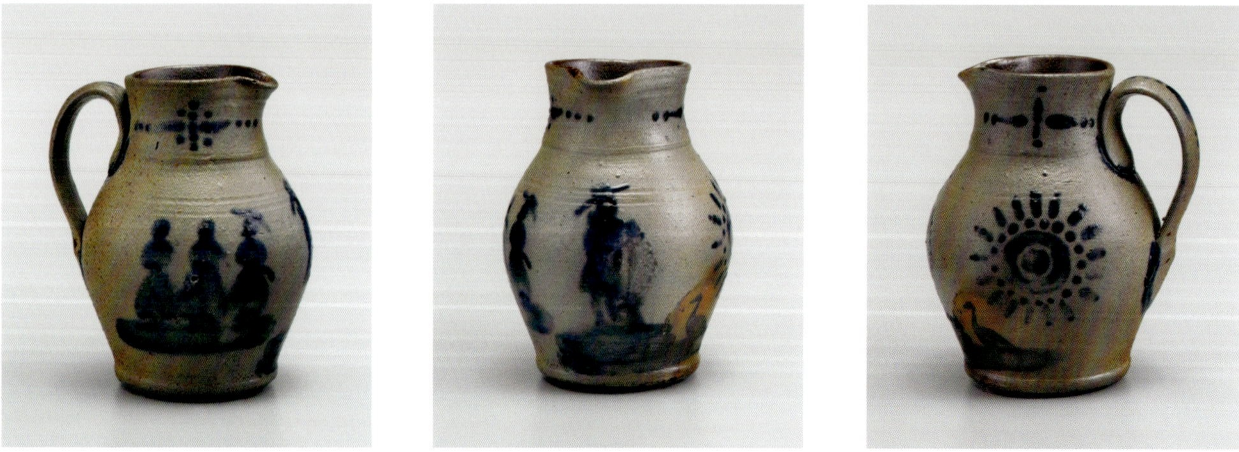

Many of the Morgantown examples show objects in the hands of the subjects. A fine example of a small pitcher, 6¼ inches tall (fig. 49), has a rounded belly, sloping shoulder, high-collared rim with burgeoning spout, and a neatly tucked foot. Beneath the spout a man wearing a wide-brimmed

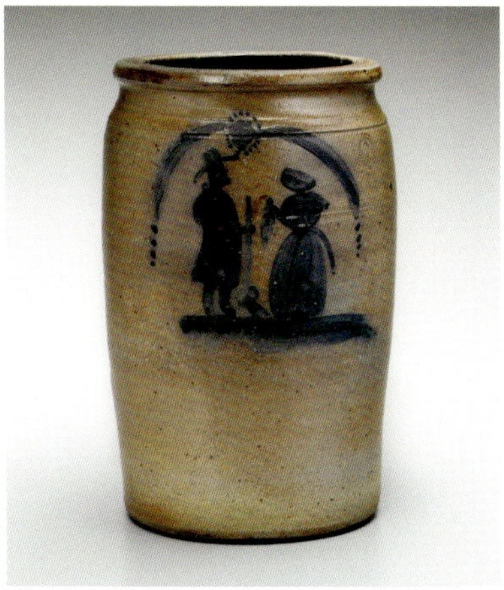

Figure 50 Jar, Morgantown, West Virginia, ca. 1860. Salt-glazed stoneware. Mark: 2 [gallon capacity stamp within a beaded circle]. (Private collection.)

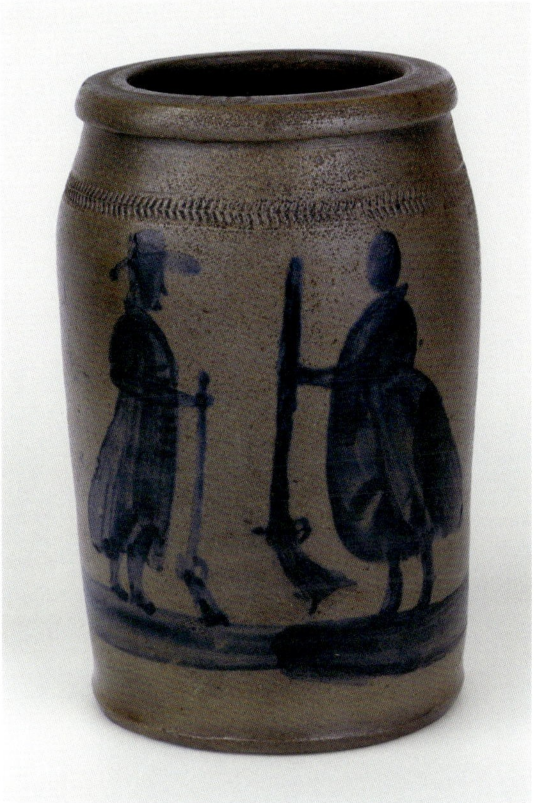

Figure 51 Jar, Morgantown, West Virginia, 1854–1865. Salt-glazed stoneware. H. 9¹³⁄₁₆". (Courtesy, Colonial Williamsburg Foundation.) This jar was found in Aurora, Preston County, West Virginia. A simple rouletted pattern with chevrons is impressed around the shoulder. The cobalt decoration shows a man and a woman, each holding a gun, facing each other. A nearly identical rendition of the figure on the left appears on the jar illustrated in fig. 50.

hat carries a long rifle. A brilliantly detailed sun radiates from one side, with a duck or goose below its rays; on the opposite side stand three women. The arrangement of the scenes appears somewhat random but may have held special significance for Thompson. Rifles appear on several other pieces (fig. 50), including a comical statement between a man and a woman on the small jar illustrated in figure 51.

The figures seen on the wares attributed to the Morgantown Pottery are generally painted in great detail. Interestingly, the decorator chose to place well-defined noses on the men, whereas the women's faces are usually featureless. Hats are placed on many of the men. The women are shown in flowing aprons and skirts, many also wearing bonnets and carrying parasols (figs. 52–54).

John W. Thompson is also credited with depicting a number of stylized animals, birds, and fish on his jars—for example a whale (fig. 55) and the stag on the enormous jar illustrated in figure 56—which are highly sought after by modern folk art collectors. Thompson's painting style shows brushstrokes that are often very broad and very thick. These overlapping layers of cobalt can trace the evolution of the composition.

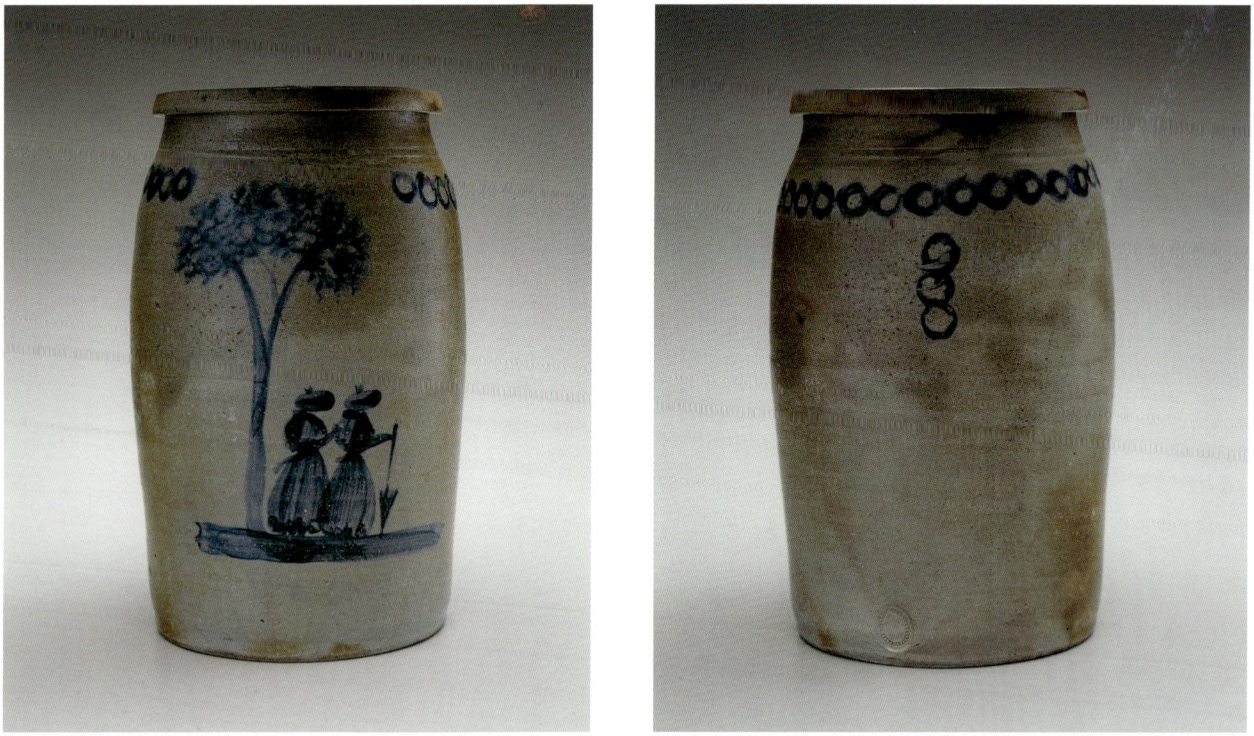

Figure 52 Jar, Morgantown, West Virginia, ca. 1860. Salt-glazed stoneware. H. 12½". Mark: 2 [gallon capacity stamp within a beaded circle]. (Private collection.) This jar was found in Upshur, West Virginia. The painted decoration shows two women, one with a parasol, under a tree. The cobalt chain decoration around the shoulder is unusual for Morgantown stoneware.

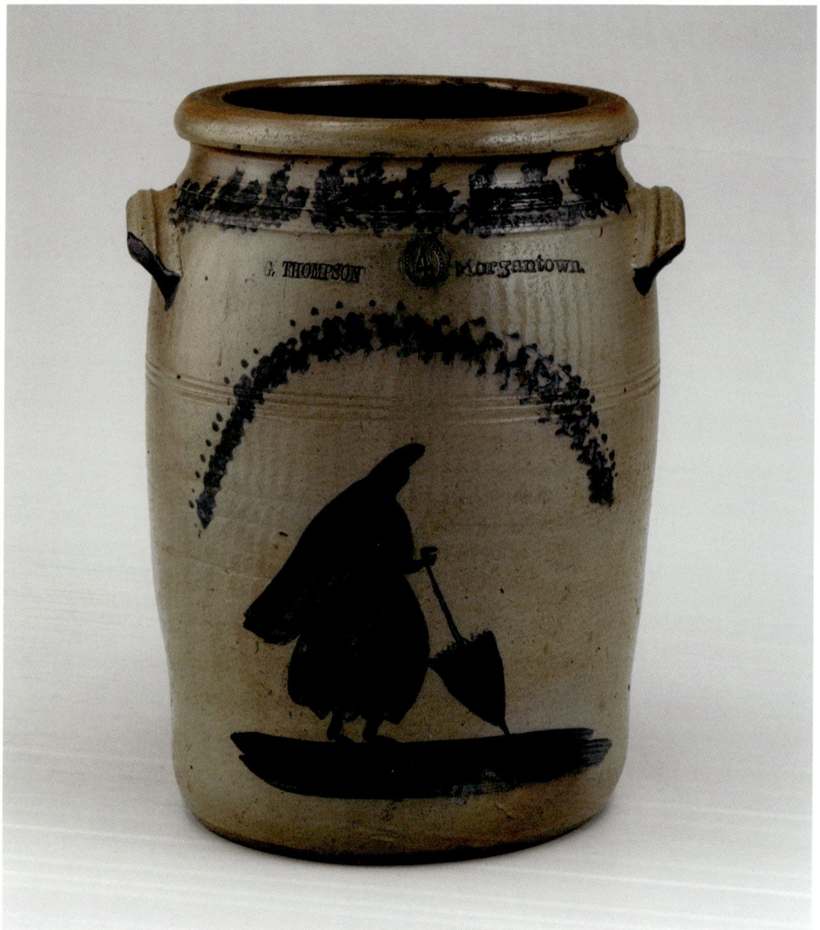

Figure 53 Jar, Morgantown, West Virginia, ca. 1860. Salt-glazed stoneware. H. 14". Marks: 4 [gallon capacity stamp within a beaded circle]; D. G. THOMPSON / Morgantown. (Private collection.) The cobalt decoration includes a sponged pattern around the rim and a sponged crescent arcing over a woman holding a parasol. The jar was found in Fairmont, Marion County, West Virginia.

Figure 54 Jar, Morgantown, West Virginia, ca. 1860. Salt-glazed stoneware. H. 13¾". Mark: 3 [gallon capacity stamp within a beaded circle]. (Private collection.) The scene illustrated here shows three women standing between two trees. The capacity stamp is filled in with cobalt.

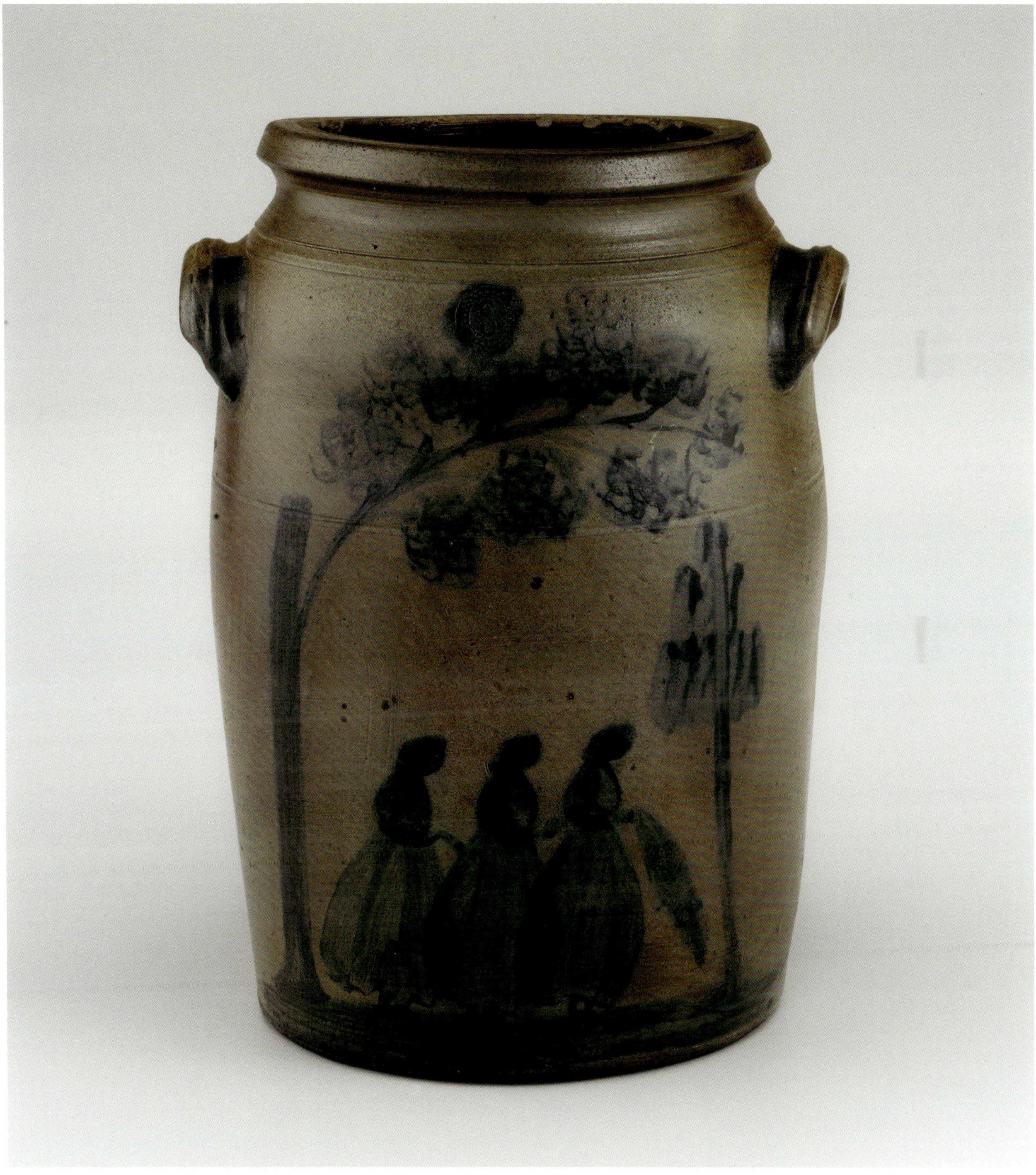

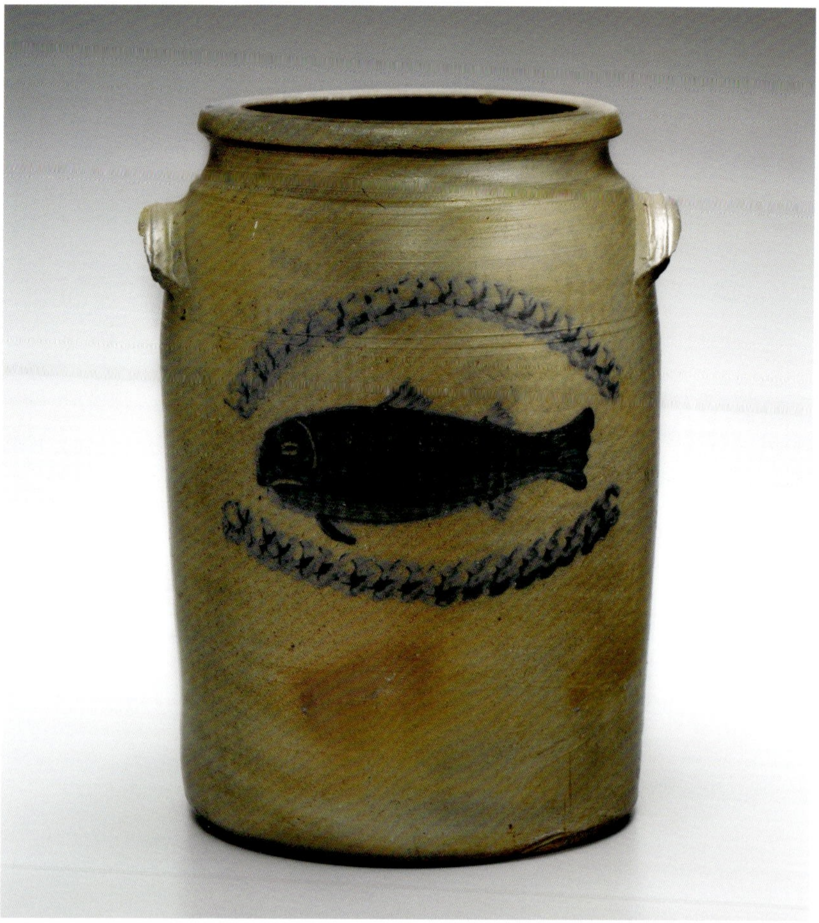

Figure 55 Jar, Morgantown, West Virginia, 1854–1860. Salt-glazed stoneware. (Private collection.) The cobalt decoration shows a whale framed within a sponged pattern.

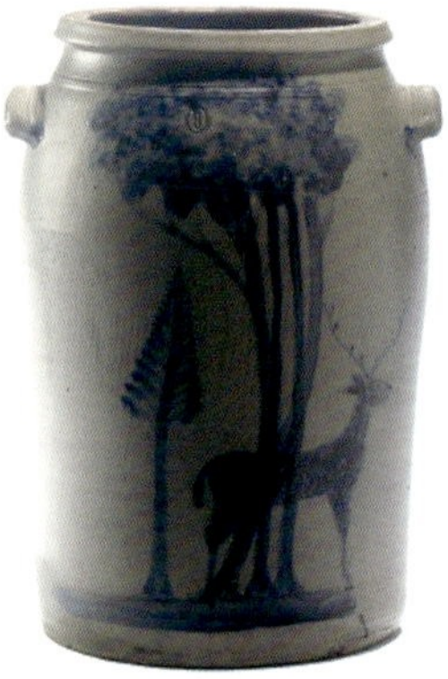

Figure 56 Jar, Morgantown, West Virginia, 1854–1863. Salt-glazed stoneware. H. 19". Mark: 8 [gallon capacity stamp within a beaded circle]. (Courtesy, Northeast Auctions.)

Figure 57 Jar, Morgantown, West Virginia, ca. 1875. Salt-glazed stoneware. H. 20½". (Duez Collection.)

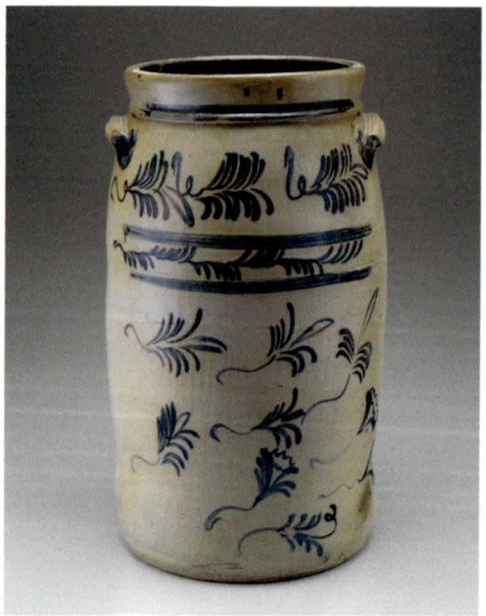

By 1870 David Greenland Thompson and his brother Frank were the only immediate members of the Thompson family listed as potting in Morgantown, with David living with Frank.[24] David reported having a "5 h.p. steam engine" (installed by James Jackson Thompson) using "415 cords of wood," paying "$300 in wages," and producing "$1900 of products" in nine months of operation.

Sometime after 1870, Frank began operation of a flour and gristmill on the same site as the pottery using the engine and discontinued work as a potter. The days of elaborately hand-decorated stoneware (fig. 57) were coming to an end in Morgantown. In 1875 David entered a three-year partnership with Robert T. Williams, who had finished an apprenticeship with Littel/Little and Company and with Hamilton and Jones of Greensboro, Pennsylvania.[25] One known example of this partnership shows Thompson's willingness to employ stenciling in the cobalt decoration. Stenciled decoration would become de rigueur for the New Geneva Pottery, of which Robert T. Williams became proprietor in 1882.[26] Stencils became increasingly popular with potters as a way to produce and decorate their wares in a less costly, more efficient manner, an adjustment necessary to keep up with the rising competition of industrialized pottery production.[27]

Despite the industrialization of the process of making stoneware, David Thompson's inventiveness still showed. Plate 7 of "An Early West Virginia

Figure 58 Flower pot, attributed to David Greenland Thompson, Morgantown, West Virginia, ca. 1885. Salt-glazed stoneware. H. 13¾". (Photo, courtesy of James and Karen Giuliani.)

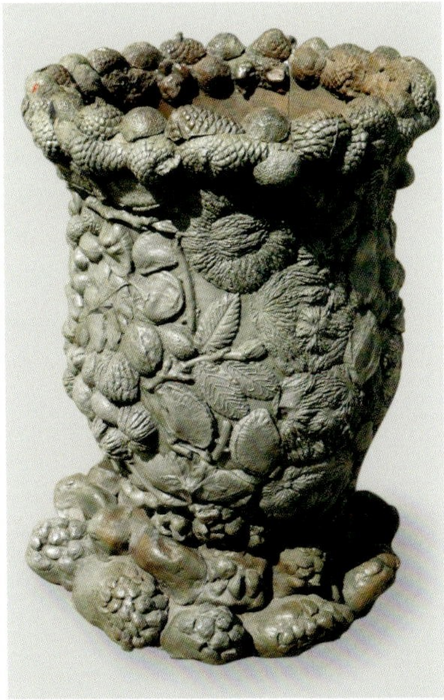

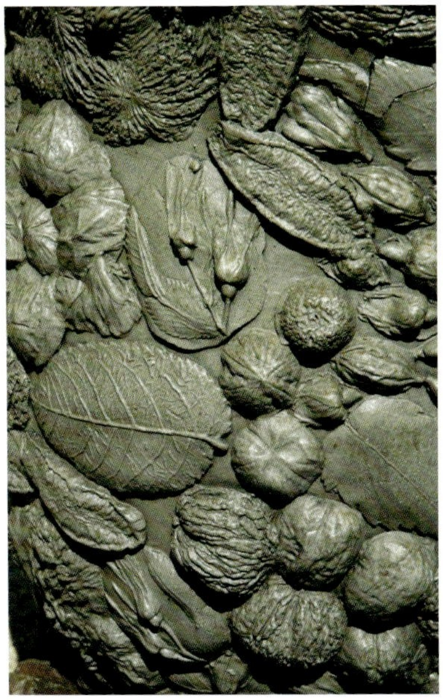

Figure 59 Vases and bowl, attributed to David Greenland Thompson, Morgantown, West Virginia, ca. 1885. Salt-glazed stoneware. H. of center vase 9¼". (Duez Collection.) Nuts, peach pits, and freshwater clams are all part of the applied decoration covering these objects.

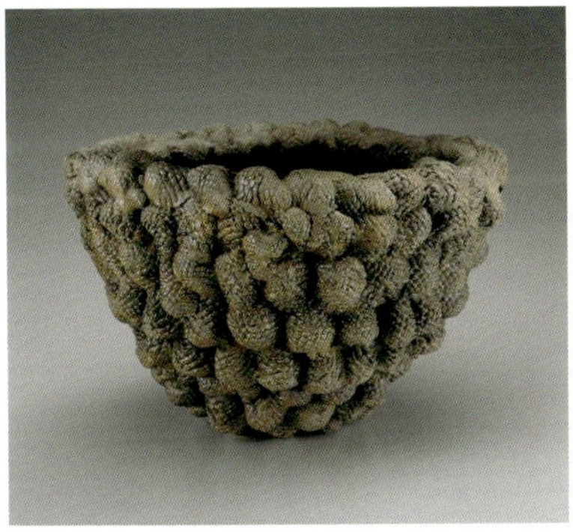

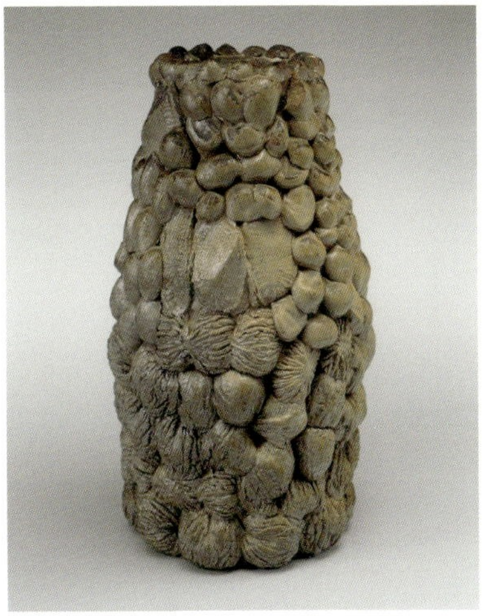

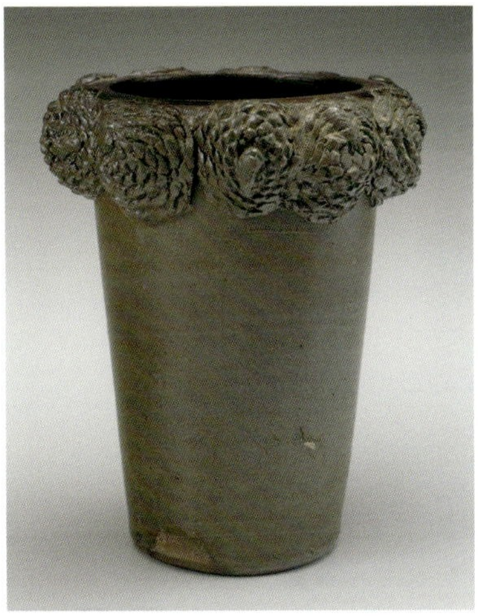

Pottery" illustrates four pieces from circa 1870–1885 having applied clay that had been pressed in molds of pine and spruce cones, nuts, and simulated flowers and leaves. In a number of examples David Thompson used what Walter Hough later described as applied creativity with a natural approach to the form (figs. 58, 59). Hough noted that these forms, which he attributed to "Greenland Thompson," were not sold but rather offered as gifts to people in the community; others were given to family members (fig. 60).[28]

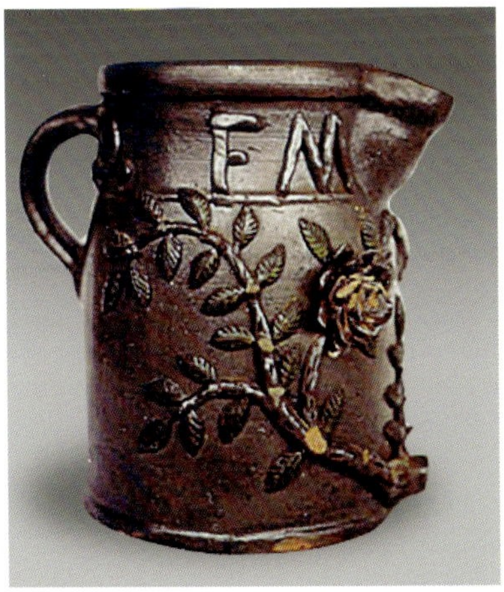

Figure 60 Pitcher, attributed to David Greenland Thompson, Morgantown, West Virginia, ca. 1885. Salt-glazed stoneware. H. 9". (Private collection; photo, Don Horvath.) The pitcher is covered with applied leaves and stems. Applied lettering around the neck spells out "F. M. Haymond," one of Thompson's relatives.

The End of an Era

When David Greenland Thompson died, in 1890, the Morgantown Pottery ceased operation.[29] After approximately ninety years of operation, the Morgantown pottery did not see a diminishing interest in their wares, nor did the creativity of the pieces or attention to detail lessen with the owner's advancing age. Yet surely it would have met the same destiny other stoneware pottery factories faced at the turn of the twentieth century. These factories, which typically employed five to thirty workers at a time, were little match for the industrial factories that were beginning to take over the stoneware market. Faced with the need to turn their entire factory of handmade pottery over to machines and molds, many potters could not afford to make the transition, having neither the space nor the workforce. Moreover, as railroads established new, faster transportation routes in the United States, the once-strategic location of these potteries—on the riverbanks—became obsolete.[30]

The history of the Morgantown Pottery transition from earthenware to stoneware is a unique, and important, example of a pottery rising to the challenge of imported stoneware in a local economy and grasping an opportunity for development in decoration and technique. The existence of the wares and tools used to make them is a rare glimpse into the history of ceramic manufacture in America.

1. The somewhat comical adventures of Jacob Foulk Jr. (Fouke, Folk, Fulk, Faulke) were first described in the 2004 article. Recent research conducted by Millie Covey Fry has revealed that Jacob Foulk moved from Morgantown to Brooke County, Virginia, where he appeared once more in the court records, as he often had in Morgantown. He appears in the 1820 census but presumably died shortly thereafter, since in 1821 his wife requested that the estate of her late husband be administered.

2. Don Horvath and Richard Duez, "The Potters and Pottery of Morgan's Town, Virginia: The Earthenware Years, Circa 1796–1854," in *Ceramics in America*, edited by Robert Hunter (Hanover, N.H.: University Press of New England for the Chipstone Foundation, 2004), p. 111. The pottery moved to the wharf in 1830, after the pottery on Lot 88 had burned.

3. Walter Hough, "An Early West Virginia Pottery," in *Report of the National Museum* (Washington, D.C.: Government Printing Office, 1899), p. 515.

4. Listed below are the only known Thompson wares bearing dates:

1854	"First 2 jar for Francis Madiera June the 2nd, 1854" (see figs. 2, 3)
1866	"JHT / BB / 1866" conical wax sealer (not illustrated)
1867	"Oct. the 8th 1867 Sarah E. Chew" milk bowl (not illustrated)
1869	"Thompson / Morgantown / 1869" preserve jar (see fig. 19)
1880	"1880 / 2" jar (not illustrated)
1885	"Rufus West / Oct 19th 1885" jar (not illustrated)

5. Monongalia Historical Society, *Sesquicentennial of Monongalia County, West Virginia* (Morgantown, W. Va.: Monongalia Historical Society, 1926).

6. Phil Schaltenbrand, *Big Ware Turners* (Bentleyville, Pa.: Westerwald Press, 2002), pp. 152–53. Schaltenbrand notes that even though the clay seam was recognized by 1810 as being usable for pottery, potters in this area continued to produce earthenware into the 1840s.

7. Professor J. C. White, "Evidence of a Great Ancient Glacial Lake in West Virginia," in Jebediah Hotchkiss, *The Virginias* (Staunton, Va.: Jebediah Hotchkiss, 1883), p. 139. White described the seam of clay located across the river from Morgantown in Pennsylvania as having "thick deposits of very fine, bluish white clay." These clays have been used extensively for the manufacture of pottery at Geneva and Greensboro, and to some extent Morgantown and Fairmont, West Virginia. White also referred to the "transported material" in Morgantown's terraces of clay and soil, possibly inferring that Morgantown had a source of earthenware when they were producing such wares. A student of his, Walter Hough, assisted in the analy-

sis of these clay seams and went on to become a curator at the Smithsonian and the author of "An Early West Virginia Pottery."

8. He was reported to have worked for the Thompsons, but his endeavor in stoneware was short-lived. Hough, "An Early West Virginia Pottery," p. 516.

9. Ibid., p. 514.

10. Ibid., p. 515.

11. Horvath and Duez, "The Potters and Pottery of Morgan's Town, Virginia," p. 120.

12. Phil Schaltenbrand, *Stoneware of Southwestern Pennsylvania* (Pittsburgh, Pa.: University of Pittsburgh Press), p. 76.

13. Ibid., p. 37.

14. Ibid., p. 76.

15. Ibid., pp. 76–78.

16. Hough, "An Early West Virginia Pottery," pp. 515–16. Hough referred to Thompson simply as "Greenland Thompson," a name used in several other references as well.

17. In most cases the handle molds are believed to have been made by impressing cast-iron hardware bin handles into a clay "master," which was then fired to become an inverse form with which to shape the lugs.

18. An image of these extrusion dies is illustrated in Horvath and Duez, "The Potters and Pottery of Morgan's Town, Virginia," p. 120, fig. 51.

19. William Ketchum, *American Stoneware* (New York: Henry Holt and Co., 1991), p. 4.

20. Schaltenbrand, *Stoneware of Southwestern Pennsylvania*, pp. 120–21. Schaltenbrand also provides figures in *Big Ware Turners*, pp. 82–83, writing that the "Isaac Hewitt enterprise in Rices Landing . . . was paying 50 cents a pound for cobalt in the 1860s but annually needed only 50 lbs to decorate approximately 90,000 pots." Although the cost of cobalt was rather high, it could be stretched or watered down to accommodate large production.

21. Schaltenbrand, *Stoneware of Southwestern Pennsylvania*, p. 79.

22. The Uniontown pottery had at least three beaded gallon-capacity stamps. They can be seen on a 3-gallon jar with two figures painted on the front; a 5-gallon Uniontown floral jar with a redundant Roman "V"; and a 20-gallon jug.

23. Schaltenbrand, *Stoneware of Southwestern Pennsylvania*, p. 79.

24. The census lists "2 males above 16 years," "1 female above the age of 16 years," and "3 lathes" (potters' wheels). Other female potters are known. For example, the 1880 census for Nicholson Township, Greene County, Pennsylvania, lists Ann Mariah Sandusky as one of the New Geneva Pottery workers (the authors thank Michael Gallis for this information). The Philadelphia City Directory for 1798 includes "Margaret Freytag, potter" at the corner of South Fifth and Small Streets. For a discussion of the position of women in the pottery studio, see Schaltenbrand, *Big Ware Turners*, p. 70.

25. Gordon C. Baker, "Robert T. Williams, New Geneva, Pennsylvania Potter," *Spinning Wheel* 34 (April 1978): 12–14.

26. Schaltenbrand, *Stoneware of Southwestern Pennsylvania*, p. 89.

27. Ketchum, *American Stoneware*, pp. 6–7.

28. Hough, "An Early West Virginia Pottery," p. 514.

29. Ibid., p. 517.

30. Schaltenbrand, *Stoneware of Southwestern Pennsylvania*, pp. 159–60.

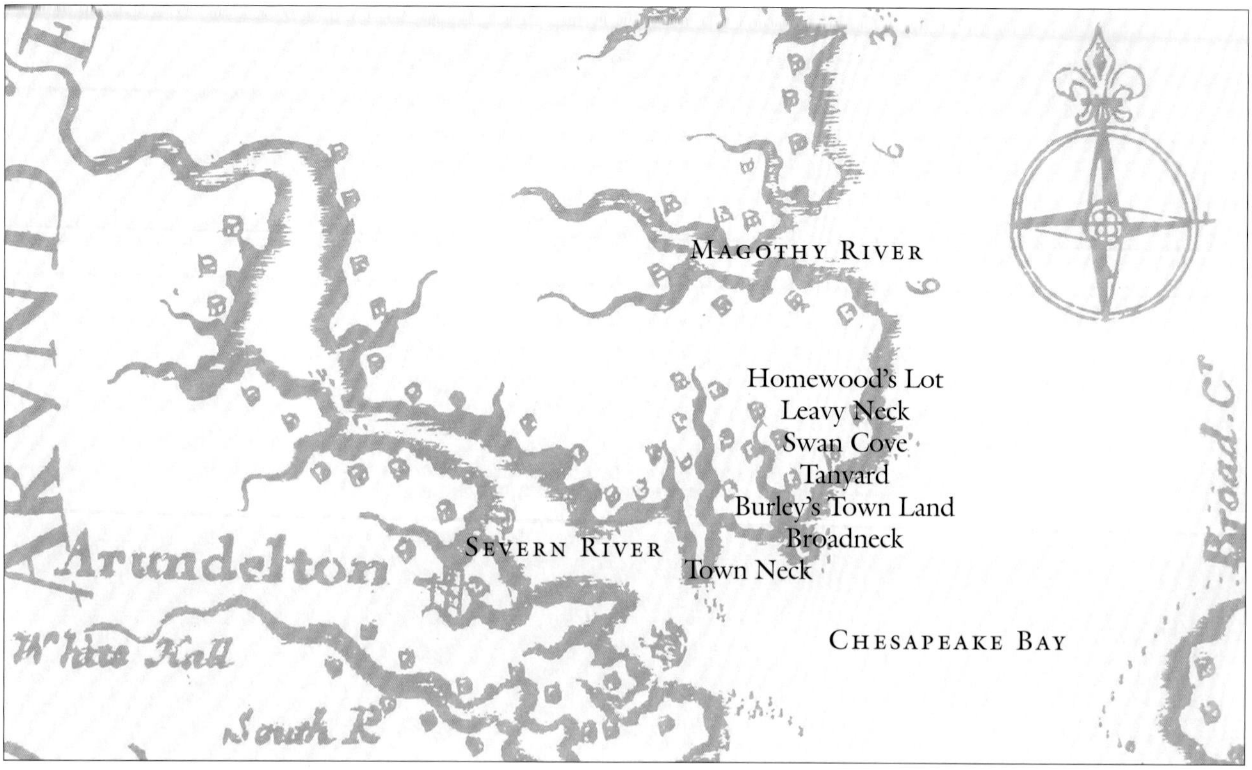

Figure 1 Providence settlement sites overlaid onto detail of *Virginia and Maryland*, map drawn by Augustine Herrmann in 1670 and published in 1673. (Private collection.)

Al Luckenbach and John E. Kille

Ceramic Treasures in Seventeenth-Century Trash: A 1660s Cellar Deposit

Figure 2 Lost Towns Project archaeologists and volunteers conducting salvage excavations at Homewood's Lot, 2002. (Photo, Shawn Sharpe, Lost Towns Project.)

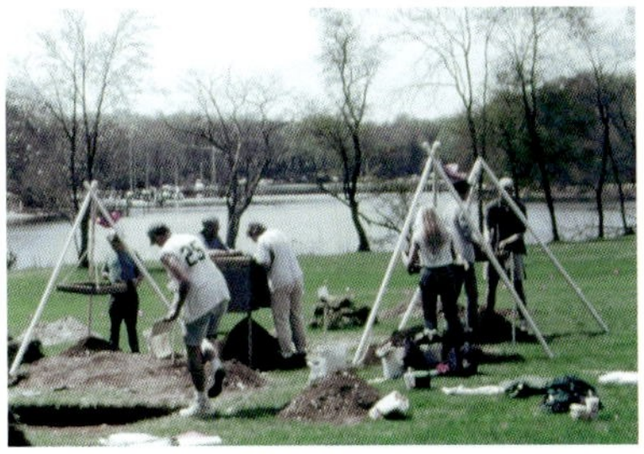

▼ ANNE ARUNDEL COUNTY'S Lost Towns Project has recovered an interesting and informative ceramic assemblage of twenty-eight vessels from a seventeenth-century homesite in present-day St. Margaret's, Maryland. Discarded circa 1665 in a cellar measuring 10 feet by 6 feet, these vessels show an unusual bias toward sophisticated tin-glazed earthenwares as well as Dutch influence.

Homewood's Lot, designated 18AN871, is located off Whitehall Creek near the Chesapeake Bay in Anne Arundel County, Maryland. The site represents the remains of the circa 1650 occupation by Quaker planter Thomas Homewood and, for another century, his descendants. It is one of eight known sites associated with the 1649 Puritan town called Providence, the first European settlement in the county (fig. 1).[1] Between 1999 and 2002 a series of archaeological salvage excavations at this location, conducted by Anne Arundel County's Lost Towns Project (fig. 2), uncovered a number of structures, including two dating from the 1650s and 1660s and two built in the early eighteenth century. The latter two dwellings show evidence of abandonment and demolition circa 1770.[2]

The earliest structure, designated Homewood's A, was an unusually small building constructed with sills resting on a highly degraded, native ironstone foundation or piers. It had an external wattle-and-daub chimney on its northern elevation and a storage cellar, designated as Feature 30. Compared with the other structures found at Homewood's Lot and other Providence sites, this heated building was tiny, measuring approximately 16 feet by 10 feet (fig. 3). The stratigraphic profile of the cellar fill displayed a readily visible alternating series of discarded trash and reddish fireplace ash (fig. 4). The discovery of a remarkable assortment of mid-seventeenth-century ceramic vessels within this tightly dated and undisturbed context provides an unusually detailed glimpse of European ceramic ware consumption in Providence.

The artifact assemblage recovered from the cellar indicates that the structure probably was built during the initial settlement, circa 1650, although it should be noted that construction debris was also found that does not appear to be related to the building of this structure. Green-glazed Dutch floor tiles, bricks, and a 1661 dated window lead suggest that Homewood A was occupied during the construction of another, more elaborate dwelling nearby (Homewood B), presumably during the early 1660s.[3]

The most obvious diagnostic artifact from the cellar is the window lead, which provides a useful alternative to dating from the ceramic assemblage.

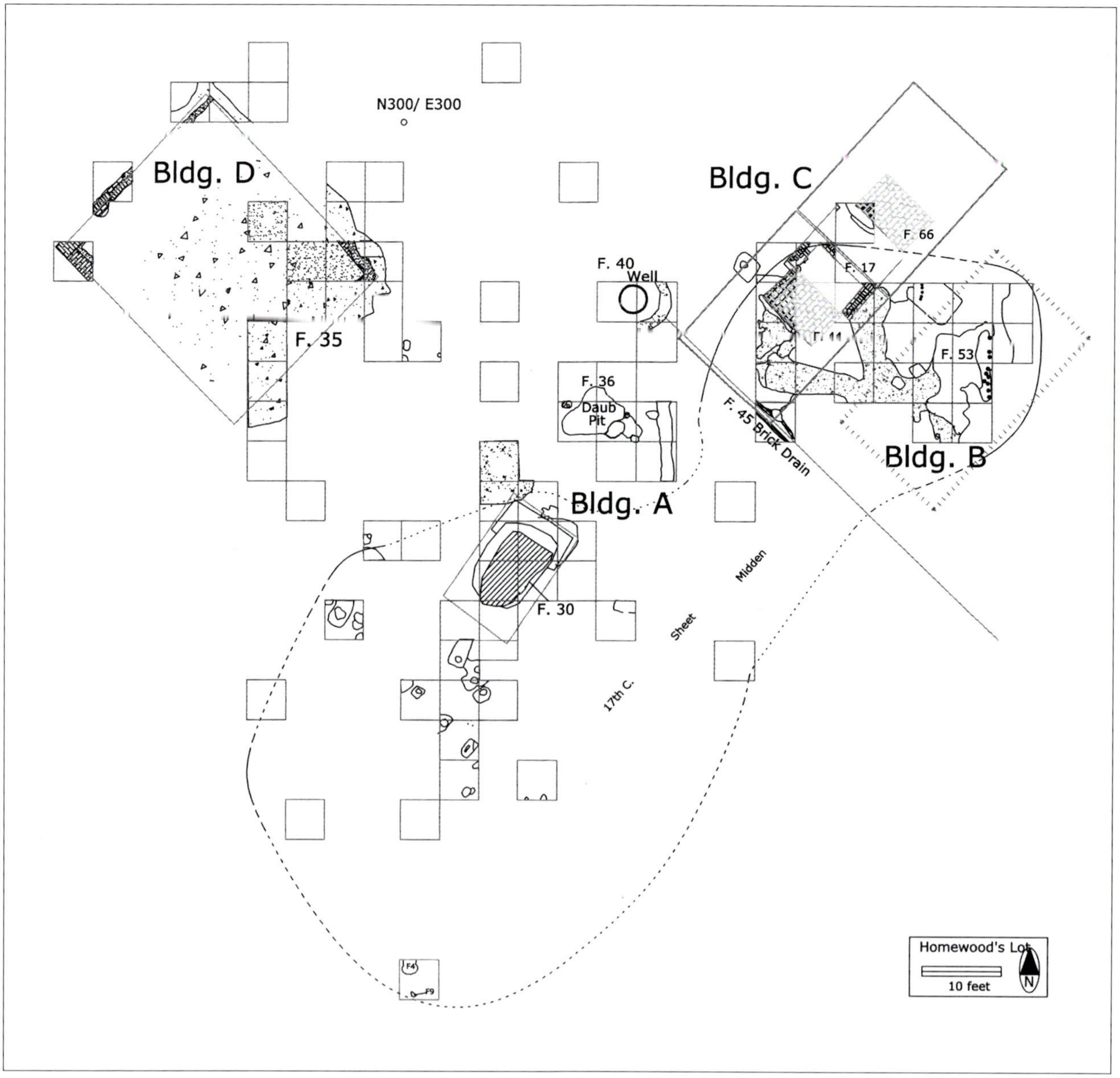

Figure 3 Archaeological plan view map of Homewood's Lot, 2004. (Courtesy, Lost Towns Project.)

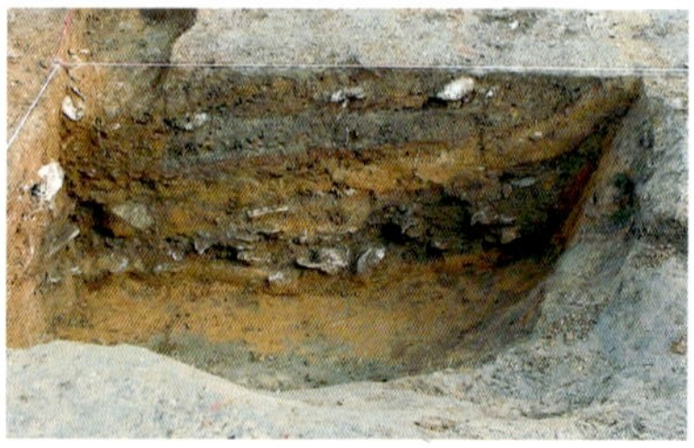

Figure 4 Archaeological profile of the cellar designated as Feature 30, Homewood's Lot, 2002. (Photo, Al Luckenbach.)

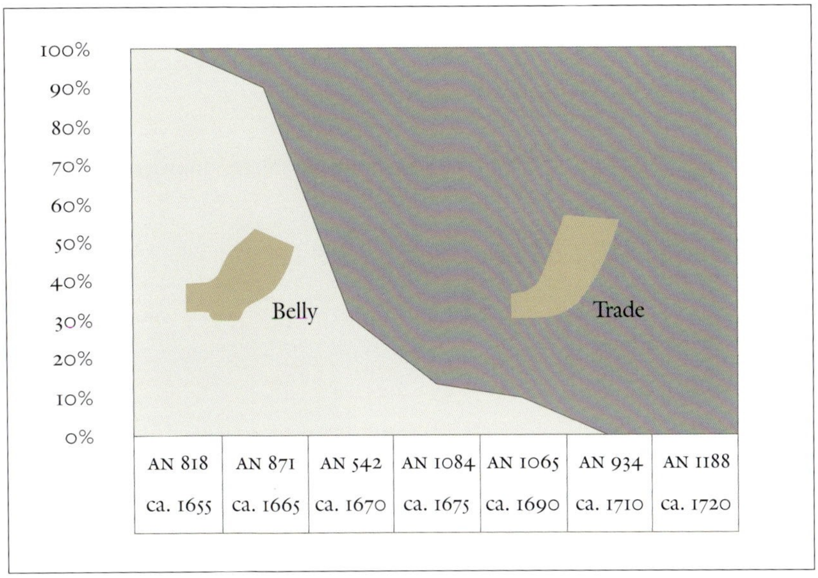

Figure 5 Chart comparing temporal ranges and percentages of belly bowl and trade pipe forms recovered from archaeological sites in Anne Arundel County, Maryland, including Homewood's Lot (18AN871).

An analysis of the bore diameters of clay tobacco pipes produced relative dates of 1654 and 1659.[4] Analysis of the pipes recovered from the plow zone soils immediately over Feature 30 produced dates of 1664 and 1666, which are in closer accord to the fill date being conjectured here.

The authors believe, however, that ceramic tobacco pipe bowl forms provide a much better diagnostic indicator for temporal placement than do stem bores. Based mainly on the ratio of belly bowls to trade pipes, analysis of pipe bowl forms suggests a mid-to-late-1660s deposit fill (fig. 5), which corresponds nicely to the 1661 dated window lead.[5]

Ceramic Vessels

The excavation of the cellar produced an unusual array of different types of ceramic vessels, including several interesting examples of white-glazed and decorated delftware, Rhenish blue-and-gray salt-glazed stoneware, and small amounts of North Devon and Staffordshire wares. For dating purposes, it is the absence of certain ceramic wares that is most significant. For example, English brown stoneware, manganese-mottled earthenware, and combed Staffordshire slipwares as well as the later white salt-glazed stonewares, all of which are common in other Providence sites, were not evident in the cellar feature. These other ceramic types are usually found in post-1680 contexts, and their absence in the cellar suggests that it had been filled before 1680.

Stoneware

Rhenish stoneware with both brown and blue-and-gray finishes is usually quite common on Providence sites, occasionally reaching 7.5 percent of

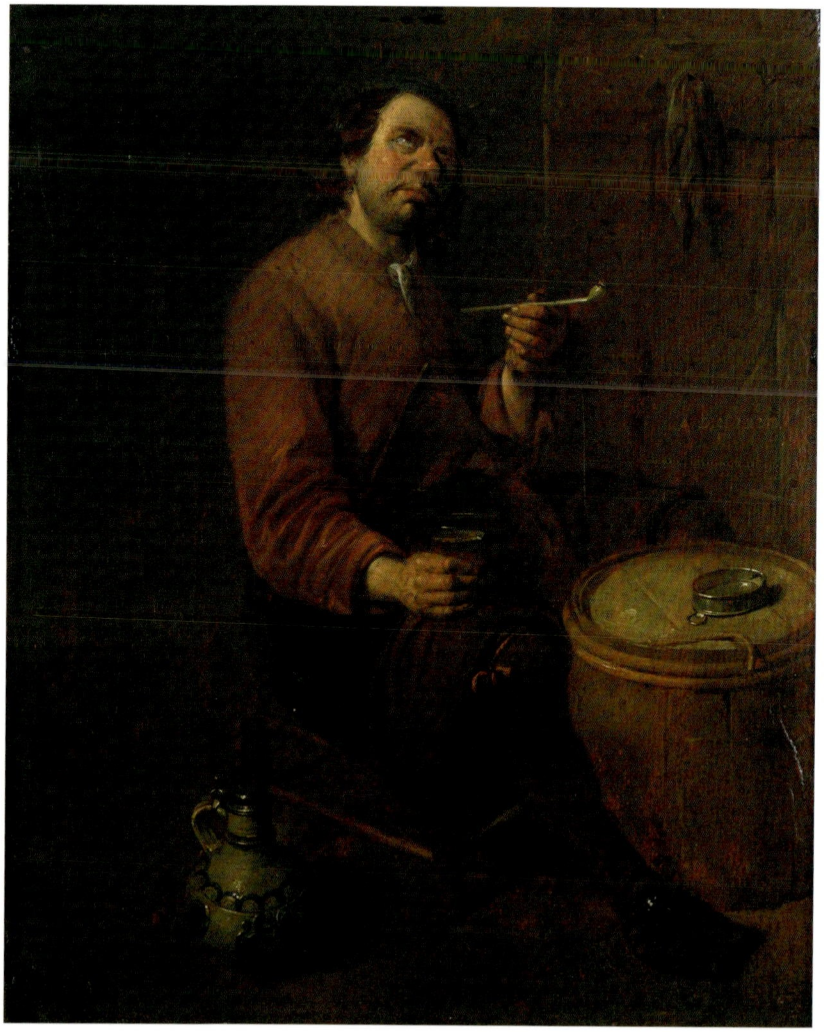

Figure 6 Arent(?) Diepraem, *A Peasant Seated Smoking*, ca. 1650. Oil on oak. 11¼" x 9". (Courtesy, The National Gallery, London.) The decorative elements on a Rhenish jug seen in the foreground of this painting bear a remarkable similarity to the fragments below, which were recovered from the shoreline of St. Mary's City, Maryland (*right*), and Feature 30, Homewood's Lot (*left*). A date of 1646 is visible (at approximately 4:00) on the medallion found at St. Mary's City. (Courtesy, Historic St. Mary's City.)

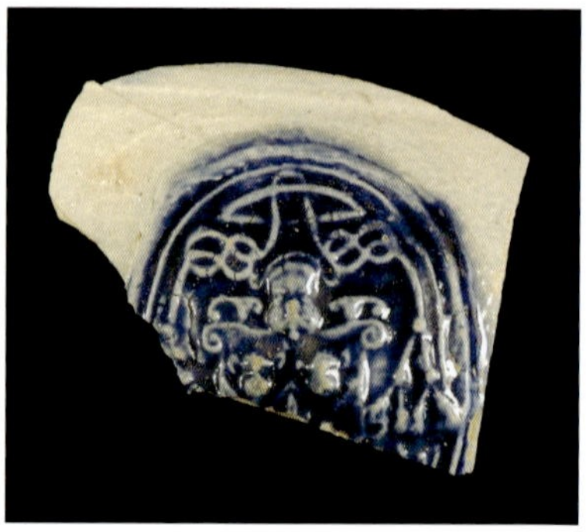

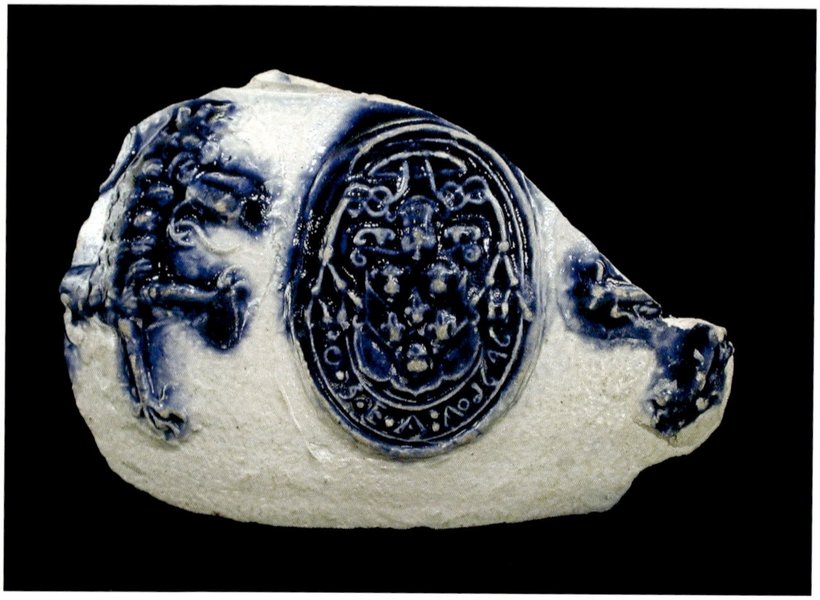

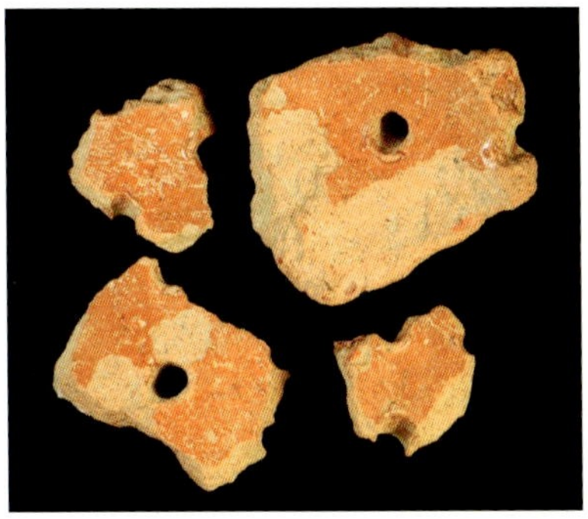

Figure 7 Detail of fragments of a colander similar to the one illustrated in fig. 8. The lead-glazed earthenware fragments are possibly Holland, ca. 1660s.

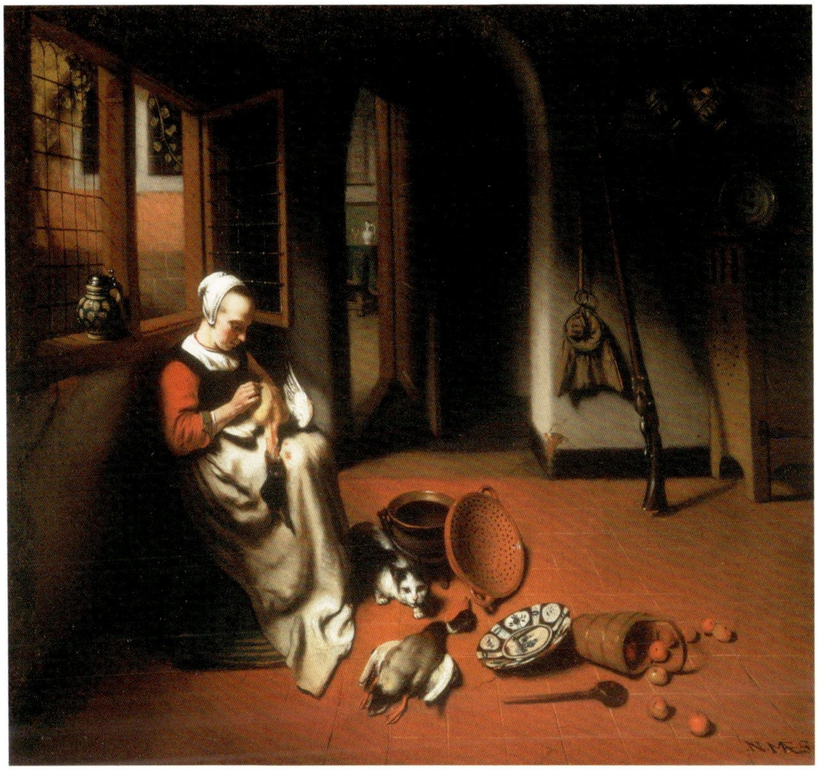

Figure 8 Nicholas Maes, *A Girl Plucking a Duck*, 1634. Oil on canvas. 23⅝" x 26". (Courtesy, Philadelphia Museum of Art.) The colander depicted in this painting provides a suitable comparison for an overlay of several fragments from a similar vessel recovered from Feature 30, Homewood's Lot.

assemblages. Yet sherds from only two vessels of this type were recovered from Feature 30: a single sherd of Rhenish brown stoneware, and thirty-one sherds of Rhenish blue-and-gray stoneware from a single jug. Among the surviving pieces of the jug, and perhaps the most interesting stoneware find, was a distinctive medallion (fig. 6). An identical medallion fragment showing the date 1646 was recovered from the shoreline of St. Mary's City, Maryland. Fortunately, this comparable example is large enough to ascertain that the vessel had alternating rampant lions; the armorial crest was attributed to an unidentified Roman Catholic cardinal.[6] Similar vessels not only appear in Dutch genre paintings of the period but also are found at sites in the New World with links to the Dutch trade.[7]

Earthenware

The cellar deposit also contained vessel fragments representing several types of earthenware, including Border ware, lead-glazed earthenware, North Devon sgraffito and gravel-tempered wares, and green-glazed floor tiles. While it is impossible to determine vessel form from the small sherds of yellow-glazed Border ware, pierced fragments from another vessel made of lead-glazed earthenware suggest it once functioned as a colander or bowl typically used for drying fruit or vegetables (fig. 7). This type of vessel is commonly depicted in seventeenth-century Dutch genre paintings (fig. 8).

Careful mending of other redware fragments with a manganese lead glaze helped to identify a bulbous, handled cup form (fig. 9). The cup is surprisingly thin-walled and delicate for what is generally considered a utilitarian ware.

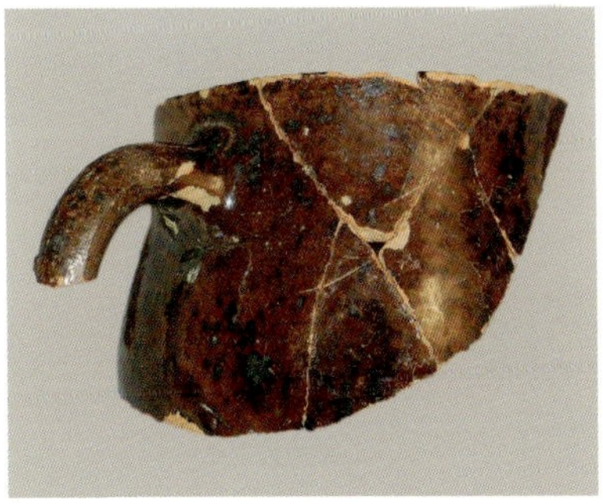

Figure 9 Handled cup fragments, England, ca. 1660. Manganese-enriched lead-glazed earthenware. (Photo, John E. Kille.)

Figure 10 Chart comparing ceramic assemblages recovered from Burle's Townland, Leavy Neck, and Homewood's Lot sites, Providence, Maryland.

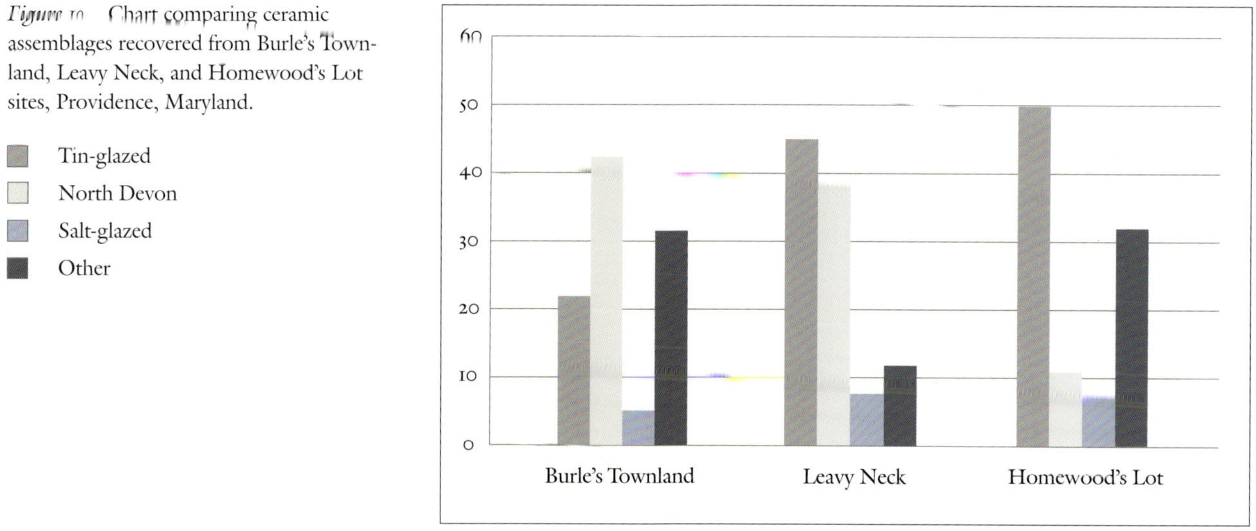

The assemblage also included sherds from two unidentified North Devon sgraffito vessels but only one example of North Devon gravel-tempered earthenware, evidenced by the partially mended rim of a milk pan with a diameter of seventeen inches. The low number of gravel-tempered vessels found is particularly unusual for a Providence settlement site, suggesting the assemblage reflects a relatively high-status household.

The ceramic assemblage from Feature 30 is also notably distinct from other Providence settlement sites that have been subjected to vessel analysis (fig. 10), and from this arise two main points. The first, already noted, is the low percentage of ceramic products from North Devon, England. In most other Providence sites the average is greater than 40 percent—and in some contexts significantly more than that. The second important point is that tin-glazed earthenware makes up fully half of the twenty-eight vessels recovered from Feature 30, the highest amount yet seen from a Providence context. Archaeologist Paul Huey notes that a similarly high percentage of delftware is an attribute often seen in archaeological assemblages from a number of seventeenth-century Dutch colonial sites in New York, notably Fort Orange.[8]

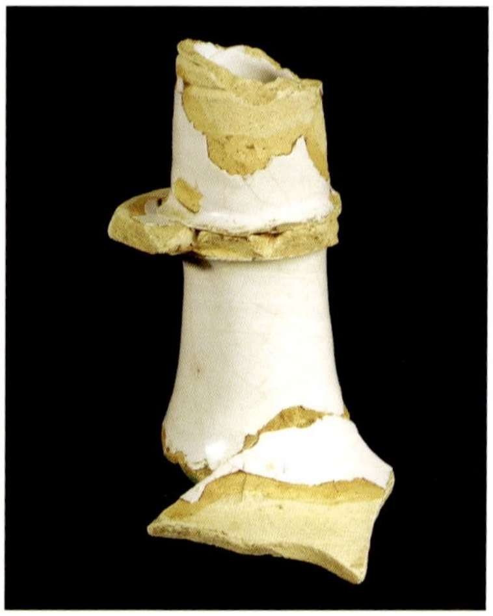

Figure 11 Mid-drip candlestick fragment, probably England, 1660s. Tin-glazed earthenware. (Photo, John E. Kille.)

Tin-glazed Earthenware

As previously noted, the cellar contained a highly unusual assortment of decorated and undecorated delftware design motifs and forms. A white-glazed delftware candlestick is perhaps the most extraordinary find, being only the second example recovered by the Lost Towns Project in more than a decade of archaeological excavations at nearly twenty seventeenth-century sites (fig. 11).[9] The candlestick's mid-drip pan and trumpet base are undoubtedly modeled after a mid-seventeenth-century metal prototype (fig. 12).

A similar candlestick in the possession of the Victoria and Albert Museum, London, which is dated 1648 and bears the arms of the Fishmongers' Company, is purported to be the earliest known dated candlestick (fig. 13). Michael Archer has noted that candlesticks from later in the seven-

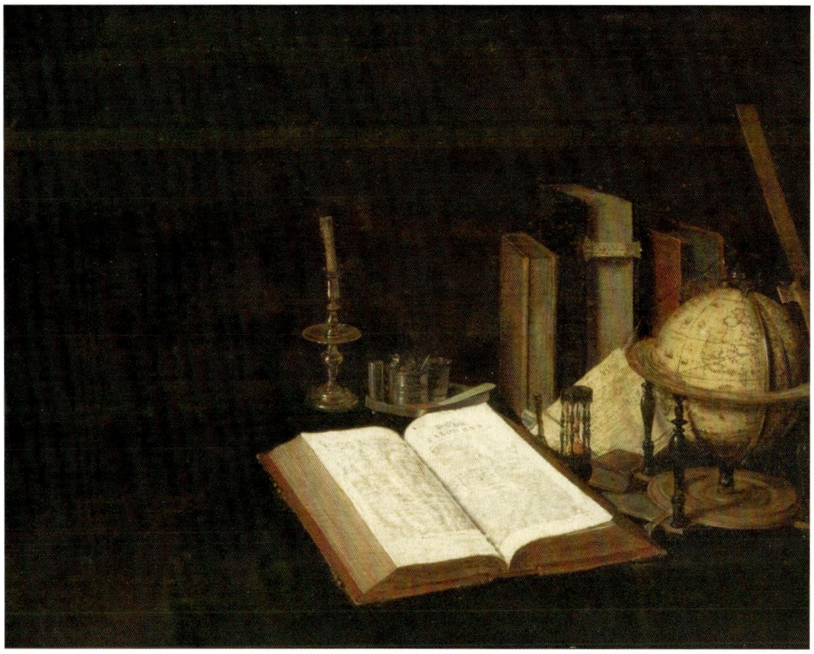

Figure 12 Jan van der Heyden (1637–1712), *Still Life with Books and a Globe*, seventeenth century, Southern Netherlands (modern Belgium). Oil on panel. 8$^{15}/_{16}$" x 10$^{5}/_{8}$". (Courtesy, Philadelphia Museum of Art.) This painting includes a candlestick typical of the seventeenth century.

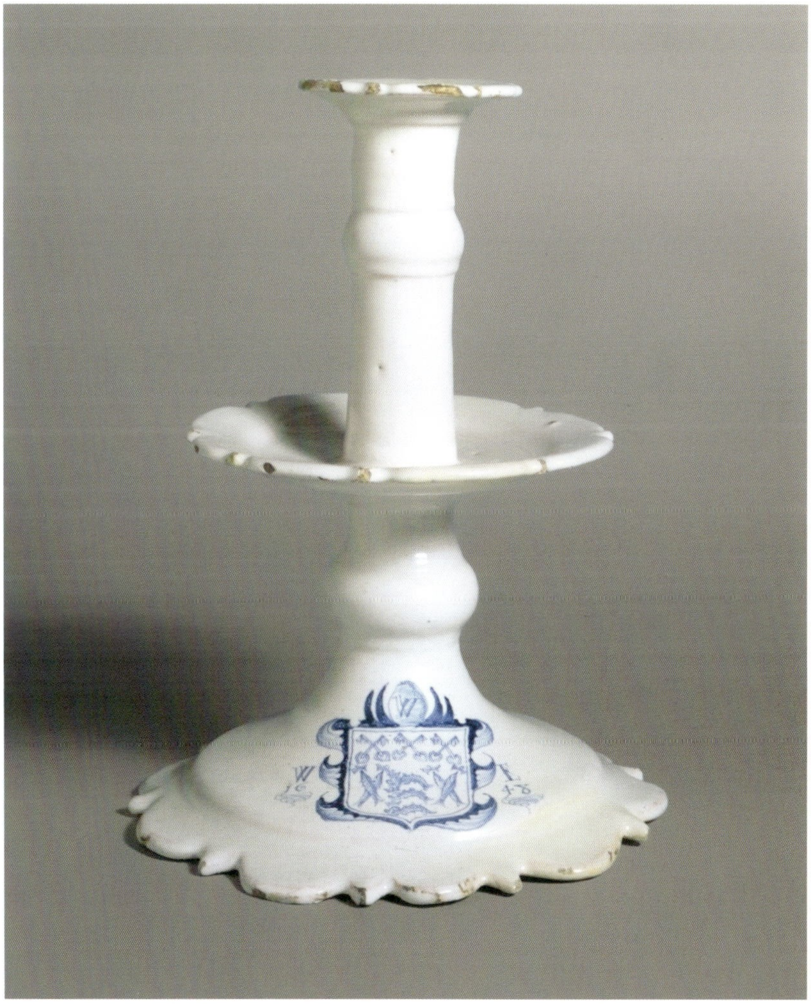

Figure 13 Candlestick, Southwark, probably Pickleherring, London, 1648. Tin-glazed earthenware. H. 10$^{3}/_{16}$". (Courtesy, Victoria and Albert Museum.) This rare example is decorated with the arms of the Fishmongers' Company and is inscribed "W / W / E / 1648."

teenth century have a simpler, straight form, which suggests that the more elaborate molded form from Homewood's Lot is closer to midcentury, as would be expected.[10]

Candlesticks of all types, including those made of delftware and more costly brass and silver, were considered a luxury in the New World during the mid-seventeenth century. Their relative scarcity is often tied to the expense associated with the manufacture of candles.[11]

The Homewood ceramic assemblage also included fragments of no fewer than three seemingly identical white-glazed delftware bowls, referred to as porringers and sometimes "bleeding bowls," with rim diameters of 5 inches (fig. 14). Porringers probably served many uses in seventeenth-

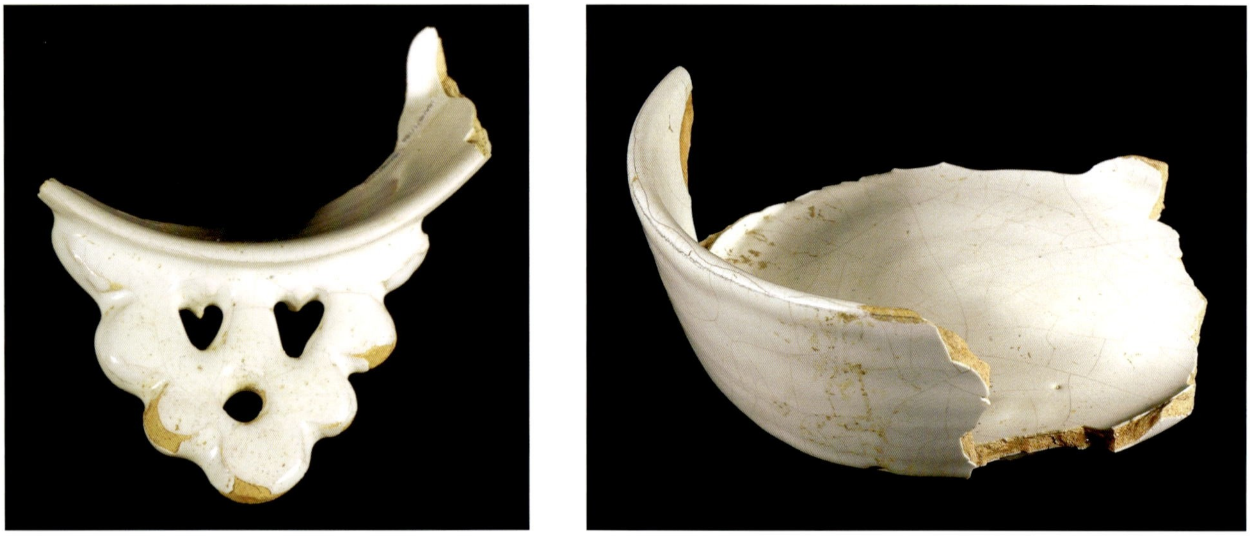

Figure 14 Handle and porringer fragments, probably England, ca. 1660s. Tin-glazed earthenware. D. of rim 5". (Photo, John E. Kille.)

century households, and their presence here (as well as a pewter example found in a nearby plow-zone context) makes it likely that they were commonly used for the consumption of food. The surviving handle of one example displays an elaborate scroll pattern with two heart-shaped holes accompanied by a third round hole. Extant delftware porringers with similar handles, albeit decorated and dated, are illustrated in *Dated English Delftware* by Louis Lipski and Michael Archer. The two vessels with the closest handles have the unfortunately broad temporal range of 1650–1730, although the latter lacks the vertical sides seen in the earlier Homewood specimens.[12]

In addition to the candlestick and porringers, partially mended rims of two undecorated white-glazed plates were also recovered from Feature 30. These vessels had rim diameters of 8½ inches. It has been suggested that the appearance of such plain white-glazed delftware in the mid-seventeenth century is tied to mass production or even the stylistic sensibilities of the

Puritan reign.[13] It has also been noted that the plain white glaze on this type of ware "focuses attention on the form itself."[14] Notwithstanding these theories and observations, it seems reasonable to assume that the manufacture of plain wares achieved a cost savings, as decorating by hand always involved another labor-intensive step in production during this period.

The Homewood deposit also included fragments of two delftware plates and a small bowl decorated with distinctive floral patterns not seen in specialized literature devoted to either English or Dutch delftware of the

Figure 15 Plate fragments, the Netherlands, ca. 1660s. Tin-glazed earthenware. D. 8½". (Photo, John E. Kille.)

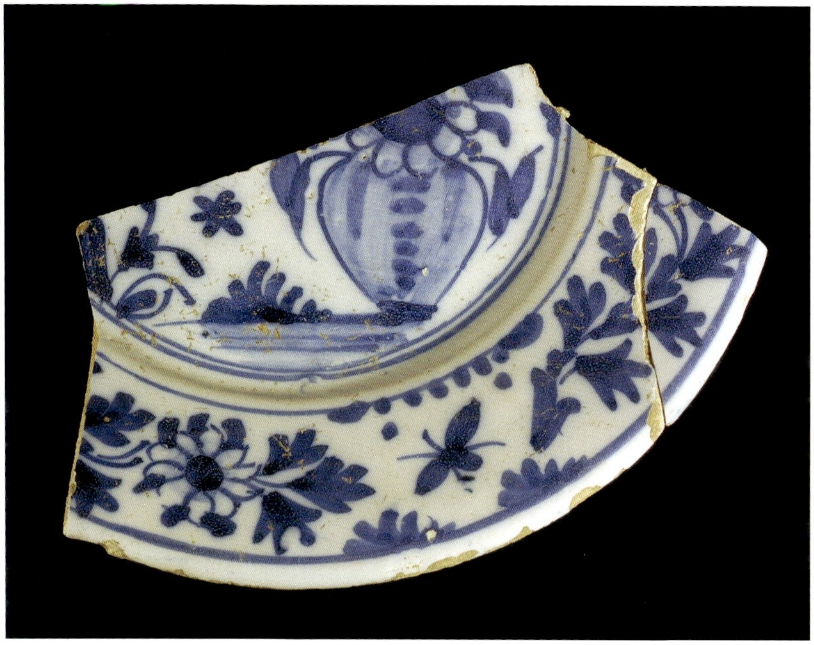

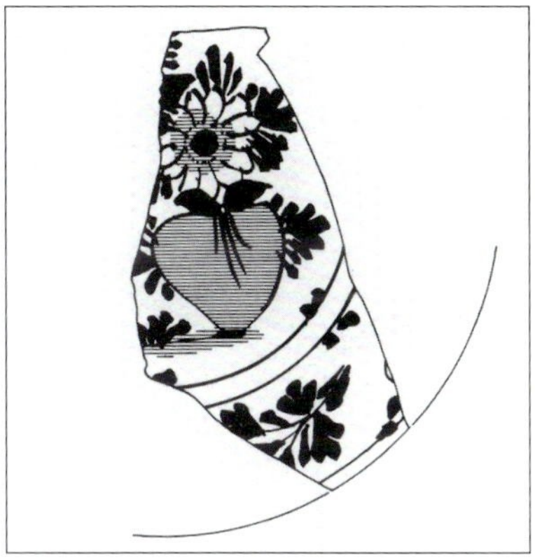

Figure 16 Drawing of a 17th-century tin-glazed earthenware plate attributed to Holland manufacture, illustrated in Sarah Jennings, *Eighteen Centuries of Pottery from Norwich*, East Anglian Archaeology Report No. 13 (Norwich, Eng.: Norwich Survey in collaboration with Norfolk Museums Service, 1981), p. 191.

seventeenth century.[15] As with the plain white examples, these plates had diameters of 8½ inches. The first plate fragment displays a central motif involving an ovoid flower vase (fig. 15). This type of design is seen on two plates excavated in Norwich, England, which are illustrated in *Eighteen Centuries of Pottery from Norwich* (fig. 16).[16] One of the two examples (number 1375) is an almost identical match; excavated from a spoil dump on London Street, it is attributed as "probably Dutch, 1640–60."

The second plate fragment from Feature 30 displays a floral border that is very similar to the design seen on a late-seventeenth-century Dutch plate, also found in Norwich, and attributed to Dutch manufacture (fig. 17).[17]

Finally, it has proved to be even more difficult to find any stylistic analogies for the fragment of a small bowl or teacup with a footring diameter of

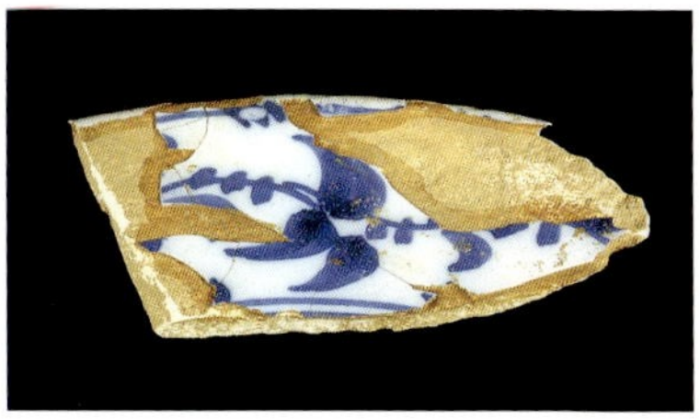

Figure 17 Plate fragment, the Netherlands, ca. 1660s. Tin-glazed earthenware. (Photo, John E. Kille.) The extrapolated diameter of the plate is 8½".

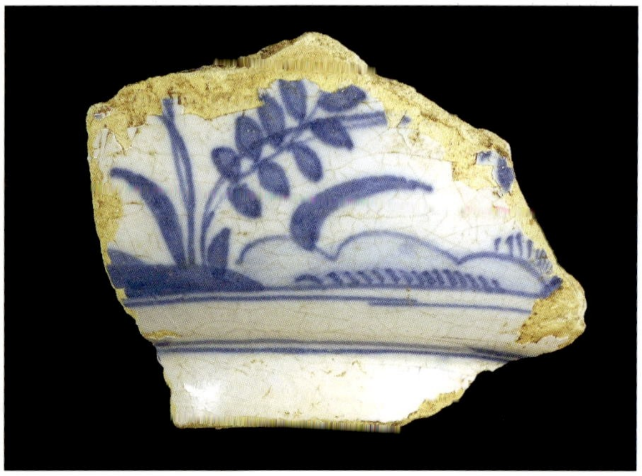

Figure 18 Bowl fragment, probably England, ca. 1660s. Tin-glazed earthenware. (Photo, John E. Kille.) The extrapolated diameter of the footring is 2½".

2½" (fig. 18). Although completely different in execution and body, a Portuguese plate from the early seventeenth century is the closest comparable design motif discovered in the literature.[18] While the Feature 30 vessel might seem more in line with English teacups from later periods, its 1660s discard date is indisputable, coming as it did from the lower levels of the intact cellar feature.

Conclusions

Homewood's Building A, the oldest of eight structures known to have existed on the property, was an unusually small, heated building. Similarly, the trash-filled storage feature beneath its floorboards was notably smaller than several others excavated at the Providence settlement. Despite this limited volume, Feature 30 proved to have an interesting and enlightening ceramic assemblage.

The tightly dated and unusual assortment of ceramics discovered in this deposit affords a fascinating glimpse of the range of different types of European wares accessible to a single household in the Chesapeake in the mid-seventeenth century. Particularly notable were the unusual composition of certain items—including low percentages of North Devon ware and utilitarian ceramics in general and a high percentage of tin-glazed earthenware, including a rare candlestick.

The high percentage of delftware has been seen in other New World contexts as an indication of Dutch influence. In fact, many types of artifacts from Holland are found at Providence sites, notably pipes and ceramics as well as basic construction materials such as pantiles, painted tiles, yellow brick, and green- and yellow-glazed floor tiles. The ceramics recovered from Feature 30 include some that are clearly Dutch in origin, some that are possibly Dutch, and some that are often portrayed as being components of Dutch trade.[19] These findings provide additional evidence of Dutch influence within the Chesapeake, at a mid-seventeenth-century Puritan settlement.[20]

1. Al Luckenbach, *Providence 1649: The History and Archaeology of Anne Arundel County, Maryland's First European Settlement* (Annapolis: Maryland State Archives; Crownsville: Maryland Historical Trust, 1995).

2. Lauren Franz and Al Luckenbach, "The Building Sequence at Homewood's Lot, Anne Arundel County, Maryland," *Maryland Archeology* 40, no. 2 (2004): 19–28.

3. Ibid.

4. David A. Gadsby and Rosemarie Callage, "Homewood's Lot through Four Generations," in *The Clay Tobacco-Pipe in Anne Arundel County, Maryland (1650–1730)*, edited by Al Luckenbach, C. Jane Cox, and John Kille (Annapolis, Md.: Anne Arundel County Trust for Preservation, 2002), p. 18. Because the diameters of pipe stems have been shown to change over time, a statistical analysis of their measurements is used to determine the dates of various archaeological sites. Formulas proposed by Lewis Binford and Lee Hanson were used in this analysis; Binford, "A New Method of Calculating Dates from Kaolin Pipe Stem Fragments," *Southeastern Archaeological Conference Newsletter* 9, no. 1 (1962): 19–21; and Hanson, "Kaolin Pipestems—Boring In on a Fallacy," *Historical Site Archaeology* 4 (1969): 2–15.

5. Al Luckenbach and James Gibb, "Dated Window Leads from Colonial Sites in Anne Arundel County, Maryland," *Maryland Archeology* 30, no. 2 (1994): 23–28 (1999 Addendum). An analysis of lead window cames has indicated that they were seldom curated for long and that their dates generally correspond with some specificity to the assumed construction dates of the associated building. See also Al Luckenbach and Shawn Sharpe, "A Seriation Analysis of 'Trade' and 'Belly Bowl' Tobacco Pipe Forms from ca. 1655 to ca. 1725," *Maryland Archeology* 43, no. 2 (2007): 28–33.

6. Silas D. Hurry, Martin E. Sullivan, Timothy B. Riordan, and Henry M. Miller, *"—once the metropolis of Maryland": The History and Archaeology of Maryland's First Capital* (St. Mary's City, Md.: Historic St. Mary's City Commission, 2001), p. 46.

7. Richard G. Schaefer, *A Typology of Seventeenth-Century Dutch Ceramics and Its Implications for American Historical Archaeology*, BAR International Series 702 (Oxford, Eng.: J. and E. Hedges, 1998), p. 134; and Charlotte Wilcoxen, *Dutch Trade and Ceramics in America in the Seventeenth Century* (Albany, N.Y.: Albany Institute of History and Art, 1987), p. 73.

8. Schaefer, *A Typology of Seventeenth-Century Dutch Ceramics*, p. 102.

9. In March 2007 the Lost Towns Project recovered a portion of a yellow-glazed Border ware candlestick from the home site of the Samuel Chew family at Fairhaven, Maryland. Samuel Chew Sr. was an original lot holder in the Town of Herrington (ca. 1660s).

10. Michael Archer, *Delftware: The Tin-Glazed Earthenware of the British Isles* (London: The Stationery Office, in association with the Victoria and Albert Museum, 1997), p. 327.

11. Louis L. Lipski and Michael Archer, *Dated English Delftware: Tin-glazed Earthenware, 1600–1800* (London: Sotheby Publications, 1984), p. 152.

12. Ibid., nos. 1230 (1650), 1237 (1698), 1240 (1727), and 1241 (1730).

13. Maryland Archaeological Conservation Laboratory's "Diagnostic Artifacts in Maryland," available online at the website of the Jefferson Patterson Park and Museum, www.jefpat.org/diagnostic/Historic_Ceramic_Web_Page/Historic%20Ware%20Descriptions/tin_glazed.htm (accessed December 7, 2010).

14. Amanda E. Lange, *Delftware at Historic Deerfield, 1600–1800* (Deerfield, Mass.: Historic Deerfield, Inc., 2001), p. 90.

15. This survey included John C. Austin, *British Delft at Williamsburg* (Williamsburg, Va.: Colonial Williamsburg Foundation, in association with Jonathan Horne Publications, 1994); Frank Britton, *London Delftware* (London: J. Horne, 1987); Dingeman Korf, *Nederlands Majolica* (Haarlem: DeHaan, 1981); Lange, *Delftware at Historic Deerfield*; Lipski and Archer, *Dated English Delftware*; Anthony Ray, *English Delftware Pottery in the Robert Hall Warren Collection, Ashmolean Museum, Oxford* (Boston: Boston Book & Art Shop, 1968); and Jan Daniël van Dam, *Delffse Porceleyne: Dutch Delftware, 1620–1850* (Amsterdam: Rijksmuseum, 2004).

16. Sarah Jennings, *Eighteen Centuries of Pottery from Norwich*, East Anglian Archaeology Report No. 13 (Norwich, Eng.: Norwich Survey in collaboration with Norfolk Museums Service, 1981), pp. 189, 191, fig. 83, nos. 1375, 1376.

17. Ibid., pp. 192–93, fig. 84, no. 1379.

18. Rafael Salinas Calado and Jan Baart, *Portuguese Faience 1660–1660* (Lisbon: Secretary of State for Culture; Amsterdam: Amsterdam Historical Museum, 1987), p. 29.

19. See Wilcoxen, *Dutch Trade and Ceramics in America*, p. 79.

20. Luckenbach, *Providence 1649*; C. Jane Cox and Al Luckenbach, "The Dutch Connection: 'Providence' within the 17th Century Chesapeake," presented at the Society for Historical Archaeology Annual Conference, Jamestown, Virginia, 2007.

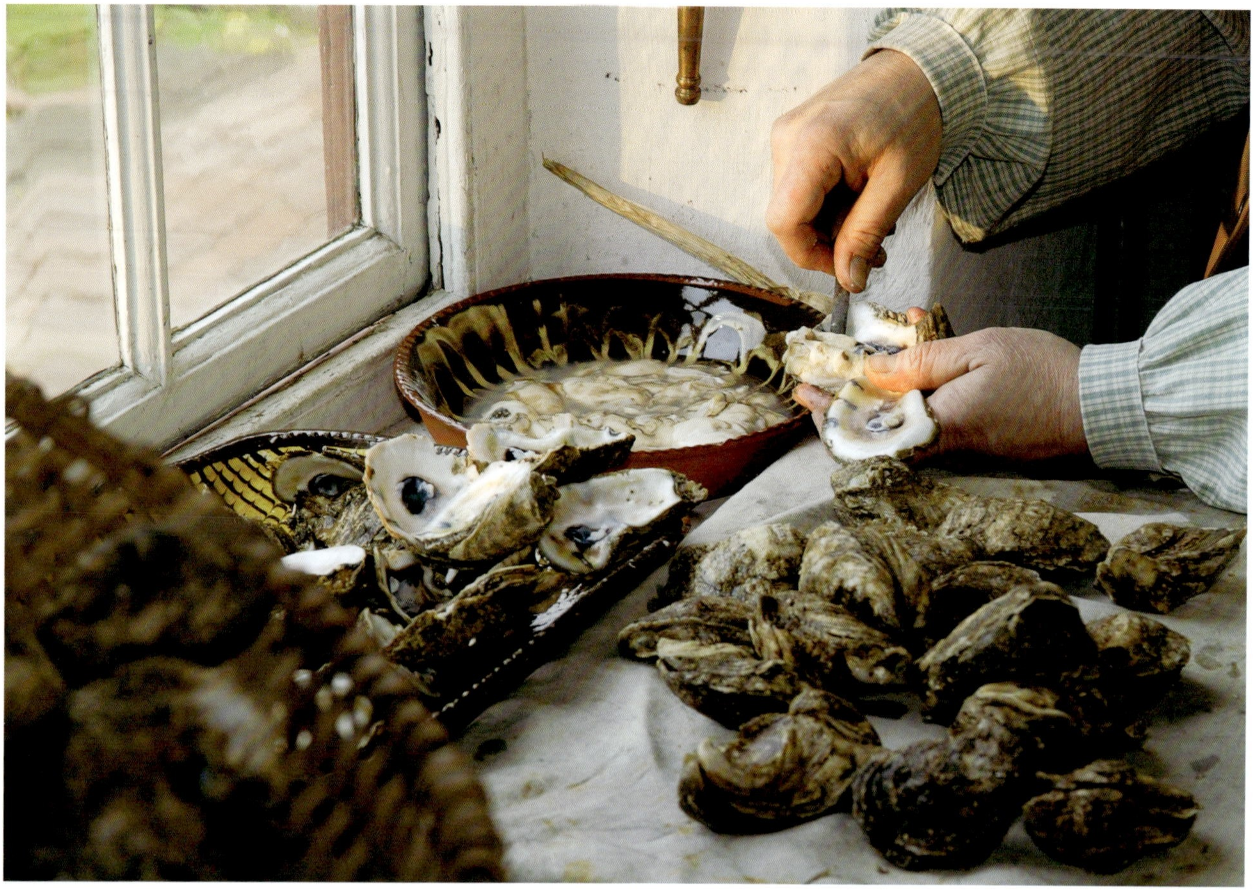

Figure 1 Shucking Virginia oysters (*Ostreidae Crassostrea virginica*) in a colonial kitchen. (Courtesy, Colonial Williamsburg Foundation.)

Ivor Noël Hume

Cap-Hole Oyster Jars: A Racial Message in the Mud, or, Shipping *Ostreidae Crassostrea virginica*

▼ IN 1967 two American adventurers prospecting for diamonds in Guyana's backcountry highlands were encouraged by local officials to abandon their project and leave the country. On their way out, the men traveled down to the mouth of the Essequibo River and came to an island variously known as Flag or Fort Island (fig. 2).[1] On it stand the ruins of an eighteenth-century Dutch fort, and in the island's muddy pits and ditches the men found large numbers of Dutch and French glass bottles which they proceeded to salvage and surreptitiously ship to the United States. The bottles were then sold to antique dealers, restoration projects, and collectors up and down the East Coast, some of the purchasers exhibiting or selling them as English—which they were not. Nevertheless, for many years thereafter the influx of these bottles and their availability at modest prices inadvertently ruined the market for genuine eighteenth-century English bottles.

When the prospectors arrived at my door, I bought a representative number and, in appreciation, the sellers gave me a brown stoneware jar of a shape I had never before encountered (fig. 3). As they did not know what it was, they said it had no marketable value.[2] I was pleased to agree with them, and the jar went into my stoneware collection to be brought out whenever a stoneware specialist happened by. One said that he thought it might have been an acoustic jar, often used by the Spanish in the arches and ceilings of churches, but no confirming documentation was forthcoming. Someone else thought it might be a container for oysters, but he, too, was

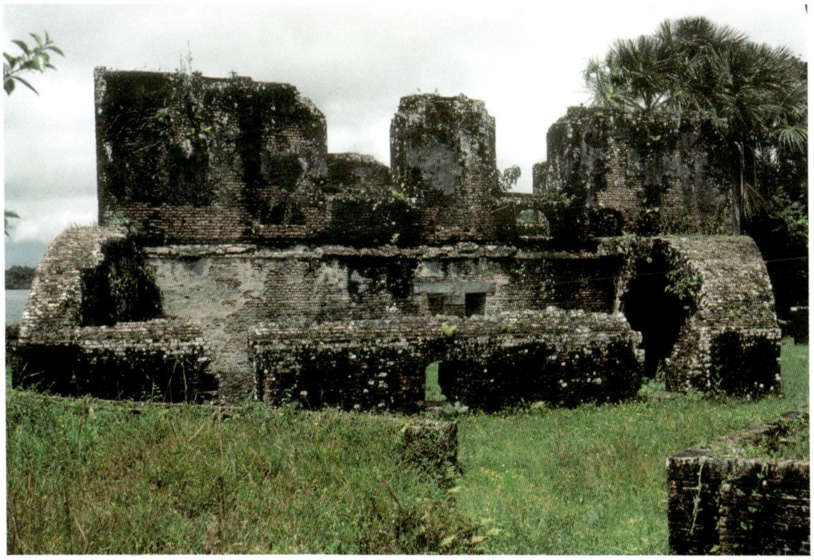

Figure 2 The brick-built Dutch colonial fort on Flag Island, Guyana, 1968. (Photo, Charles H. Kable.)

unable to prove it. Consequently, the jar went onto a back shelf and was almost forgotten until I received a letter of inquiry from Susan Williams of Springfield, Virginia. She enclosed a photograph of three Dutch wine bottles and a stoneware jar almost identical to mine, found in 1968 by her brother buried in mud beside the river at Fort Island (fig. 4). She asked, "How old might it be and what was it used for?"

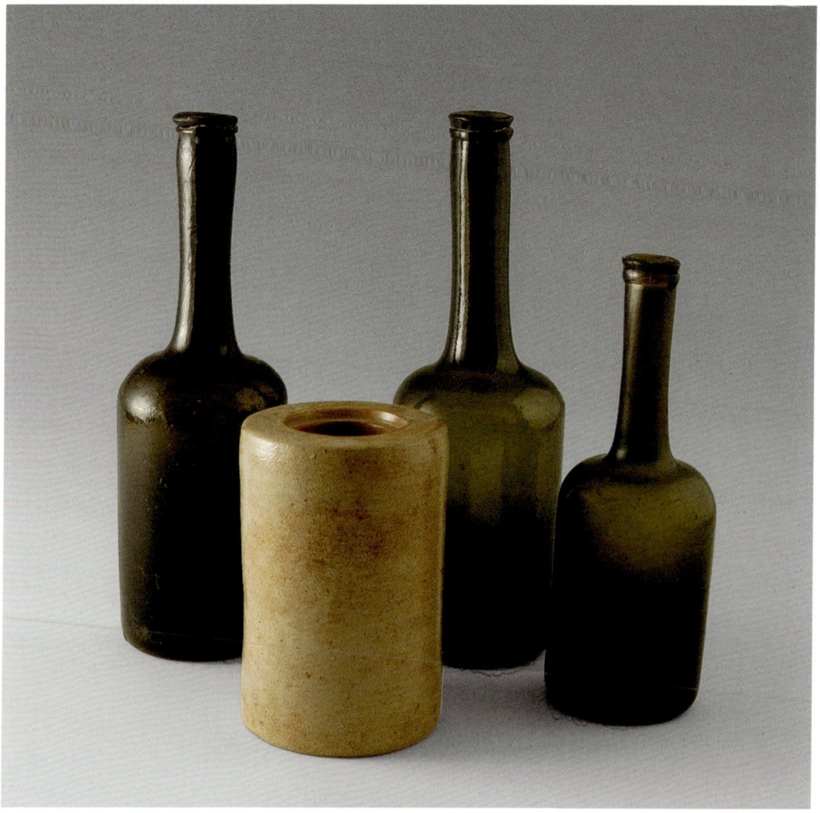

Figure 3 Dutch glass bottles and an enigmatic brown stoneware jar found on Flag Island by diamond prospectors. H. of jar 5¾". (Author's collection; photo, Robert Hunter.)

Although Susan was unaware of it, she had the answer and I did not. Her cylindrical stoneware jar is stamp-inscribed "DANIEL JOHNSON / N^{O} 24 / LUMBER STREET / N.YORK" (fig. 5). When I told her that the town was almost certainly New York and not York in England, she immediately set to work searching the city directories and soon found that in Manhattan in 1796 there were two Lumber Streets as well as a Daniel Johnston living at 1 Lumber Street, his trade given as "oysterman."[3] Assuming that Johnston was a misspelling of Johnson, that seemed to settle the matter—until it was found that the 1799 directory lists both Daniel Johnson and Daniel Johnston. The former is listed at an unnumbered address on Lumber Street; the latter at 276 Water Street and his trade given as "oysterstand."[4] Taking this information at face value one might conclude that Johnston was not an oyster fisherman but the proprietor of a "small often open-air structure for a small retail business."[5] However, in the 1800 directory Johnson is listed at both 24 Lumber Street and 276 Water Street. Thus, it seems his primary address is confirmed, while the Water Street number puts him in the same building as Daniel Johnston. It makes much more sense, therefore, to con-

Figure 4 Dutch and English bottles recovered from Flag Island along with the brown stoneware oyster jar illustrated in fig. 5, which bears an impressed inscription for Manhattan oysterman Daniel Johnson. (Courtesy, Susan Williams; photo, Robert Hunter.)

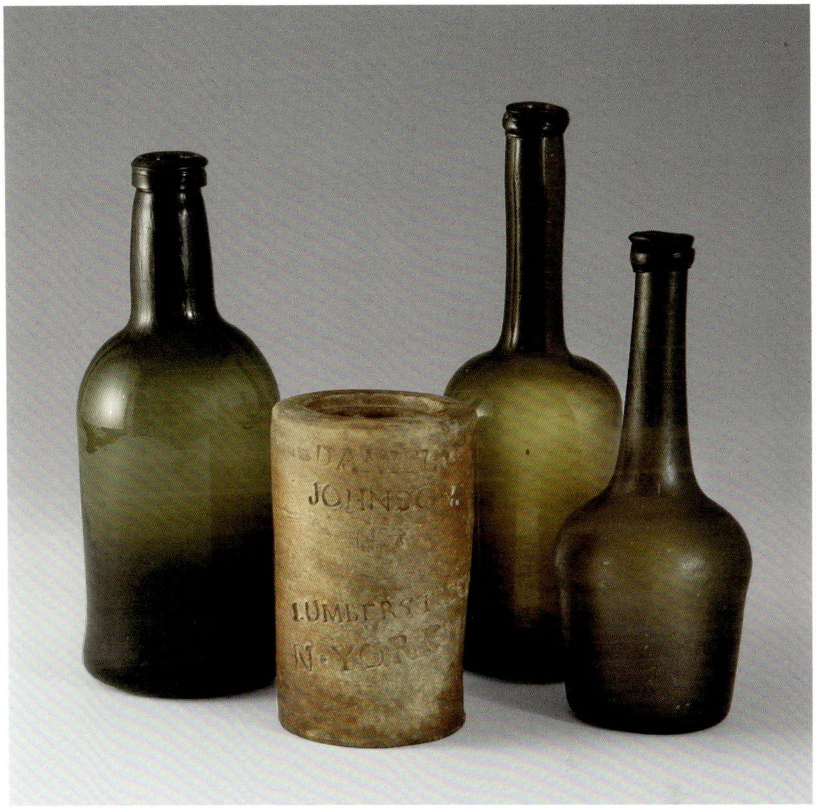

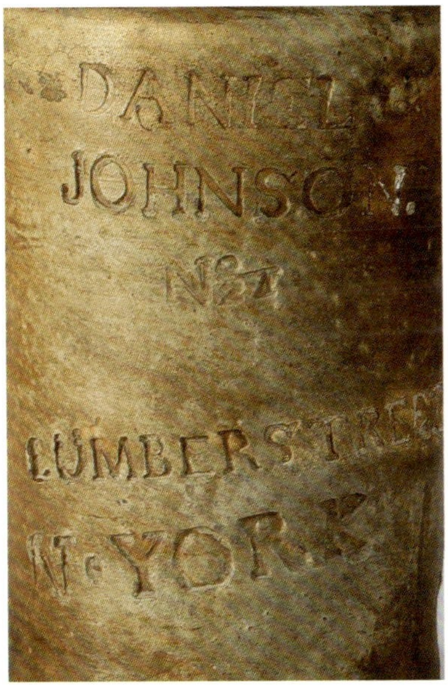

Figure 5 Detail of the oyster jar illustrated in fig. 4, attributed to Thomas Commeraw, New York, New York, 1800–1805. Salt-glazed stoneware. H. 5¾". Impressed on side: DANIEL/JOHNSON/NO. 24 LUMBER STREET/N. YORK.

clude that the name is misspelled in both the 1796 and 1799 directories, that Johnson and Johnston are one and the same, and that he lived on Lumber Street and had his fresh oyster stand at 276 Water Street.[6]

At the turn of the nineteenth century, Manhattan was rapidly expanding along the East River in the direction of Corlears Hook (figs. 6, 7). Why Johnson's address was unnumbered in the 1799 directory and given as No. 24 a year later is unknown, but taking street-name changes into account, it is probable that Johnson's jar was made no later than 1804.

The 1800 directory puts Johnson where his jar says he belongs, but in the 1804–1805 directory he is listed at both 24 Lombard Street (its name before becoming Lumber) and 106 Cherry Street.[7] The transition from Lumber to Lombard—and then Lombardy—was almost certainly due to problems of pronunciation. To New York oystermen who had never heard of the medieval Lombards, Lumber was easier to say and remember. Although the 1804 *New-York Register* showed Daniel Johnson on Lombard Street, its name was not officially changed to Lombardy until 1809.[8] A city directory (Low's) of 1807 warned that "care should be taken to understand the

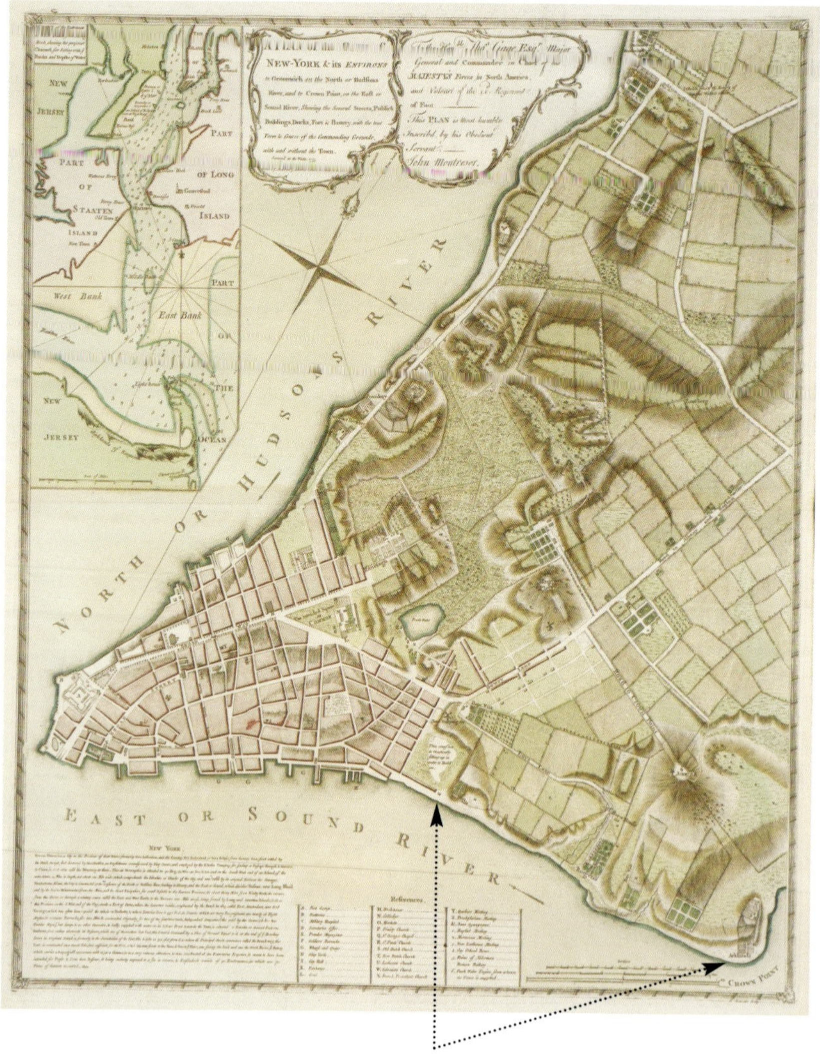

Figure 6 John Montresor, *A Plan of the City of New-York & Its Environs to Greenwich, on the North or Hudsons River, and to Crown Point, on the East or Sound River, Shewing the Several Streets, Publick Buildings, Docks, Fort & Battery, with the True Form & Course of the Commanding Grounds, with and without the Town*, London, England, 1767 (this copy reissued in 1775), 25⅜" x 20⅝". (Courtesy, Colonial Williamsburg Foundation.) Daniel Johnson's workplace would have faced the East River within the area marked "This overflow is continually filling up in order to Build on." In the lower right corner of the map, the property marked "Acklands" is the approximate site of Thomas Commeraw's later stoneware factory. By 1800 the curve identified by the British as Crown Point would return to its old Dutch name of Corlears Hook.

difference of the spelling of Lombardy and Lumber, as the want of attention to this has led many strangers astray." To this one can only add "Amen."

However, the thought serves as a useful segue to the Manhattan Trinity Church records, which show that on June 9, 1793, a Daniel Johnson married Phillis Dickson,[9] and that on August 10, 1805, a Phillis Johnson married John Edwards, whose race was cited as Negro. If, as seems likely, this was the same Phillis who twelve years earlier had married Daniel, it follows that he, too, was an African-American. But was he the Daniel of the oyster jar? There had been yellow fever outbreaks in New York in 1802 and 1805. The distinguished African-American scholar Ralph Ellison cited an example from the 4th Ward: "Out of the 458 blacks, living in 10 cellars, 33 were sick, of whom 14 died, while out of 120 whites, living immediately over their heads in the apartments of the same houses, *not one* even had the fever."[10]

It is important to note that the United States Federal Census was designed to identify the white population and so was divided into columns that recorded the ages of both "Free White Males" and "Free White Females." At the end of each line was space for "All Other Free Persons" and for "Slaves." To be a free black householder was an anomaly calling for an unspecified citation. In 1800 a Daniel Johnston living in Ward 7 was identified by name as "a Black" who answered for four more in the "other Free Persons" column. Ten years later there was a Daniel Johnston, "a Black" in Ward 4 who had five "free persons" in his household or business. The increase from four to five strongly suggests that this was the same person—that is, Johnson of the jars who had moved from one ward to the other. That implies, too, that Johnson was relatively prosperous and employed four and then five African-Americans to work his boat and shuck his oysters.

On another page of the 7th Ward Census, property occupier (or owner?) Thomas Commeraw also bore the suffix "a Black." It is significant, however, that he was the only individual so identified in a page of forty names. More on him anon.

Only from the census records can we be sure of a New Yorker's race; but although the more frequent directories do not so state, their inclusion of laborers, maids, cartmen, coachmen, and the like renders it safe to assume that many were people of color. By 1810 there were 7,470 free blacks in New York City, and social historian Shane White has stated that the presence of blacks "was most noticeable in the oyster trade, which they dominated."[11] In 1810 the official census listed twenty-seven oystermen, of whom sixteen were black. Five years earlier *Longworth's Trade Directory* listed only fifteen, among them "Johnson Daniel" and "Johnson John."[12] That increase in the number of oystermen belies the evidence that the dumping of waste into New York Harbor had been ruining the oystering business.[13]

Figure 7 John Bachmann, *Bird's Eye View of New York and Environs*, 1865. 31" x 46". Published by Kimmel & Forster, New York. Corlears Hook is being pointed to by the steamship emerging from the inlet at right. (Courtesy, Library of Congress, Geography and Map Division.)

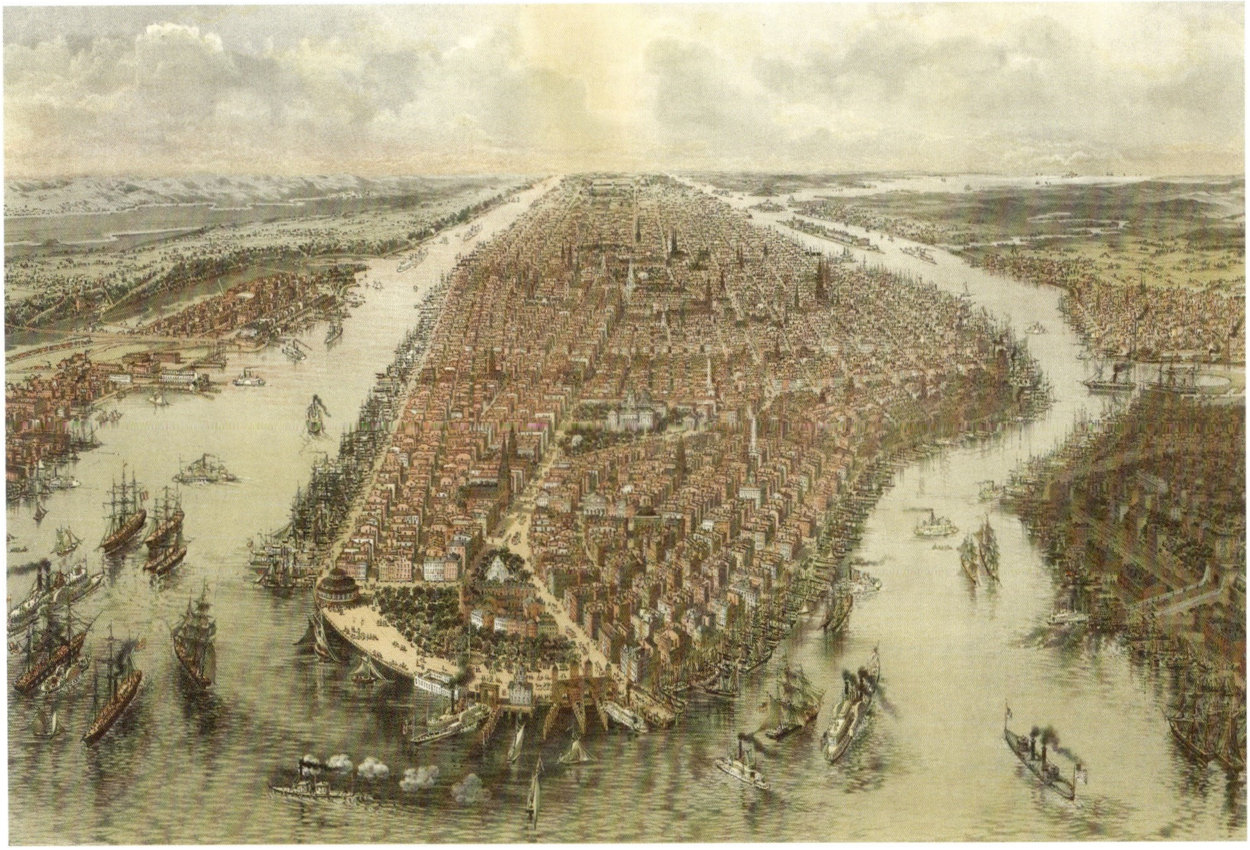

The New York ward system had been set up in 1683, the 4th Ward being one of the original six. By 1800 there were twenty-two wards; the 7th was located east of Catherine Street, an area subject to rapid expansion in the first decades of the new century. That may explain why Daniel was at 1 Lumber in 1796 (the first dwelling built?) and Lumber—with no number—three years later. He may have been an early beneficiary of that urban growth. By 1804 his home address had been changed to 24 Lombard Street, and he had left his oyster stand at 276 Water Street and gone to 106 Cherry Street. The directory for 1810 lists only the Cherry Street address. If, as seems likely, the "stand" had been Johnson's, it may be that he was dead and his well-established trade was being continued there in his name.

With Johnson located and possibly dead by the summer of 1805, the next question focuses on his jars and where they were made.[14] Were they English or American? As ceramics made in England for the American market were sometimes inscribed, it might be that a New York importer placed a custom order with an English manufacturer on Johnson's behalf. In 1763, for example, James Gilliland was established on nearby Wall Street as an importer of china, glass, and earthenware. However, one doubts that a briney-handed oysterman would have patronized a high-end importer. It is far more likely that he would have gone to the edge of the city to an earthenware potter like Jonathan Durell, who in 1775 had advertised that he made "butter, water, pickle and oyster pots, porringers and milkpans. . . ."[15]

Durell's linking of pickle and oyster pots indicates that one shape could be used for either—or both. However, another Durell ad published two years earlier suggested that pickles and oyster pots were not necessarily related. There he was selling "butter, water, pickle, oyster and chamber pots."[16] The moral, therefore, is that one should not read too much into the placement of a comma.

A much earlier advertisement, printed in England in 1708, lists "right Ground Oysters pickled to be sold by the Pot for 2s there being 25 large oysters in the Pot."[17] By the mid-eighteenth century Swedish traveler Peter Kalm visited New York and described two methods for pickling oysters, the simplest as follows: "They are taken out of the shells fried in butter, put into a glass or earthen vessel with melted butter over them, so they are fully covered with it and no air can get to them. Oysters prepared in this manner have likewise an agreeable taste, and are exported to the West Indies and other parts."[18] But buttered or in brine, by the mid-eighteenth century pickled oysters were being shipped from New York to the West Indies. Fresh oysters, by contrast, had a relatively short shelf life and were sent in bags and barrels to Boston and elsewhere on the American mainland.

In 1774 mercenary army captain John Gabriel Stedman, then stationed in Suriname in neighboring Guyana, wrote in his journal that he had received from "Mrs. Gordon 3 bottles English muskat, a cask of Boston biscuits, and a small keg of oysters."[19] He did not say whether they were fresh or pickled, but since he mentioned a keg one must assume that he meant a small wooden barrel and not a stoneware jar.

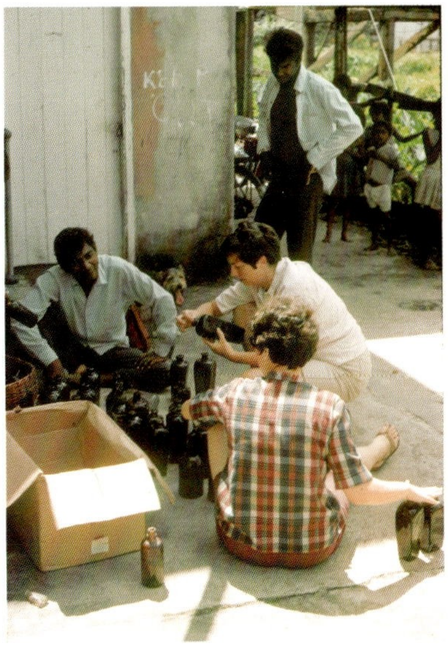

Figure 8 Members of the Kable family bargaining for bottles in Guyana, 1968. (Photo, Charles H. Kable.)

Jars like Johnson's were of a distinctive shape, being wide-mouthed and having a recessed collarlike orifice, presumably to take a flush-fitting stoneware lid. Most of the bottle and jar "treasures" from the mud around Fort Island were salvaged by local children, and it is highly unlikely that anyone told them that these flat discs had any value and were worth retrieving.

Susan Williams recalls her sister-in-law telling her that most of the bottles were recovered by children probing the mud with sticks. After discovering that foreigners would pay real money for the glass bottles the children or their parents started selling them to tourists in Georgetown (fig. 8). When the locals heard that saleable bottles were being found on the island, they hurried there to buy more. How many oyster jars were recovered and sold before the trove ran out is anybody's guess. Two more were auctioned by Crocker Farm, Inc., in November 2005, one bearing the inscription "DANIEL / JOHNSON. AND CO. 24 / LUMBER STREET / N·YORK," the other "BROWN / WATER STREET / N·YORK" (fig. 9). Unfortunately, for confidentiality reasons the auctioneers declined to reveal who bought them (and, more important, who sold them),[20] but it is reasonable to deduce that these, too, came from the mud of Fort Island, the inference being that they were part of a single shipment. To support that thesis it is necessary to determine whether Brown was in business prior to 1804. He was. In 1799 he was an oysterman at 15 Ann Street, although by 1810 he had moved to 128 Water Street and 48 Nassau Street.[21] As previously noted, Daniel Johnson/Johnston was operating on Lumber Street as early as 1796, and soon thereafter (ca. 1800) he was buying jars bearing his name as the sole proprietor. The "AND CO." marked on the second Johnson jar suggests an expansion of his business and that the two jars are of successive dates, probably between circa 1800 and 1805. Although such business designations are

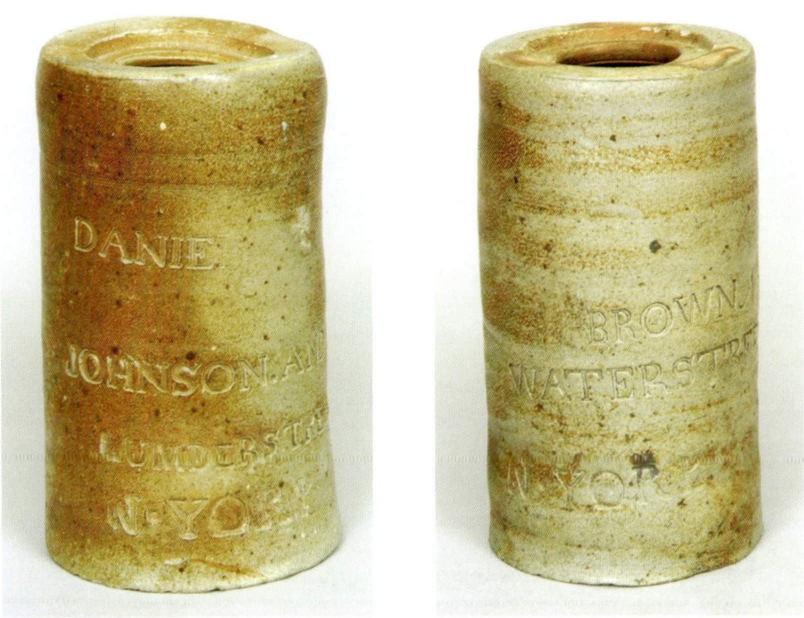

Figure 9 Oysters jars, attributed to Thomas Commeraw, ca. 1800–1805. Salt-glazed stoneware. H. $5\frac{3}{4}$". Left jar impressed on side, DANIEL JOHNSON AND CO. 24/LUMBER STREET/N·YORK.; right jar impressed on side, BROWN/WATER STREET/N·YORK. (Courtesy, Brandt Zipp, The Thomas Commeraw Project/Crocker Farm, Inc.; photos, Luke Zipp.) The jar on the right was made for William Brown, who was still in business as an oysterman ca. 1810. The history of these jars is unknown, but it is reasonable to suppose that both came from the Guyana source, possibly part of a combined shipment.

recorded as early as the mid-eighteenth century, they were rare until well into the nineteenth. Johnson's usage, therefore, implies that he was literate and a possessor of more education than one might expect of an oysterman.

The only recent book devoted to oyster containers is Jim and Vivian Karsnitz's *Oyster Cans*, whose title means what it says; the book focuses almost exclusively on the tin-plated cylinders known first as canisters. The authors attributed the introduction of food-preserving containers to a Frenchman, Nicholas Appert, who in 1809 used glass bottles. By 1840, because glass containers were so easily broken, we are told that they were replaced by cylindrical tin cans whose entry and dispensing apertures could be sealed with solder. Known as "cap-hole cans," they so closely resemble Johnson's stoneware cylinders there is every reason to contend that in America the design evolved directly from ceramic to metal without the intrusion of glass.[22] Indeed, the Guyana brown stoneware cylinders can claim to be the earliest examples of oyster jars yet recorded, be they in stoneware, glass, or tin.[23]

The previously cited Jonathan Durell was a relative newcomer. The German stoneware-potting families of Crolius and Remmey had been in business on the edge of town as early as 1728 and continued through several generations.[24] In 1800, still in the area known as Pot Baker's Hill, John Remmey II sold his premises to his brother John Remmey III, and John Crolius sold his to his son Clarkson. There is no direct proof that either factory made Daniel Johnson's jars, but there is some evidence of a connection, albeit slight. By the end of the eighteenth century these two families are known to have marked their better products using block stamps assembled from printer's type. Clarkson Crolius stamped his "C. CROLIUS / MANUFACTURER / MANHATTAN-WELLS / NEW YORK"; on many surviving examples John Remmey used the same address, but he occasionally marked his wares "J. REMMEY / MANHATTAN-WELLS / N.YORK."[25] The city abbreviation, therefore, parallels the usage found on Johnson's oyster jar. But jumping to conclusions can be a dangerous sport. Another stoneware potter, Thomas Commeraw, long thought also to be of European descent, was working in New York City between 1793 and 1812. He, too, stamped his name and business address on his wares: "COMMERAWS / STONEWARE / CORLEARS / HOOK" (fig. 10).[26]

As it turns out, Commeraw was a prominent figure in the black congregation known as the African Church in the City of New-York. On June 2, 1809, he led the singing at a service heralding the abolition of the slave trade.[27] In the 1800 census he was listed as "Thomas Commeraw, Black."[28] There is little doubt that he was a free African-American, one of many who inhabited lower Manhattan at the beginning of the nineteenth century. His name, possibly an anglicized version of Commerau, suggests that he was of French origin. If he was, he might have been among the hundreds of French refugees who fled from Santo Domingo (Haiti) at the time of Pierre Toussaint's great slave rebellion of 1791.

Corlears Hook, as Commeraw would have known it, offered a small patch of high ground amid a marsh which, in the precolonial centuries was

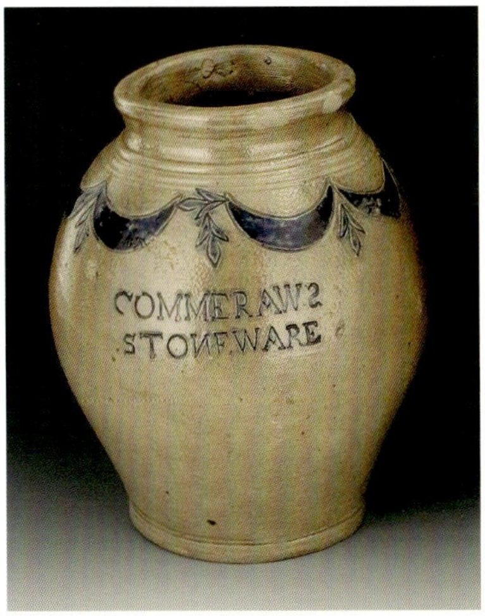

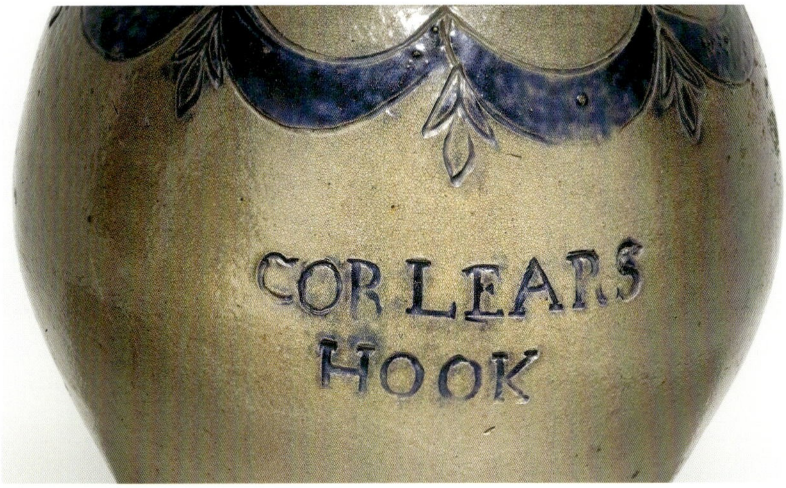

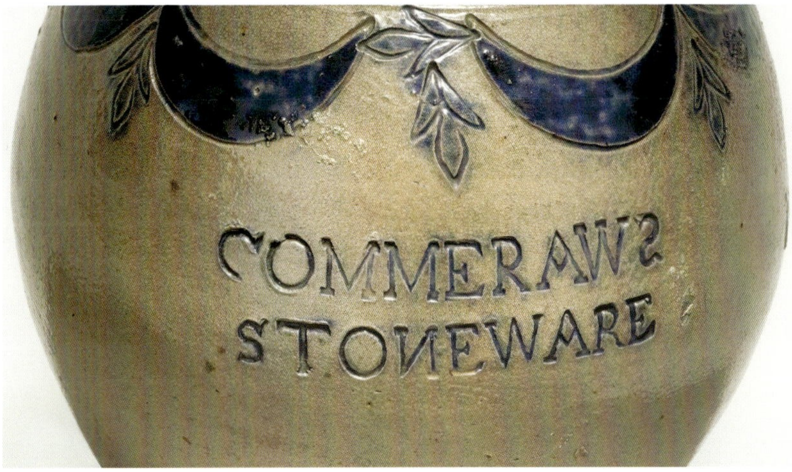

Figure 10 Storage jar, Thomas Commeraw, New York, New York, 1797–1798. Salt-glazed stoneware. H. 8 15/16". Impressed on one side, COMMERAWS/STONEWARE; impressed on the other side, CORLEARS/HOOK. (Courtesy, Colonial Williamsburg Foundation.)

inhabited by the Lenape Indians. Through the seventeenth and eighteenth centuries the marsh was gradually filled in with trash from slaughterhouses and boatyards. However, a map of 1766–1767 shows the high area at the hook—labeled Ackland—laid out and occupied.[29] By 1789 there were three unidentified buildings on the site (see fig. 6).[30] In 1803 a plan was developed by architects Mangin and Goerck to straighten out the East River shoreline to turn the hook into a corner.[31] The plan was rejected, but another of 1811 by surveyor John Randel Jr. proposed a similar straightening of the shoreline.[32] The Manhattan City Commissioners adopted the Randel plan, and through the 1820s the topography of New York was drastically changed, the hills behind the hook being flattened and their soil dumped as fill into ponds and marshes. It is likely that the site of Commeraw's factory had long been buried by the time that construction of the Corlears Hook Park was completed in 1905.

Most of Commeraw's marked wares are jugs of various sizes and usually include the word STONEWARE with the "N" in reverse, although the "N" in N. York is correctly applied. No reversed "N" is present on the Johnson oyster jars. That potentially negative evidence is refuted by the fact that Commeraw was using block-set inscriptions to promote his one-off wares, whereas Johnson's, being a special order, would have had the letters set individually and thus avoid Commeraw's signature mistake.[33] Commeraw's named wares are cobalt-decorated and include ovoid devices that some call "bow knobs" but others see as "clam shells." Either way, Thomas Commeraw's Corlears Hook factory products are readily identifiable.

Before it became a park, Corlears Hook was a promontory at the north end of Water and Cherry Streets, much closer to Daniel Johnson's business than to the Crolius/Remmey kilns, and the waterfront was ideally located for Johnson to transport his jars by water or on a wagon down Cherry Street.

At some date prior to January 1906, the New York State Museum acquired a Daniel Johnson oyster jar recorded as "having the shape of a tobacco or snuff jar" (fig. 11).[34] Its condition prompted the cataloger to speculate that Daniel Johnson himself might have been a potter. The publication date of the museum's acquisition suggests that the jar was found in 1905. It may be no coincidence, then, that construction of the Corlears Hook Park was completed in that year.

In sum, it is reasonable to deduce that at some date between 1800 and 1804 one or more Commeraw craftsmen impressed N·YORK on the oyster

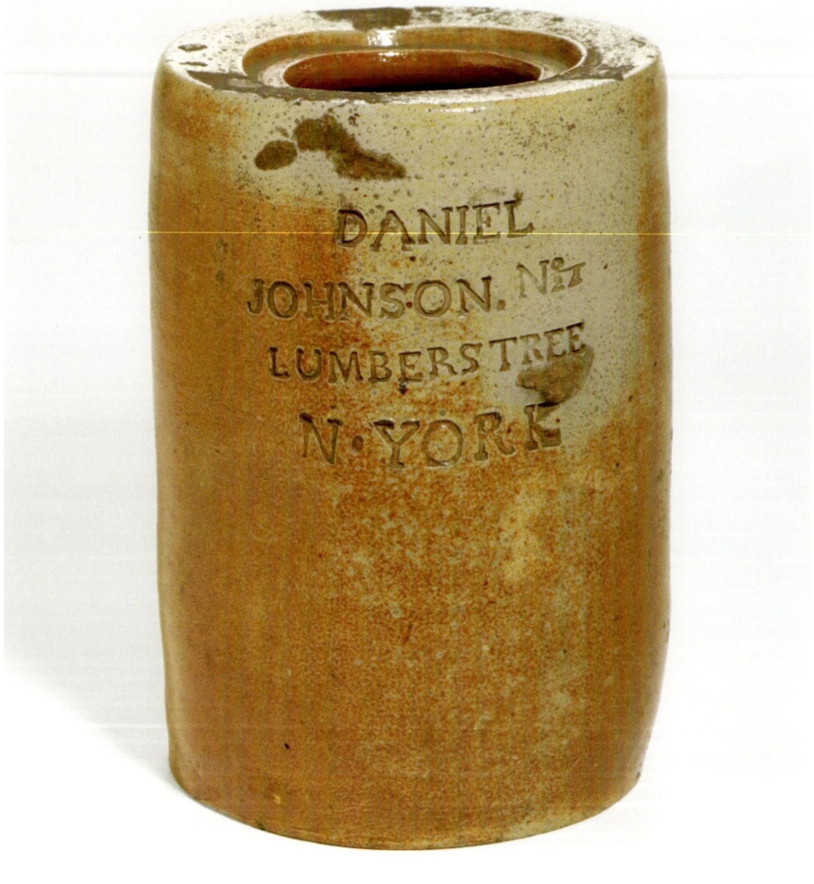

Figure 11 Oyster jar, Thomas Commeraw, New York, New York, 1800–1805. Salt-glazed stoneware. H. 5¾". Impressed on side, DANIEL/ JOHNSON.N°24/LUMBERSTREE/ N·YORK. (Courtesy, New York State Museum.) This jar from Thomas Commeraw's Corlears Hook pottery was acquired by the New York State Museum in 1905, the same year that Corlears Hook Park was laid out.

jar that in 1967 was retrieved from Guyana's Fort Island. With Daniel Johnson's race established, it seems likely that he ordered his jars from a factory where his needs were most readily understood. If true, the oyster jars are historically important evidence of free-black entrepreneurship in the last days of slavery in New York City.

Throughout the nineteenth century, ceramic jars continued to be sold as shipping containers for oysters, but as the Karsnitzes' book *Oyster Cans* demonstrates, "cap-hole" cans were losing favor against heavily black-glazed stoneware made in either Maine or New Hampshire.[35] These had straight necks and were sealed with corks, their principal characteristic being an internal black glaze that extended over the neck and down to a point below the shoulder. Beneath were printed the names of proprietors such as the C. C. Porter Fish Company of Maine or E. T. Wilson of Farmington, New Hampshire. A similar black glaze covers both the inside and outside of a jar of the "cap-hole" variety that might be a transitional example from the second quarter of the nineteenth century (fig. 12).[36]

Another jar from New England claiming to be an oyster jar was imported from England; it is Bristol-glazed and has bead-and-reel rouletting at the shoulder and a square-cut rim (fig. 13). It carries the impressed mark of Stephen Green's Imperial Pottery in Lambeth and therefore dates prior to 1858.[37] Similar jars from other London-area factories were described as being covered for "pickling and preserving." Several patents were registered for the airtight sealing of such jars. The illustrated example found in a Virginia field was patented by Alfred Singer of Vauxhall on

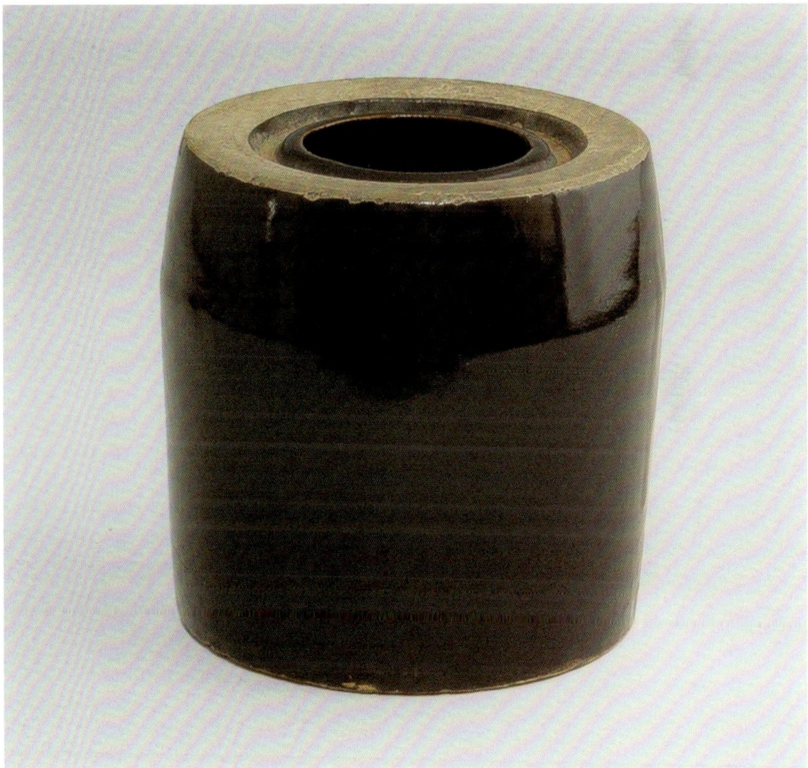

Figure 12 Oyster jar, American, probably New England, mid-nineteenth century. Black-glazed stoneware. H. 6". (Author's collection; photo, Robert Hunter.) This jar has a recessed lid opening, similar to those of the earlier Commeraw jars.

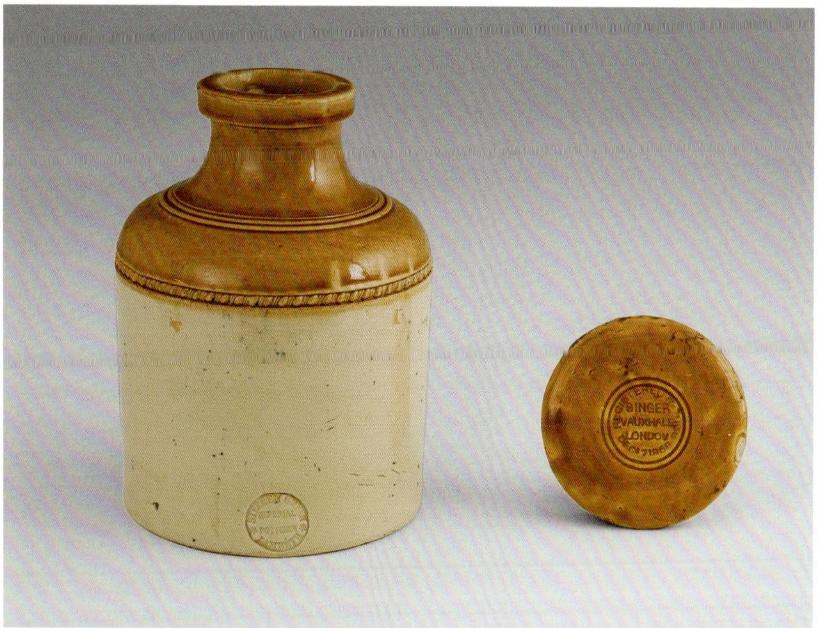

Figure 13 Oyster jar and a lid, London, England, mid-nineteenth century. Bristol-glazed stoneware. H. 6¼". Impressed on jar, IMPERIAL/POTTERIES · STEPHEN GREEN · LAMBETH; impressed on lid, SINGER/VAUXHALL/LONDON REGISTERED N^{O}.4138 DEC 17 1858. D. 3⅛". (Author's collection; photo, Robert Hunter.) The jar was found in New England and the lid in a Virginia field.

December 17, 1858 (fig. 13, right). Designed to twist and lock, the lid fitted a wider version of the Stephen Green jar.[38] In 1873 Doulton and Watts of Lambeth were selling similar shapes as "Air-tight Covered Jars." Doulton's nearby competitor James Stiff was offering comparable jars and illustrated them with a metal clamp over the lids. In neither case did the manufacturers define these jars as containers for oysters.[39] However, an example in *Oyster Cans* is print-identified as being made for the Café Royal's Oyster Bar at 17 West Register Street in Edinburgh.[40] Other oyster jars in white stoneware and with less readily sealed mouths were made for one Thomas Anderson of Glasgow in the late nineteenth century.[41]

By the mid-nineteenth century, brown stoneware cap-hole oyster jars like those made for Daniel Johnson were being copied in tin. An intact crate of the cylindrical cans was found in the wreck of the steamboat *Arabia*, which sank in the Missouri River in 1856. The crate bore the stenciled information "2 Doz 1^{lb} Cans/ FRESH / COVE OYSTERS/ FROM / PRICE & LITTIC / BALTIMORE" (fig. 14).[42]

The decline of the specifically defined cap-hole oyster jar forced potteries to seek a wider market by advertising these jars as all-purpose containers.

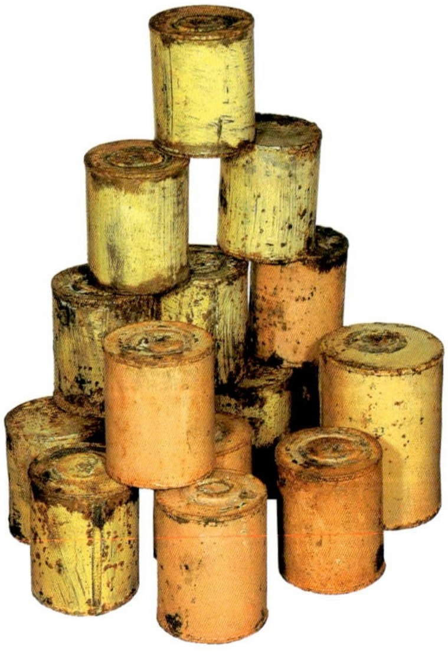

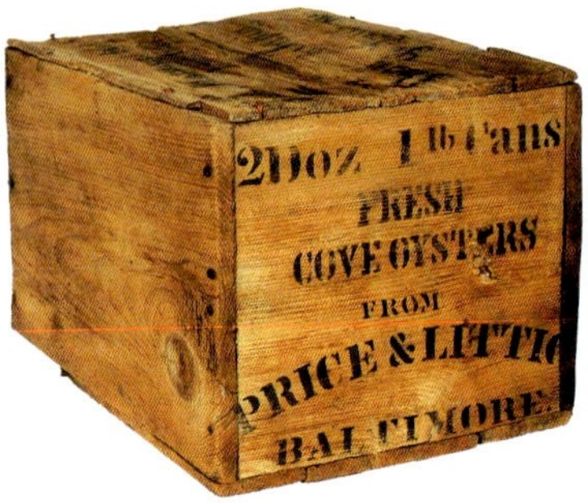

Figure 14 Baltimore wooden shipping crate and its tinplate oyster cans from the 1856 wreck of the steamboat *Arabia* in the Missouri River. Known as cap-hole cans, these vessels were sealed with solder. (Courtesy, David Hawley, *Arabia* Steamboat Museum.)

Consequently, unless the jars carry printed labels, their evolution through the latter years of the nineteenth century is hard to define. There is no doubt, however, that African-American Thomas Commeraw's stonewares were at the fountainhead of an ongoing oyster distribution industry.

ACKNOWLEDGMENTS I am indebted to Susan Williams for bringing her jar to my attention and for leading the charge in the research that led to this article. I am grateful, too, to Brandt Zipp of Crocker Farm Stoneware Auction for sharing his study of Thomas Commeraw and his Corlears Hook factory, and to Dr. Shane White in Sydney, Australia, for his help sorting out elusive, nineteenth-century Manhattan addresses. Finally, my thanks to Christine Annan at Edinburgh's Café Royal Oyster Bar.

1. Guyana had been a Dutch colony from 1616 until ceded to Britain in 1814, whereupon it was renamed British Guiana. It secured its independence in 1966. The neighboring Dutch Guiana (Suriname) gained independence in 1975.

2. In 2005 an American auction house sold two at $3,500 each.

3. One was in the 3rd Ward on the Hudson River (North River) side. The other was in the 7th Ward on the East River. Immediately above his listing was a Richard Johnston, who was cited as a potter on Magazine Street.

4. *Longworth's American Almanack, New-York Register and City Directory* (New York, 1799), pp. 262–63.

5. *Merriam-Webster's Collegiate Dictionary, Tenth Edition* (2001), s.v. "stand."

6. New York historians have long recognized that the "responsibility for public record-keeping often fell to people with relatively little formal education." Gilbert Tauber, comp., "A Guide to Former Street Names in Manhattan" (2005), available online at www.oldstreets.com (accessed July 29, 2011). Ceramic historian Brandt Zipp independently came to the same conclusion after locating the Lumber Street addresses of the people flanking Johnston in the 1800 Federal Census.

7. Previously one of the two Lumber Streets, this one was earlier known as Rutgers Street and subsequently (in 1831) changed to Monroe Street.

8. Tauber, comp., "Guide to Former Street Names in Manhattan." Lombardy became Monroe Street in 1831. The phonetically induced connection is amplified by the 1810 directory listing Ephraim Whitlock at No. 32 of nearby Oliver Street and identifying him as "inspector of lumber." The *New-York Directory* for 1810 still had people living on Lumber Street.

9. Trinity Church Parish Registers, June 9, 1793, and August 10, 1805. The minister who performed both ceremonies was Benjamin Moore, who five years later found himself in trouble for officiating at the marriage of a slave. Shane White, *Somewhat More Independent: The End of Slavery in New York City, 1770–1810* (Athens: University of Georgia Press, 1991), p. 89, citing *Daily Advertiser*, July 25, 1798.

10. Ralph Ellison, *Invisible Man* (New York: Random House, 1952), p. 9, citing a contemporary medical report.

11. White, *Somewhat More Independent*, pp. 161, p. 236 n. 34.

12. The 1804 directory has a John Johnson, oysterman, living at 92 Warren Street in Ward 5, which could indicate that Daniel and John were not related. (It should be remembered that while census records identified African-Americans, the city directories did not.)

13. Mark Kurlansky, *The Big Oyster: History on the Half Shell* (New York: Random House Trade Paperbacks, 2007), p. 121.

14. That assumption may have been scuttled by *Longworth's Directory*, which lists Daniel Johnson as still alive at 106 Cherry Street in 1810, having been at both at 106 Cherry and 24 Lombard in 1804.

15. Rita Susswein Gottesman, ed., *The Arts and Crafts in New York: Advertisements and News Items from New York City Newspapers*, vol. 1, *1726–1776* (New York: Printed for the New-York Historical Society, 1938), p. 85.

16. *The New-York Gazette and Weekly Mercury*, March 15, 1773.

17. Nancy Cox and Karin Dannehl, eds., *Dictionary of Traded Goods and Commodities, 1550–1820* (University of Wolverhampton, 2007), www.british-history.ac.uk/source.aspx?pubid=739 (accessed July 28, 2011).

18. Kurlansky, *The Big Oyster*, p. 73, citing Peter Kalm, *Travels in North America*, vol. 1, 1747–1748, edited by Adolph B. Benson (New York: Dover Publications, 1966).

19. *The Journal of John Gabriel Stedman, 1744–1797, Soldier and Author, Including an Authentic Account of His Expedition to Suranam in 1772*, edited by Stanbury Thompson (London: Mitre Press, 1962), p. 147.

20. Auction house owner Brandt Zipp recalls that both jars came together from a dealer in Canada.

21. Ann Street and Nassau Street were in Ward 2 but adjacent to Ward 4.

22. Jim Karsnitz and Vivian Karsnitz, *Oyster Cans* (Atglen, Pa.: Schiffer, 1993), p. 11 (Southern Can Co. ad, 1914). On September 6, 1856, the *Arabia* sank in the Missouri River and was discovered and archaeologically salvaged in 1988. Among the thousands of objects recovered was a crate of tin oyster cans with shapes similar to that of the Guyana jars. David Hawley, *The Treasures of the Steamboat Arabia* (Kansas City, Mo.: Arabia Steamboat Museum, 1995), p. 58.

23. Karsnitz and Karsnitz, *Oyster Cans*, p. 150. In 1922 the William Heyser Company of Baltimore was offering oysters in both "small Cans" and "Glass Jars."

24. Meta F. Janowitz, "New York City Stonewares from the African Burial Ground," *Ceramics in America*, edited by Robert Hunter (Easthampton, Mass.: Antique Collectors' Club for the Chipstone Foundation, 2008), pp. 41–66.

25. Donald Blake Webster, *Decorated Stoneware Pottery of North America* (Rutland, Vt.: Charles E. Tuttle Co., 1971), p. 48, fig. 27.

26. Ibid., p. 217. The name "Corlear's Hook" dates back to the Dutch creation of Manhattan, but during the British occupation it was known as Crown Point.

27. The order of service and the lead speaker's address are in the archives of the New-York Historical Society.

28. 1800 U.S. Federal Census Record, New York Ward 7, Roll 23, p. 905, image 286. A probable relative was John Commeraw, who was married to Zerviah Cook and listed by the New England Historic Genealogy Society in 1817 as being "of colour."

29. Bernard Razner's map of 1766–1767. See Paul E. Cohen and Robert T. Augustyn, *Manhattan in Maps, 1527–1995* (New York: Rizzoli International Publications, 1997), p. 71.

30. *New-York Directory and Register*, 1789. Cohen and Augustyn, *Manhattan in Maps*, p. 93.

31. Joseph François Mangin was an emigrant Frenchman. His partner was Casimir Goerck.

32. A copy of the Randel map is in the New York Public Library; another is in the Library of Congress.

33. There was no such error in the lettering of the BROWN / WATER STREET N.YORK jar.

34. *59th Annual Report of the New York State Museum*, 1906, pp. 28–29. The full text of this report is available online through Google Books, s.v. "59th Annual Report of the New York State Museum" (accessed July 28, 2011).

35. Karsnitz and Karsnitz, *Oyster Cans*, pp. 122–23.

36. This jar was purchased from a California dealer who could not identify it.

37. In that year Green retired and sold his business to the Fulham Pottery.

38. Ivor Noël Hume, *If These Pots Could Talk: Collecting 2,000 Years of British Household Pottery* (Hanover, N.H.: University Press of New England for the Chipstone Foundation, 2001), pp. 323–25.

39. Chris Green, *John Dwight's Fulham Pottery: Excavations 1971–79*, Archaeological Report No. 6 (London: English Heritage, 1999), pp. 361–68.

40. The Café Royal's oyster bar opened in Edinburgh in 1826 and is still renowned for its oysters both Rockefeller and Kilpatrick. It moved to its present location in 1863. According to Christine Annan at the Café Royal, the address changed from 17 to 19 West Register Street about eight years ago.

41. Karsnitz and Karsnitz, *Oyster Cans*, p. 123.

42. Hawley, *The Treasures from the Steamboat Arabia*, p. 56.

New Discoveries

Introduction

▼ AS LONG AS new archaeological sites are excavated and old archaeological collections are analyzed, new ceramic discoveries will appear. What we find out about them is significant, because whether they were used to furnish homes or public buildings, to store or prepare food in the kitchen or dairy, or placed before eager diners, historic ceramics are as varied as the people they served. These objects, always easily broken, rapidly changed in form, decoration, and ware, yet they communicate their purpose to us by revealing who made them, and when and why they were made. And, based on where they are found, they even tell us about those who bought, used, and discarded them. From locations as diverse as Trinity College in Cambridge, England, and Monterey, California, when it was the capital of the Mexican Republic province, all of this year's New Discoveries are archaeological finds, and they pose stimulating questions and provide intriguing glimpses into the past.

Try to imagine what Robert Cotton at Jamestown was thinking as he pressed pipe clay into a Powhatan basket in 1608–1610, or David Drake, the legendary Edgefield potter, as he signed the bottom of a stoneware jug as a freedman on January 25, 1868. Consider the specialized uses of three vessels on diverse British sites in rural colonial Virginia: a circa 1613–1625 Normandy stoneware drug jar, a circa 1700 delftware salt cellar, and an eighteenth-century Langerwehe stoneware butter churn. Was punch improved when strained through a refined stoneware vessel at the Chew Mansion in Maryland in the 1700s? Did hungry Trinity College students in the eighteenth century even notice that the plates from which they ate bore their cook's name? How did ordinary terracotta flowerpots express the scientific knowledge and wealth of the powerful and elite in Federal-period Massachusetts? In the 1820s, what ceramics were fit for the table of French royalty living in New Jersey, or a Scottish emigrant and his family in rural New York?

Ceramic vessels can literally speak to us, as does a circa 1792–1820 creamware jug that commemorates a British Royal Navy hospital. But why was it found at Fort Johnson, South Carolina, a place built to defend against the Royal Navy? A whiteware cup found in California surely delighted its young owner in the 1830s, and one can only imagine the disappointment when it fell, with no means of repair. Each object has its own story, if we take the time to read.

Merry Abbitt Outlaw
xkv8rs@aol.com

Anthony Butera Jr.

Whately Teapots in the Western Catskills

▼ DURING RESTORATION of a circa 1820 stone house in rural New York State, an earthenware teapot, likely made at the circa 1821–1833 Thomas Crafts teapot factory in Whately, Massachusetts, was discovered buried next to the hearth (fig. 1).[1] The house was situated in Delaware County, a remote area of the western Catskills that was settled in the late eighteenth and early nineteenth centuries. Andrew and Helen Cowan and an unknown number of children emigrated from Selkirk, Scotland, in 1820 and built this house.[2] Ceramic sherds recovered from the site's surface are typical of the era. These include a variety of dipped wares, shell-edged earthenwares, painted and printed pearlwares, and American stoneware. A small amount of American red earthenware was also found.[3]

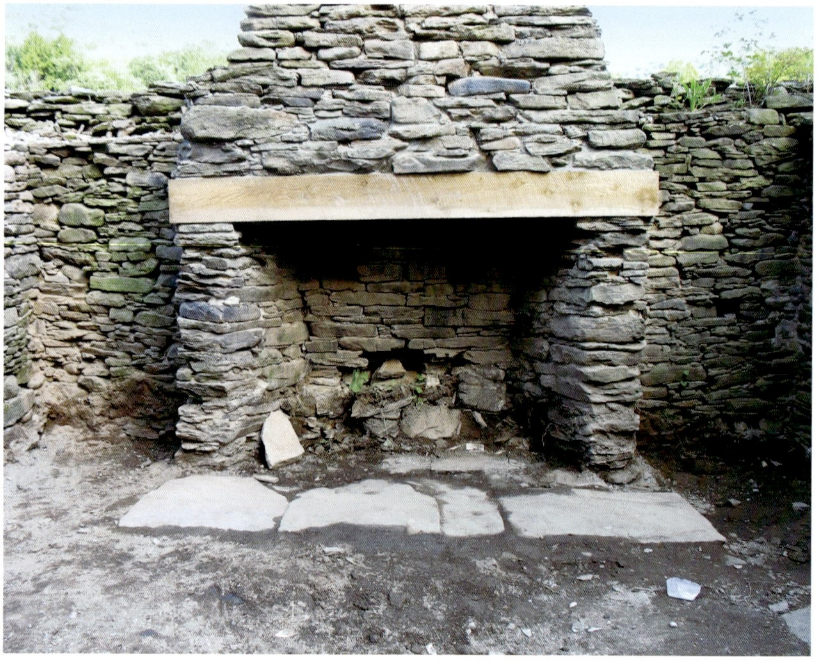

Figure 1 The original fireplace, hearth, and walls of the ca. 1820 Cowan stone house in Delaware County, New York. (Photo, Anthony Butera Jr.) It is believed that the house burned ca. 1900. The teapot was found buried on the right side of the hearth when a temporary support was built in order to replace the rotting lintel. It was probably sitting on a shelf that was supported by the chimney cheek and the adjacent wall. Charred wooden supports are still in place.

Based on its brilliant metallic black glaze and resemblance to extant examples, the teapot (fig. 2) has been attributed to the Thomas Crafts factory.[4] Its faceted eight-sided spout relates to a faceted eight-sided spout sherd found in the culvert of the house next to the Cowan house, and matching spouts have been recovered recently from the Crafts manufactory site.[5] The thick glaze on each example makes it difficult to determine how the spout was formed. In 1832 Crafts reported that he made 2,080 dozen

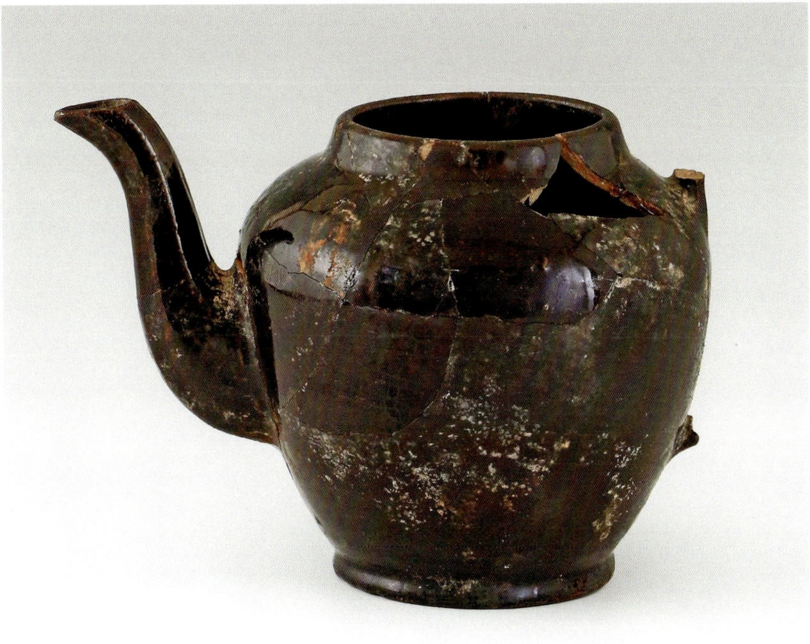

Figure 2 Teapot, Thomas Crafts, Whately, Massachusetts, ca. 1821–1833. H. 5½". (Photo, Robert Hunter.) This vessel was found on the right side of the hearth illustrated in fig. 1. A lid from a Thomas Crafts teapot in the author's collection fits this example perfectly.

teapots annually and sold them for one dollar per dozen.[6] Since the market was being flooded with these cheap teapots, it is not surprising to find evidence of two of his teapots in this remote location.

There are probably fewer than two dozen extant examples of Thomas Crafts teapots, making this specimen quite important. The original owners would be surprised to see this modest, everyday object to be of such interest to future generations.

Anthony Butera Jr., 112 Tuthill Street, Port Jefferson, New York 11777
erawder@optonline.net

ACKNOWLEDGMENTS I extend thanks to Ralph Cowan, David Graci, and Shiela Kohler from the Pocumtuck Memorial Association, and especially to Diana and Garry Stradling for their enthusiastic encouragement.

1. Brian Cullity, ed., *Slipped and Glazed: Regional American Redware*, exh. cat. (Sandwich, Mass.: Heritage Plantation of Sandwich, 1991), p. 29.

2. Andrew Cowan's great-grandnephew Ralph Cowan, who still resides in Delaware County, provided this information; personal communication, August 2009 and November 2010.

3. *History of Delaware County, N.Y.* (New York: W. W. Munsell & Co., 1880), p. 53, which offers an account of what was brought by the county's settlers: ". . . some cups and saucers, and a few plates, all of the 'shell-edged' variety, now nearly out of existence. . . ."

4. Lura Woodside Watkins, *Early New England Potters and Their Wares* (Cambridge, Mass.: Harvard University Press, 1950), p. 104. Watkins describes the teapots as "of common redware with a remarkably brilliant, almost metallic glaze." Diana and J. Garrison Stradling (personal communication) gave the teapot an attribution to Whately, Massachusetts, in January 2010.

5. David Graci, an advocational archaeologist, has recovered matching spouts and other types of spouts from the Crafts pottery site. A visit to his house in October 2010 provided the identification of these vessels.

6. Watkins, *Early New England Potters*, p. 104.

Beverly A. Straube

Incorporating the Other: A Seventeenth-Century Virginia Indian Basket Rendered in Clay

▼ MANY OF THE European elite in the early modern period enjoyed assembling collections of natural and artificial oddities from around the world. Natural history specimens, including live animals, were collected along with precious man-made objects and displayed in house zoos and in Wunderkammern, or cabinets of curiosity. These collections conveyed to others their owners' knowledge, status, and refinement. They were also reflective of political and mercantile interests and exemplified the "need to control human and natural resources in the New World, and . . . the search for things that could bring profits to the Old World."[1]

In 1608 newly arrived Jamestown colonist Francis Perkins sent to friends in England some exotic New World gifts, among them four pairs of turtledoves, two Indian-made "pots of our ordinary earth," a corncob ("ear of the native wheat"), and twelve pounds of sassafras "to use in medicines or between linens." Perkins promised to send "better things" on the next ship, including some little animals "with skins of fine fur" if he were able to acquire any.[2] King James I of England was eager to acquire some of the Virginia flying squirrels he had heard about, and London antiquarian Sir Walter Cope's cabinet of curiosities is documented as having contained many New World objects, including an Indian canoe and "flies that glow at night in Virginia."[3]

In seventeenth-century Virginia, the culture of collecting was practiced not only by the English but also by the Native Americans, who were eager to acquire the exotic objects—copper, iron tools, glass mirrors, and beads—the colonists offered in exchange for food. Just as the Virginia Company of London attempted to gain control of Virginia's resources, paramount chief Powhatan of the Powhatan tribe tried to command access to the English artifacts that were used as tribute goods in his stratified society. On one occasion, for instance, he was led to believe by the English that certain blue glass beads were "composed of a most rare substance of the coulour of the skyes, and not to be worne but by the greatest kings in the world." According to Captain John Smith, this made Powhatan "halfe madde to be the owner of such strange Jewells."[4]

The Jamestown colonists, preoccupied with trying to locate or produce commodities of value for the London investors, were largely dependent on the Indians for food. Corn, which made up a large percentage of the native diet, was the most common foodstuff obtained through trade, and it was purchased from the Indians by the basketful.[5] The baskets, made by the native women of cornhusks, tree bark, grass, and other fibers, did not have a rigid framework but were flexible containers. This is indicated in a 1609 trading encounter when, after the English had worked out the quantity of "copper and beads and other things" to exchange for corn, they discovered that the Indian warriors were craftily pushing up the bottoms of their baskets with their hands "so that the less corn might fill them."[6]

One of these Indian baskets, probably acquired through trade with the Powhatan, was used by an English colonist to form a pit-fired clay pot that was found by archaeologists in a circa 1610 context within James Fort (fig. 1). Robert Cotton, who arrived at Jamestown in April 1608, is believed

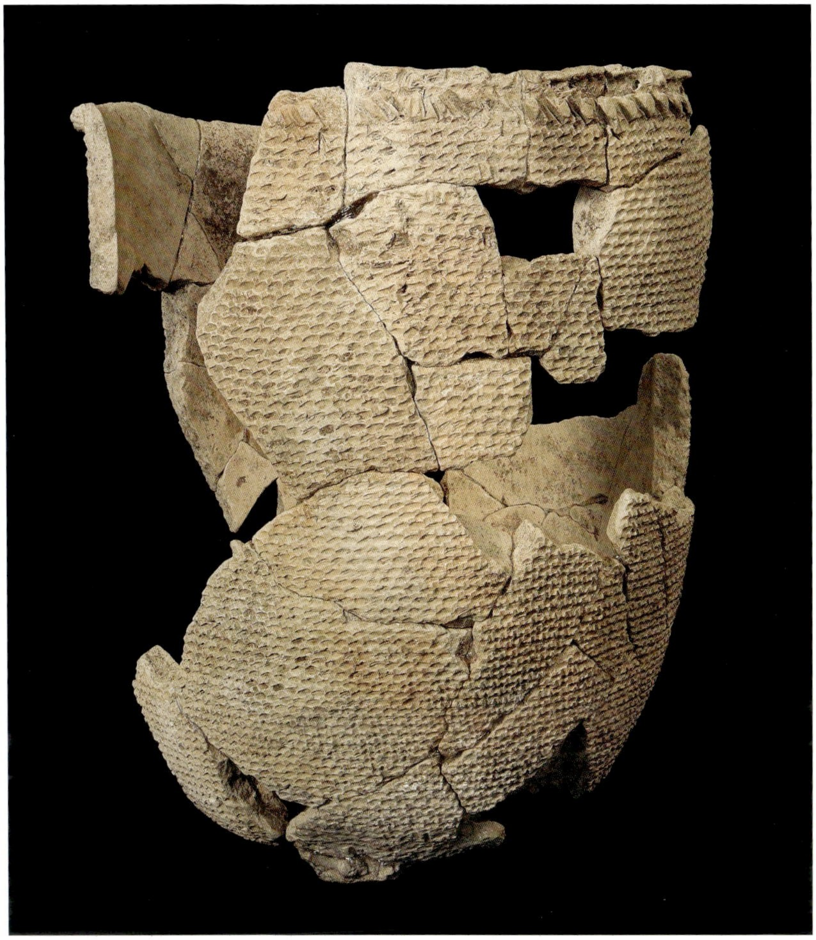

Figure 1 Basket pot, attributed to Robert Cotton, Jamestown, Virginia, 1608–1611. Unglazed earthenware. H. 10". (Courtesy, Preservation Virginia; photo, Michael D. Lavin.)

to be the source of the pot for several reasons. He was a tobacco pipe maker whose Virginia-clay pipes are numerous in the fort's earliest contexts, he is considered the only colonist working with clay at this early date, and the pot's clay matrix appears identical to that of his pipes.[7]

The pot had been shaped by pressing layers of clay into the basket's interior, resulting in walls nearly ¾-inch thick. Cotton then fired the clay pot over low heat, which burned off the basket material while leaving its impressions on the pot's exterior. The basket's fibers produced a clean imprint in the clay with no fibrous offshoots, leading basket maker and historical interpreter Sharon Walls to conjecture that the fibers were produced from the leaves of *Yucca filamentosa*, a native plant used by the Virginia Indians for its durability and strength.[8]

The pot was produced to function as a traditional English form, as evidenced by the addition of clay pads on its base to enable the vessel to sit upright, unlike the native pots, which had a cone-shaped base and were

Figure 2 (*Left*) Modeling clay impression of the rim of the basket pot illustrated in fig. 1 alongside (*right*) the actual rim sherd, showing the pattern of twined fiber.

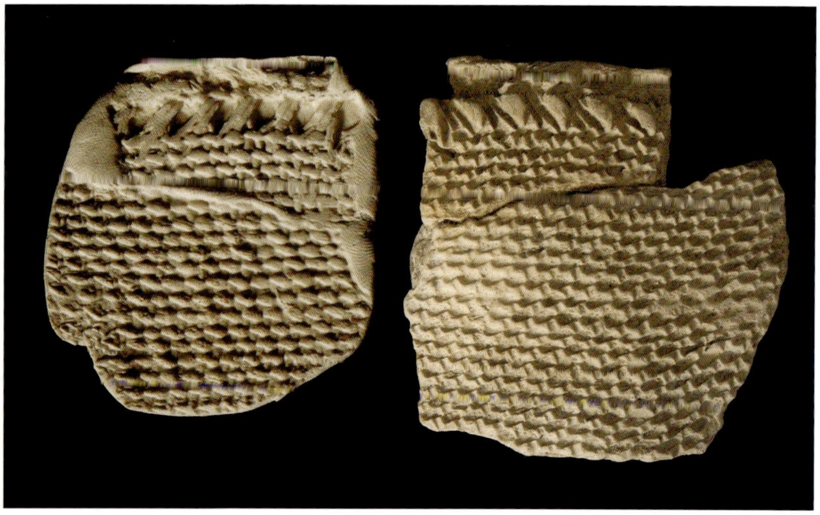

propped in the hearth using wood and stone cobbles. As the modeling clay impression of the vessel shows clearly (fig. 2), the basket used for the mold was a flexible container made of finely woven twined fiber. In twining, the wefts are worked in pairs, one going over the warp and the other under; as the wefts are woven, they are twisted, producing a distinctive diagonal pattern. The rim of the basket was finished off by pulling two warps from the left over one from the right and then braiding the fibers together.

Cotton's "basket pot" could represent a quick and easy solution for producing a needed clay pot, but it could also reflect an Englishman's appreciation for Powhatan basketry and his attempts to preserve it. Either way, the unique pot provides us with a rare opportunity to examine the fiber artistry of a native Virginia woman from four hundred years ago, as she fashioned a functional object from the natural world.

Beverly A. Straube, FSA, Senior Archaeological Curator, Jamestown Rediscovery Project, Historic Jamestowne, bly@preservationvirginia.org

1. Antonio Barrera, "Local Herbs, Global Medicines: Commerce, Knowledge and Commodities in Spanish America," in *Merchants and Marvels: Commerce, Science, and Art in Early Modern Europe*, edited by Pamela H. Smith and Paula Findlen (New York: Routledge, 2002), p. 165.

2. A New World plant, sassafras was particularly valued in Europe as a treatment for syphilis. Philip L. Barbour, ed., *The Jamestown Voyages under the First Charter, 1606–1609*, 2 vols. (Cambridge: Cambridge University Press for the Hakluyt Society, 1969), 1:161.

3. Ibid., 2:288. Cope's objects were noted in 1599 and probably were acquired from Sir Walter Raleigh's 1585–1587 colonizing attempts in modern-day North Carolina. See Thomas Platter, *Thomas Platter's Travels in England, 1599*, translated by Clare Williams (London: J. Cape, 1937), pp. 172–73.

4. Philip L. Barbour, ed., *The Complete Works of Captain John Smith (1680–1631),* 3 vols. (Chapel Hill: University of North Carolina Press for the Institute of Early American History and Culture, 1986), 2:156.

5. Historian Kathleen M. Brown has estimated that corn provided about 75 percent of the calories consumed by the Indians living in the Coastal Plain. See Brown, *Good Wives, Nasty Wenches, and Anxious Patriarchs: Gender, Race, and Power in Colonial Virginia* (Chapel Hill: University of North Carolina Press for the Institute of Early American History and Culture, 1996), p. 46.

6. Edward Wright Haile, ed., *Jamestown Narratives: Eyewitness Accounts of the Virginia Colony, the First Decade, 1607–1617* (Champlain, Va.: RoundHouse, 1998), p. 484.

7. William M. Kelso and Beverly A. Straube, *Jamestown Rediscovery, 1994–2004* (Richmond, Va.: Association for the Preservation of Virginia Antiquities, 2004), pp. 163–66.

8. Sharon Walls, personal communication, 2006.

Bruno Fajal and Taft Kiser

A Domfront (Normandy) Stoneware Pot in Jacobean Virginia

▼ VIRGINIA'S RAREST archaeological sites are those dating to its first decades as an English colony. These early contexts can contain such exotic ceramics as Ming porcelain, Algonquin tobacco pipes, and Montelupo tin-glazed earthenware, but one of the most exceptional new discoveries is a small French-made ointment pot. From a circa 1613–1625 context at Shirley Plantation in Charles City County, the pot was made near Domfront, Normandy, Orne, and is the only piece of Normandy stoneware that has been found in Virginia (figs. 1, 2).[1]

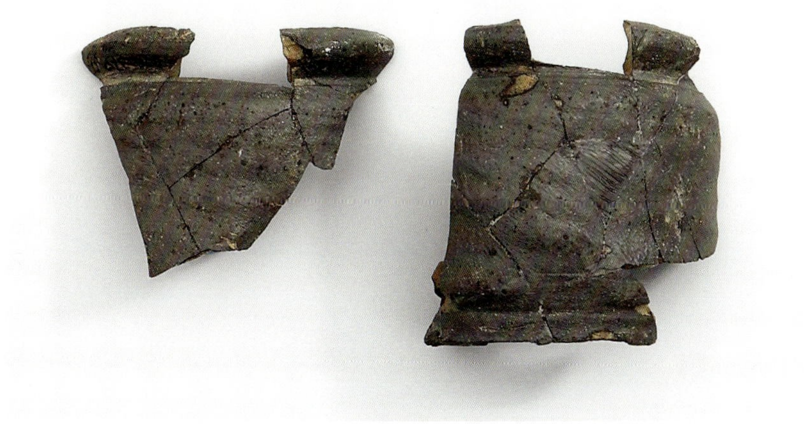

Figure 1 Drug jar, stoneware, Domfront, France, ca. 1613–1625. H. 1¾". (Courtesy, Virginia Department of Historic Resources; photo, Robert Hunter.) Recovered at site 44CC370, Charles City County, Virginia, in 2002.

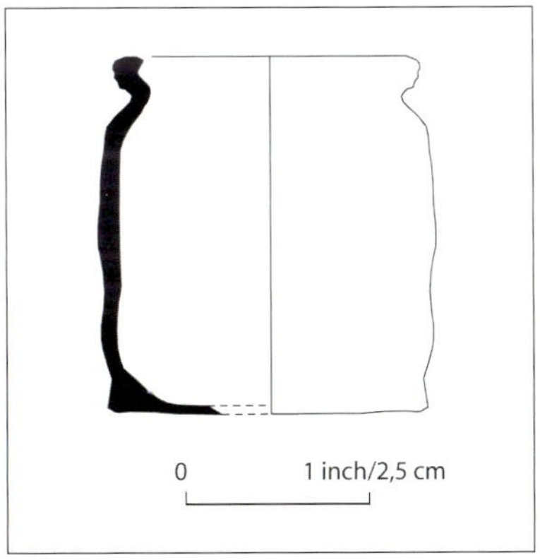

Figure 2 Profile illustration of the Domfront drug jar illustrated in fig. 1. (Illustration by Bruno Fajal and Taft Kiser.)

In Normandy, distinctive stonewares were produced around Domfront (southwest Orne and southeast Manche), in west-southwest Bayeux (Calvados), in the north of the Cotentin (Manche), and in Martincamp-Bully, near Neuchâtel-en-Bray (Seine-Maritime). The Shirley vessel is a wide-mouthed cylindrical "drug jar" that probably held medicinal salve for people or livestock.[2] At Domfront this form was produced from the early postmedieval period until the early twentieth century, when the local stoneware industry, around Ger (Manche), fired its last kilns. Made of a kaolin-type clay known as Saint-Gilles-des-Marais, which becomes semivitrified at 1,250–1,300°C., the surface is gray when unglazed and the fired clay body is pale brown. This clay has been used by potters since at least the thirteenth century and is found exclusively in a limited area west of Domfront.[3]

French pottery is rare in seventeenth-century Virginia archaeological contexts; most common is the high-fired earthenware known as Martincamp, named for a village in Normandy.[4] Stoneware jars from Beauvais have been excavated from early contexts at Jamestown and a circa 1620–1635 context at Jordan's Point in Prince George County. And although earthenware from Saintonge in southern France has not been recovered from Jamestown's early contexts, a vessel of this ware has been found in a circa 1618–1630 context at Flowerdew Hundred, in Prince George County, and in post-1630 contexts at Henrico County's Curles Plantation and the Walter Aston site in Charles City County.[5]

Elsewhere in North America, Normandy stoneware has been found on French sites in such contexts as circa 1562–1563 in South Carolina, the 1560s in Florida, and 1604–1763 in Canada and Maine.[6] It has not been found at Maine's Fort Pentagoet, probably because the settlement relied on the more southern port of La Rochelle for supplies.[7]

The Domfront pot was excavated at the earliest known European structure on Shirley. In 1613 Virginia's principal mariner, Samuel Argall, destroyed two settlements in French Arcadia, and the prizes he carried to Virginia included prisoners (who were transported to England in June 1614), food, clothing, horses, and tools.[8] The 1625 Muster lists eight people who were on Argall's ship in 1613 and 1614; half of them were at Shirley or neighboring Jordan's Point, where the two Beauvais jars were excavated. This suggests an alluring interpretation as Argall's loot, but Virginians had frequent contact with French vessels in the northern fisheries and French workers were sent to the colony to establish vineyards and produce silk.[9]

Bruno Fajal, CNRS, CRAHAM-UMR 6273, University of Caen,
14032 Caen cedex, France, bruno.fajal@unicaen.fr

Taft Kiser, Cultural Resources, Inc., Richmond, Virginia
tkiser@culturalresources.net

1. Bruno Fajal, "Ger 1988: Rapport de fouille de sauvetage urgent," unpublished report, Centre de Recherches Archéologiques Médiévales, Caen, 1988.

2. Stéphen Chauvet, *La céramique Bas-Normande ancienne* (Mortain [Manche]: Éditions du Mortainais, 1950).

3. Daniel Dufournier and Bruno Fajal, "L'apparition du grès dans la région domfrontaise, premières observations," in *La céramique du XIe au XVIe siècle en Normandie, Beauvaisis, Ile-de-France*, Cahiers Techniques du GRHIS No. 2, edited by Xavier Delestre and Anne-Marie Flambard Héricher ([Mont-Saint-Aignan]: Publications de l'Université de Rouen, 1995), pp. 73–80; and Jean-Pierre Lautridou, "Les argiles fini-tertiaires de Saint-Gilles-des-Marais (Domfrontais, Orne)," *Bulletin de la Société Linnéenne de Normandie* 118 (2002): 39–41.

4. For an online overview of Martincamp flasks (the only form found in Virginia), go to www.preservationvirginia.org/rediscovery/page.php?page_id=321.

5. According to Jamestown Rediscovery curator Beverly Straube, no Saintonge earthenware has been found in Jamestown's early contexts (personal communication, 2010); the vessel at Flowerdew Hundred was identified by the late John G. Hurst (personal communication, 1997).

6. Jean-Pierre Chrestien, Daniel Dufournier, Françoise Niellon, and Marcel Moussette, *Le site de l'habitation de Champlain à Québec: Étude de la collection archéologique (1976–1980)* (Quebec: Government of Quebec, 1987); David M. Brewer and Elizabeth Horvath, *In Search of Lost Frenchmen: Archaeological Excavations at Canaveral NS* (Washington, D.C.: National Park Service, U.S. Department of the Interior, 2008), available online at www.nps.gov/archeology/sites/npsites/canaveral.htm (accessed December 20, 2010); and Jean-Pierre Chrestien and Daniel Dufournier, "French Stoneware in North-Eastern North America," in *Trade and Discovery: The Scientific Study of Artefacts from Post-Medieval Europe and Beyond*, British Museum

Occasional Paper 109, edited by Duncan R. Hook and David R. M. Gaimster (London: British Museum, 1995), pp. 91–103.

7. Alaric Faulkner and Gretchen Faulkner, *The French at Pentagoet, 1635–1674: An Archaeological Portrait of the Acadian Frontier*, Occasional Publications in Maine Archaeology No. 5 (Augusta: Maine Historic Preservation Commission, 1987), p. 211.

8. *Dictionary of National Biography*, edited by Sir Leslie Stephen, 63 vols. (New York: Macmillan, 1885), 2:79.

9. Edward Wright Haile, ed., *Jamestown Narratives: Eyewitness Accounts of the Virginia Colony, the First Decade, 1606–1617* (Champlain, Va.: RoundHouse, 1998), pp. 786–87, 824–26; Faulkner and Faulkner, *The French at Pentagoet*, p. 14; Martha W. McCartney, *Virginia Immigrants and Adventurers, 1607–1635: A Biographical Dictionary* (Baltimore, Md.: Genealogical Publishing Co., 2007), pp. 56–57, 209, 213, 216–17, 377–78, 452, 618, 749–50; *Adventurers of Purse and Person: Virginia, 1607–1624/5*, edited by Virginia M. Meyer and John Frederick Dorman, 3rd ed. (Richmond: Order of the First Families of Virginia, 1607–1624/5, 1987), p. 29; and Susan Myra Kingsbury, ed., *The Records of the Virginia Company of London*, 4 vols. (Washington, D.C.: Government Printing Office, 1906–1935), 1:466, 3:242–43.

Glenn Farris

A Vision of a Technological Wonder in 1830s California

▼ IMAGINE you are a young person in 1830s California, which at that time was a province of the Mexican Republic. One day your father brings home a wonderful ceramic drinking cup that had been delivered from the other side of the world by a merchant ship. On the side is depicted a remarkable machine, a steam engine pulling a line of cars full of passengers (figs. 1, 2). Such a cup—the equivalent of a Buck Rogers spaceship mug of my youth—was discovered during archaeological excavations conducted in 1974 in Monterey, which had been the capital of the California province.

The cup came from the backyard of American merchant and sea captain John Rogers Cooper, who built his home in Monterey circa 1832 for his wife, Encarnación Vallejo, and their six children. Their first three children, Anita (1828–1912), John Rogers Henry (1830–1899), and Rogerio (1838–1873),

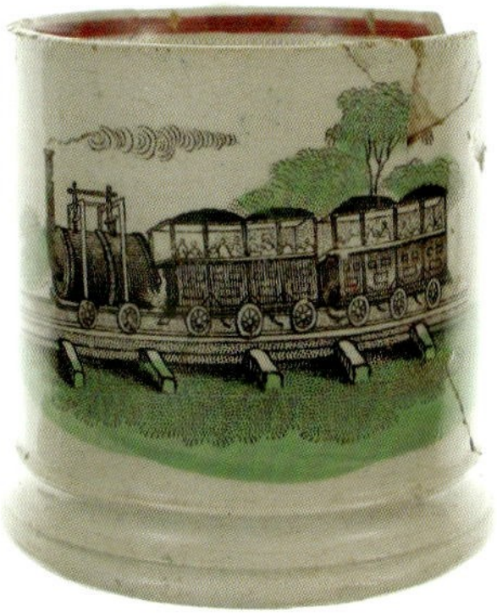

Figure 1 Cup, England, ca. 1835–1840. Whiteware. H. 2⅝". (Courtesy, California State Parks; photos, Lydia Reed-Carpenter.)

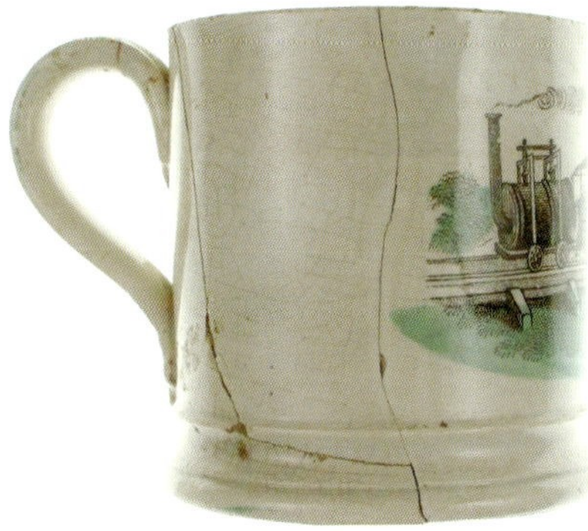

Figure 2 Side view of the cup illustrated in fig. 1.

would have been young in the 1830s and early 1840s. Whether this cup was presented to any of these children is not known, but the image certainly would have appealed to them. The family ultimately moved from Monterey to San Francisco in 1864, and for whatever reason the cup was left behind.

The English-made cup features a black transfer-printed image of a very early English steam-engine train that dates to 1820–1830, according to Dieter Hopkins, Curator of Collections at the National Railway Museum of York, England.[1] The cup itself was probably made between 1835 and 1840, although there is no mark on it to know for sure. It is decorated in pink with a horizontal stripe below the interior rim and a vertical stripe on the handle exterior; a swath of overglaze green highlights the grass and trees. Given its size and decoration, this cup would surely have been, in Noël Riley's words, a "gift for good children," although I have yet to find a parallel.[2]

Glenn Farris, 508 Second Street, Suite 108, Davis, California 95616
gfarris@omsoft.com

1. Hopkins noted that the locomotive was distinctive in its "large, cylindrical boiler with direct drive from the cylinders on top of the boiler via a grass-hopper motion direct to the four driving wheels. This was the common arrangement on locomotives up to the 1820s and on conservative designs up to circa 1830." Personal correspondence with Dieter Hopkins, December 9, 2002.

2. Noël Riley, *Gifts for Good Children: The History of Children's China*, 2 vols. (Somerset, Eng.: Richard Dennis, 1991–1996), vol. 1, *1790–1890*.

Carl Steen

The Jolly Young Waterman at Fort Johnson, Charleston, South Carolina

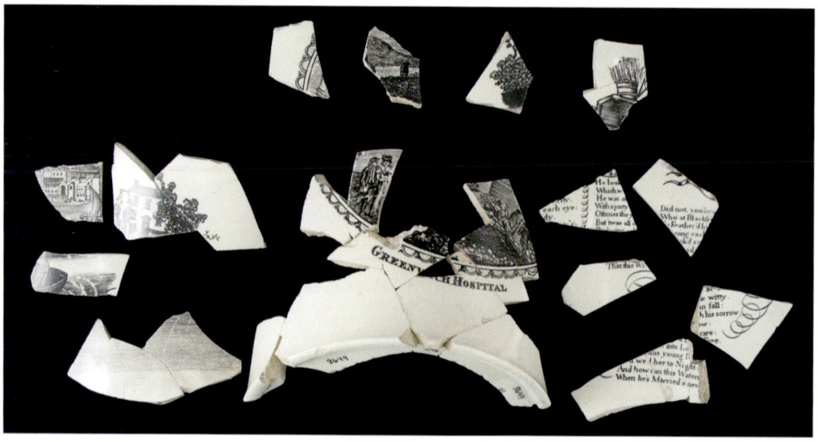

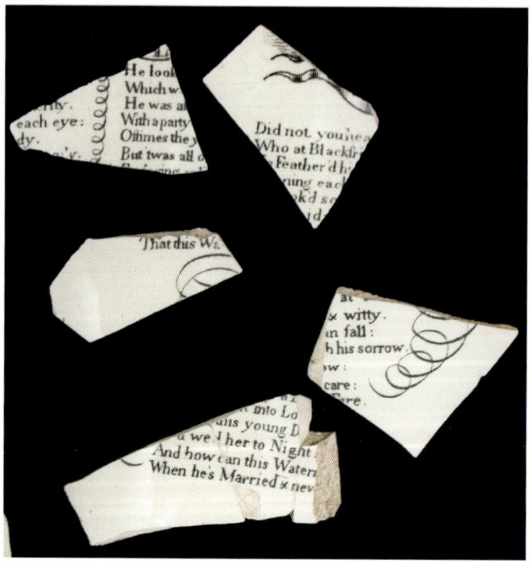

Figure 1 Jug fragments, probably Liverpool, England, 1792–1800. Creamware. H. of base sherd 3". (Photo, Carl Steen.) Over its glaze the vessel is decorated with a black transfer-printed motif labeled "GREENWICH HOSPITAL" on one side and, on the other, the lyrics from a musical written in 1774 by Charles Dibdin.

▼ IN EXCAVATIONS at Fort Johnson, on Charleston Harbor in South Carolina, sherds of a creamware pitcher with a black transfer-printed overglaze decoration were found in a midden deposit dating to the late eighteenth–early nineteenth century. Many of the prints used on such vessels were commemorative, celebrating heroic figures and great events, such as Lord Nelson and the British victory at Trafalgar. Creamware vessels like this are not unfamiliar finds in deposits dating between the 1780s and 1820s, but this is the most complete example found at Fort Johnson, and is one of the few archaeological examples whose commemorative scene and a seemingly unrelated verse can be identified.

Without the Internet, and the Google search engine specifically, this identification likely would not have been made. Many pieces survive, but they did not form a cohesive picture. Two of the first sherds found were marked ". . . ICH HOSPI . . ." (fig. 1), and a Google search revealed that Greenwich Hospital was founded in 1694 as the Royal Naval Hospital and served that purpose until 1869.[1] Numerous hospital images were also available online, and a mug with the identical print helped prove that all the sherds came from the same vessel (figs. 2, 3).

More frustrating were the sherds with writing on them, as all had to be retrieved from the collection and crossmended. In the absence of a visible title or full sentence, the phrase "And wed her to Night" seemed unique enough to qualify for a Google search, which indicated it is a portion of a

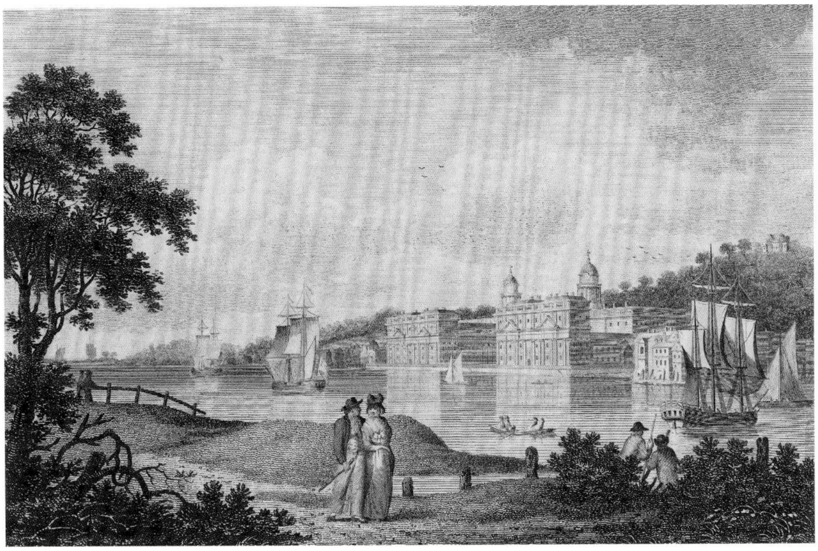

Figure 2 Thomas Nicholson, William Walker, and John Walker, *Greenwich Hospital*, London, England, 1792. Etching and engraving. 5¾" x 7¾". (Courtesy, National Maritime Museum, London.) This print is the source for the transfer-printed image on the jug fragments illustrated in fig. 1, and thus dates the vessel to 1792 or later.

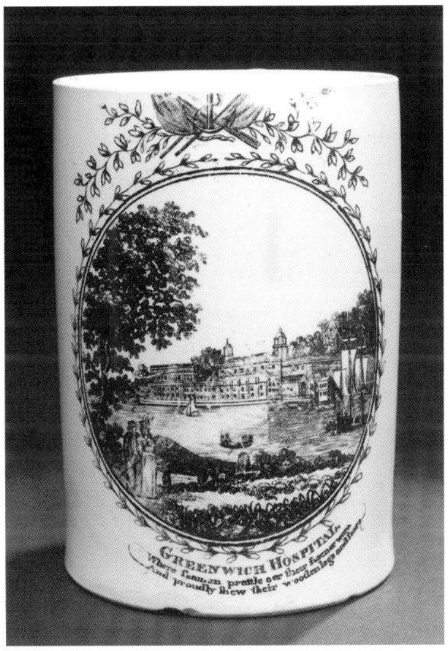

Figure 3 Mug, probably Liverpool, England, 1792–1800. Creamware. H. 6⅛". (Courtesy, National Maritime Museum, London.) The print on this vessel has been recorded on commemorative mugs and jugs, and proved helpful in the identification of the fragmented jug from Fort Johnson. The transfer-printed motif of London's Greenwich Hospital is based on the engraving illustrated in fig. 2.

song composed by an Englishman named Charles Dibdin (b. ca. 1745).[2] The son of a parish clerk in Southampton, Dibdin chose music over the priesthood (against his parents' wishes) and moved to London in 1760, at age fifteen. He became a singer and actor, and within two years he had written his first production. Dibdin continued to write music for his own works as well as for such theatrical luminaries as David Garrick, a highly influential English actor, playwright, and producer. Dibdin later became famous for his war songs, which were financed by Lord Pitt and the Royal Navy and used to stir national spirit against France at the turn of the nineteenth century.

In 1774 Dibdin wrote the musical *The Waterman*, from which our song was taken, the title referring to the men who transported people along and across Britain's rivers on small boats, or wherries.[3] "And wed her to Night" turns out not to be a dark metaphor but simply a phrase whose meaning—"And wed her tonight"—was clouded by the syntax of another era.

The message conveyed by this artifact is multifaceted. A jug commemorating a naval hospital and watermen in general is certainly appropriate. When not defending Charleston against military enemies, Fort Johnson and its hospital served as a quarantine facility and the wharf a public landing for the people of James Island, many of whom in the early nineteenth century traveled to the city in boats manned by watermen.

Yet in the early nineteenth century tension was high among the British, French, and Americans, and in 1807–1808 these tensions led to federal renovations of the fortifications at Fort Johnson that sealed the midden deposit in which the sherds were found. London's borough of Greenwich is home to more than just the naval hospital, however; it is also the site of the naval observatory and has numerous military connections. In that sense, it is a bit odd that a vessel commemorating the royal navy would be found at the place built to defend against it.

One of the interesting and enjoyable things about archaeology are the little-traveled paths that one encounters. To go from a creamware sherd in the sand to a night at the theater in 1774 London is a long journey through time and space. The technology that makes this possible has been available for only a short while—and one can only imagine what is in store for future researchers.

Carl Steen, President, Diachronic Research Foundation, P. O. Box 50394 Columbia, South Carolina 29250. carl.steen@gmail.com

1. http://en.wikipedia.org/wiki/Greenwich_Hospital_%28London%29#Greenwich_Hospital.

2. http://en.wikipedia.org/wiki/Charles_Dibdin. Dibdin reputedly wrote more than fourteen hundred songs, thirty dramatic plays, and three novels. Today there is a memorial to Dibdin, complete with statue, at the Royal Hospital in Greenwich.

3. Joseph Woodfall Ebsworth, ed., *The Bagford Ballards: Illustrating the Last Years of the Stuarts, Volume 1* (Hertford: Stephen Austin and Sons, 1878), p. 254. Dibdin's full verse of "The Waterman's Delight," from act 1 of *The Waterman* (1774), was likely influenced by a ballad entitled "The Waterman's Delight; or the Fair Maid," written a century earlier.

Carl Steen

The Last Dave Pot

▼ DAVID DRAKE, the legendary Edgefield potter who signed his name "Dave," is renowned for producing large-capacity alkaline-glazed stoneware vessels and, especially, for those he incised with compelling rhymed couplets. Born a slave circa 1800, Dave marked the earliest of his known dated vessels in 1829 and the last in 1864.[1] An exciting new discovery in Dave's chronicles was revealed during a recent visit to Miles Mill stoneware manufactory near the Gopher Branch Heritage Preserve of the South Carolina Department of Natural Resources. Miles Mill, one of the potteries belonging to Lewis Miles, opened in 1867 one-half mile from his kiln at Stoney Bluff.[2]

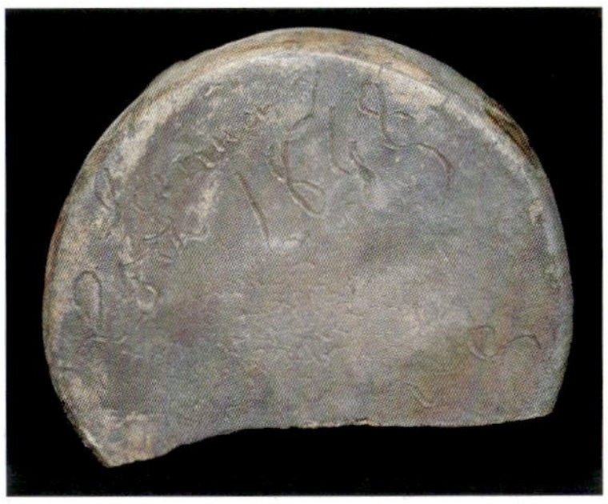

Figure 1 Storage jar base fragment, attributed to David Drake, Miles Mill Pottery, Edgefield, South Carolina, 1868. Alkaline-glazed stoneware. Incised in script: "January / 25th 1868" above the partial word ". . . ve." (Private collection; photo, Carl Steen.)

While clearing debris from the mill's channel, the property owner found a large sherd from a cylindrical slip-glazed jug (fig. 1), the base of which is inscribed "January / 25th 1868" above the partial word ". . . ve." Although no Dave vessels from Miles Mill have been previously identified, this example provides proof that he worked there, because the letters clearly share handwriting elements seen on his many signed pieces.[3] Like all of Dave's known pots, this example bears a date that includes month, day, and year.[4] Unlike all of his other signed vessels, however, this one is signed on the bottom. Most significantly, the date makes it the only known vessel Dave signed as a freeman—as an employee rather than a slave.

Carl Steen, Diachronic Research Foundation, Columbia, South Carolina
carl.steen@gmail.com

1. Arthur F. Goldberg and James P. Witkowski, "Beneath His Magic Touch: The Dated Vessels of the African-American Slave Potter Dave," in *Ceramics in America*, edited by Robert Hunter (Hanover, N.H.: University Press of New England for the Chipstone Foundation, 2006), pp. 67–70. See also Jill Beute Koverman, "Searching for Messages in Clay: What Do We Really Know about the Poetic Potter, Dave?" in *I made this jar . . . : The Life and Works of the Enslaved African-American Potter, Dave*, edited by Koverman, exh. cat. (Columbia: McKissick Museum, University of South Carolina, 1998), pp. 24–25.
2. Leonard Todd, *Carolina Clay: The Life and Legend of the Slave Potter Dave* (New York: W. W. Norton & Co., 2008), p. 172.
3. Personal communication with Steve Ferrell and Jill Beute Koverman, 2010.
4. Goldberg and Witkowski, "Beneath His Magic Touch," p. 75.

Angelika Ruth Kuettner

A Dish from Your Alma Mater?

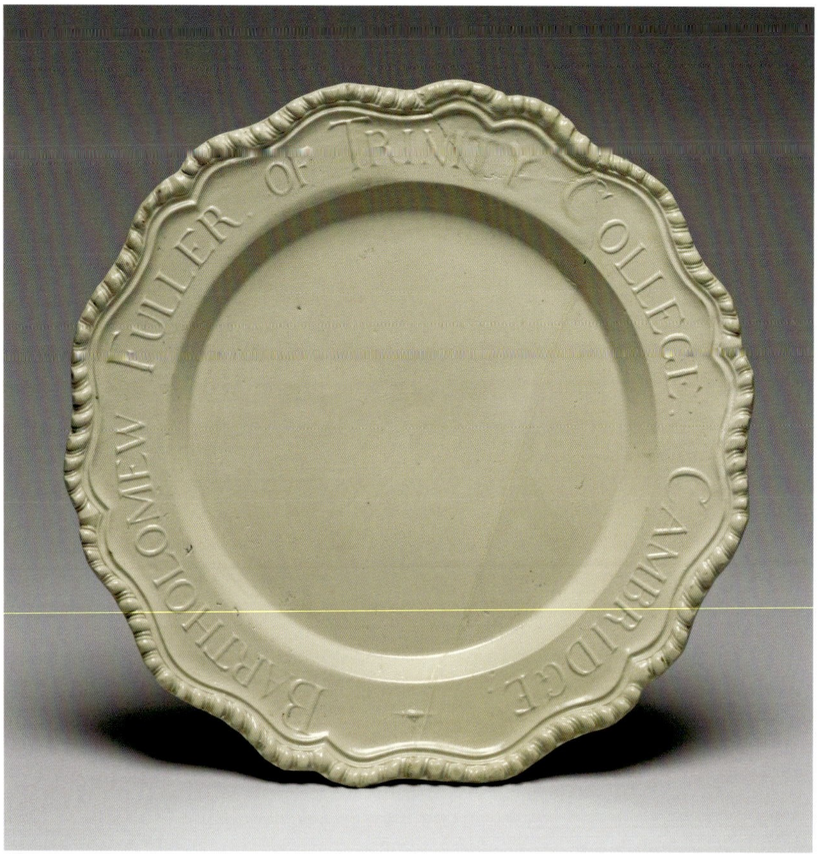

Figure 1 Plate, probably Staffordshire, England, ca. 1765–1770. White salt-glazed stoneware. D. 9¼". (Courtesy, The Fitzwilliam Museum, Cambridge.) While commemorative salt-glazed dishes with molded inscriptions for "The King of Prussia" and "William Pitt" are well known and were produced in large quantities, the "Bartholomew Fuller" plate is the only known example of a previously undocumented custom order for Trinity College.

▼ BOARDING SCHOOLS and institutions of higher learning often boast sets of china distinguished by the school's name or crest. Some even use the decorative motifs to delineate the wares of one dining hall from another within the same university. Yale University, for example, furnishes eleven of its dining halls with unique sets of dishes. The student who takes meals in the Georgian Palladian setting of one college eats from Fitzhugh-inspired patterned dinnerware, whereas the student who feasts in the Elizabethan setting of another college dining hall eats off plates bordered with grotesque faces. Other sets of dishes used on the campus spell out the name of the hall in which they are used. The College of William and Mary, the second oldest school in the United States, currently uses monogrammed dishes only at special functions, such as those held in the President's House. But in the recent past the student body ate from dishes that were stamped with the name of the school.

The use of "branded" dining wares is not a new fashion. It has its roots in a long tradition of embellishing dishes and flatware with the names of schools in order to prevent the wares from being stolen and resold. While nineteenth- and twentieth-century ceramic collegiate dining wares are prevalent, surviving examples from the eighteenth century are virtually nonexistent. The exception is the newly rediscovered white salt-glazed stoneware plate illustrated for the first time in figure 1.

The general form of the plate, with its molded gadrooned border, is typical of those manufactured by dozens of British potteries during the 1760s.

However, this special-ordered example proclaims in high-molded relief "BARTHOLOMEW FULLER OF TRINITY COLLEGE CAMBRIDGE." It is one of a handful of molded patterns that have surfaced since my catalog of border patterns was published in *Salt-glazed Stoneware in Early America* by Janine E. Skerry and Suzanne Findlen Hood.[1]

The dish was found in 1908, beneath a staircase, during renovations at Trinity College's Old Court at Cambridge University in England. It is mentioned in Bernard Rackham's 1935 *Catalogue of the Glaisher Collection of Pottery and Porcelain in the Fitzwilliam Museum Cambridge* but is not pictured and the reference is ambiguous about whether the inscription is molded or painted. An online search for "plates" in the Fitzwilliam Museum's collection database and my subsequent visit to the museum confirmed the existence of this rare molded example of white salt-glazed stoneware.

Bartholomew Fuller served as the cook for the Trinity College student body from at least the early 1760s. In 1764 Fuller became a Common Councilman to the Cambridge municipal corporation, which included the university and the surrounding community. The position would have elevated Fuller's status in the community. Perhaps due to the demands of his new civic post, he took on an apprentice cook that same year. Fuller served as Common Councilman while maintaining his role as cook until his death in April 1770. During his tenure he acquired thirty-two acres of arable land in the parish of Little Wilbraham, conveniently located six miles from Cambridge.

Now exhibited in the Fitzwilliam Museum, this plate is apparently the sole survivor of Bartholomew Fuller's culinary tenure at Trinity College. Since the special mold for the embossed inscription on the rim of this custom order would have required a considerable effort to produce, many multiples were certainly made. It is plausible that there are other dishes lurking in the homes of the ancestors of Cambridge's eighteenth-century alums.

ACKNOWLEDGMENTS Special thanks to the catering departments and the archivists of Cambridge University, The College of William and Mary, and Yale University. Information on Bartholomew Fuller was gathered from the Cambridge University archives, the Fitzwilliam Museum, and the 17th–18th Century Burney Collection Newspapers, British Library.

Angelika Ruth Kuettner, Associate Registrar for Collections Documentation and Imaging, Colonial Williamsburg Foundation, akuettner@cwf.org

1. Angelika Ruth Kuettner, "White Salt-glazed Stoneware Plate Patterns," in Janine E. Skerry and Suzanne Findlen Hood, *Salt-glazed Stoneware in Early America* (Williamsburg, Va.: Colonial Williamsburg Foundation, 2009), pp. 226–39.

Christa M. Beranek

and

Rita A. DeForest

Planting Pots from Gore Place, Waltham, Massachusetts

▼ EXCAVATIONS at an early-nineteenth-century greenhouse at Gore Place in Waltham, Massachusetts, yielded 2,083 terracotta planting pot sherds, a common yet little studied artifact type.[1] Gore Place was one of a number of federal-period country seats in the greater Boston area, home to Christopher Gore, a Massachusetts governor and United States senator, and his wife, Rebecca. A politician by profession, Gore invested a great amount of energy in his farm, as did many of his peers. Many federalist merchants and political leaders owned country properties and joined the Massachusetts Society for Promoting Agriculture (MSPA), an elite organization whose members were interested in scientific agriculture, such as attempting to improve crop yields through selective propagation or soil enrichment or importing new livestock and plant breeds.[2]

Excavations in 2008 by the Fiske Center for Archaeological Research at the University of Massachusetts, Boston uncovered some of the remains of the Gores' 1806 greenhouse, including the planting pot fragments. Keeping potted plants in greenhouses allowed residents of colder climates to grow species that were "rare, foreign, and tender."[3] In Massachusetts, this allowed people to cultivate exotic flowers and citrus trees.[4] The Gores' greenhouse seems to have been a relatively lavish structure designed to be seen by visitors; it was located along the entrance drive to the property, and excavations recovered fragments of marble floor tile, suggesting that the interior was a well-finished space. The greenhouse stood until at least 1841, but seems to have fallen out of use during the late 1830s or early 1840s during the period when Theodore Lyman Jr. owned the property (1834–1838) or during the occupation of John Singleton Copley Greene (1838–1856).

Description of the Collection

All of the planting pots from Gore Place are unglazed earthenware, reddish yellow or yellowish red, with no inclusions visible to the naked eye. At least 149 pots are represented, ranging in diameter from about 13⁄16" to 11 13⁄16".

There are six basic rim profiles with significant variation within each one (fig. 1). Twelve of the identified pots are decorated, eleven with non-

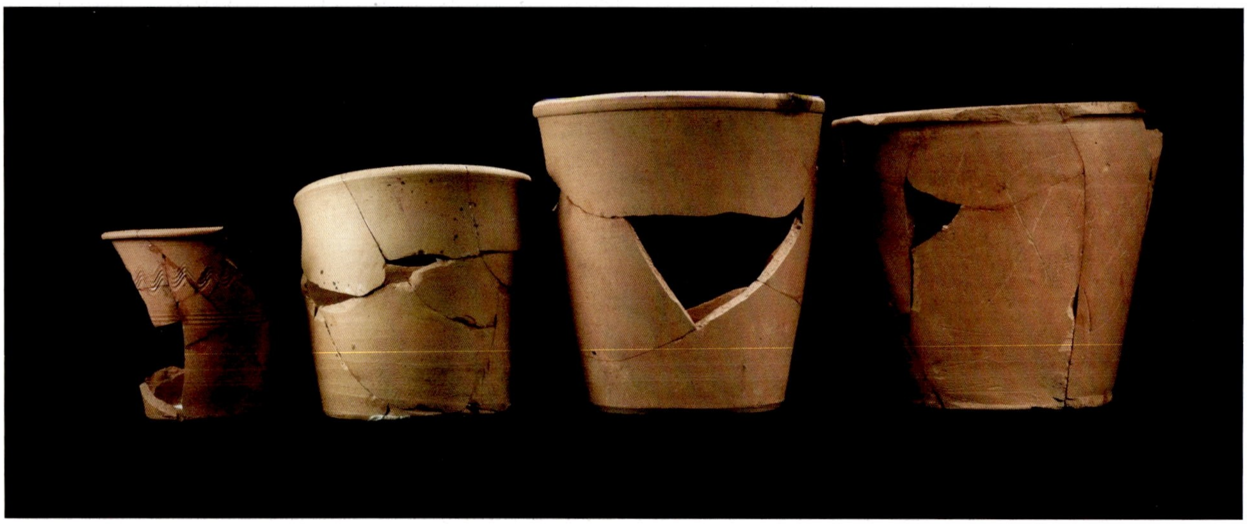

Figure 1 Flowerpots, probably American, ca. late 1830s–early 1840s. Unglazed earthenware. H. of tallest 7". (Photos, Melody Henkel.) The pot on the far left is decorated; the one on the far right is incised.

Figure 2 Flowerpot, probably American, ca. late 1830s–early 1840s. Unglazed earthenware. H. $10\frac{3}{16}$". The decoration on this vessel is the most common type on the decorated sherds in the collection.

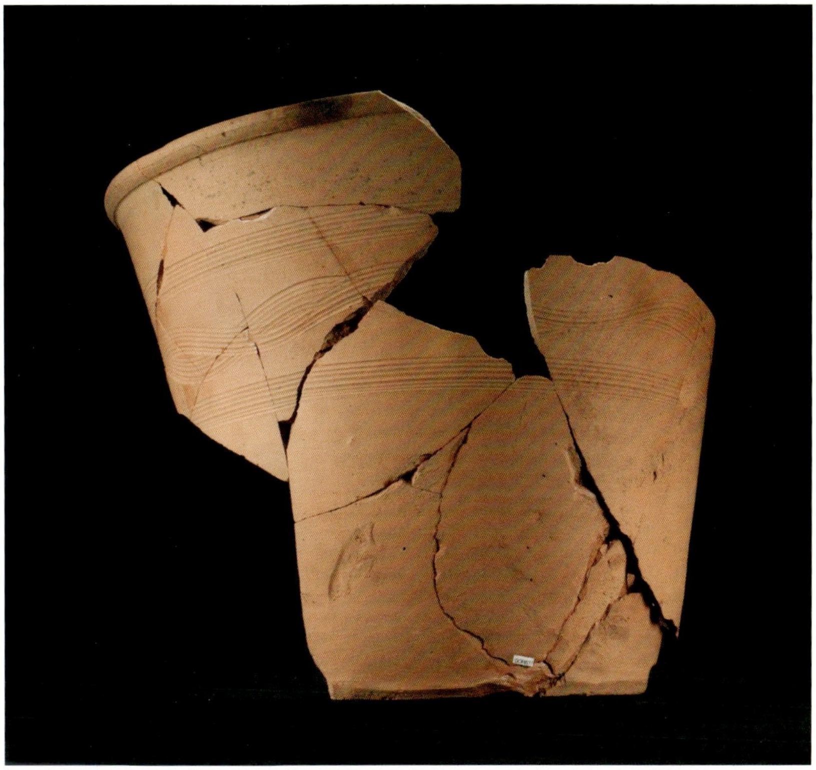

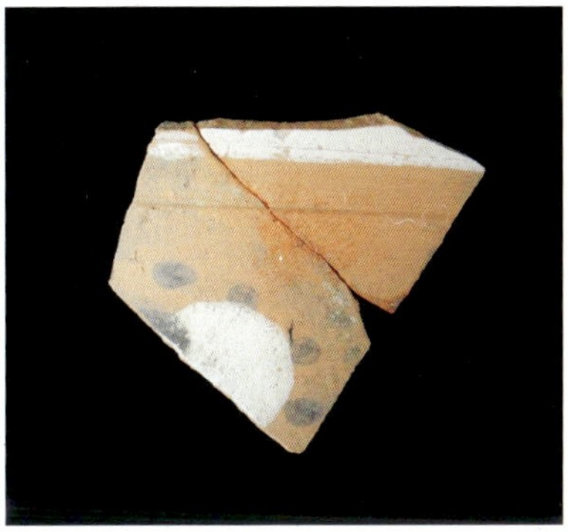

Figure 3 Flowerpot fragment, probably American, ca. late 1830s–early 1840s. Unglazed earthenware. H. $1\frac{5}{8}$". This is the sole example of slip decoration in the collection.

identical tooled or engine-turned bands of straight and/or wavy lines (fig. 2) and one with texturing seemingly created by sponging the exterior.

Additional body sherds are also decorated, most with similar tooled patterns, although one is painted with white and black slip (fig. 3). Another bears letters, numbers, and symbols scratched onto the exterior, presumably during its use-life (fig. 4).

Fifty-nine distinct bases were identified, and of the thirty-two that extend to the center all have a central drainage hole, mostly about $\frac{13}{16}$" in diameter but ranging from about $\frac{3}{16}$" to $1\frac{5}{8}$". No drainage holes on the lower body were found as is sometimes seen on other flowerpots from this period.

Like the eighteenth-century planting pots found at Mount Vernon in Virginia, these do not have footrings.[5] No terracotta saucers, or plant trays, were found. Sherds found in the debris suggest refined earthenware plates may have served this purpose, although some contemporary authors advised against using saucers in the greenhouse because they inhibited drainage and held stagnant water.[6]

Manufacture

All of the pots from Gore Place were wheel thrown, as they predate the 1861 invention of the flowerpot molding machine by William Linton, and thus display individual variations in rim profile, decoration, and diameter. The similarity of the decoration on the eleven tool-decorated pots suggests that they all came from one workshop, but the manufacturer has not been identified. Other scholars have noted the longevity of the general flowerpot form and the difficulty of dating or sourcing individual examples.[7] It is

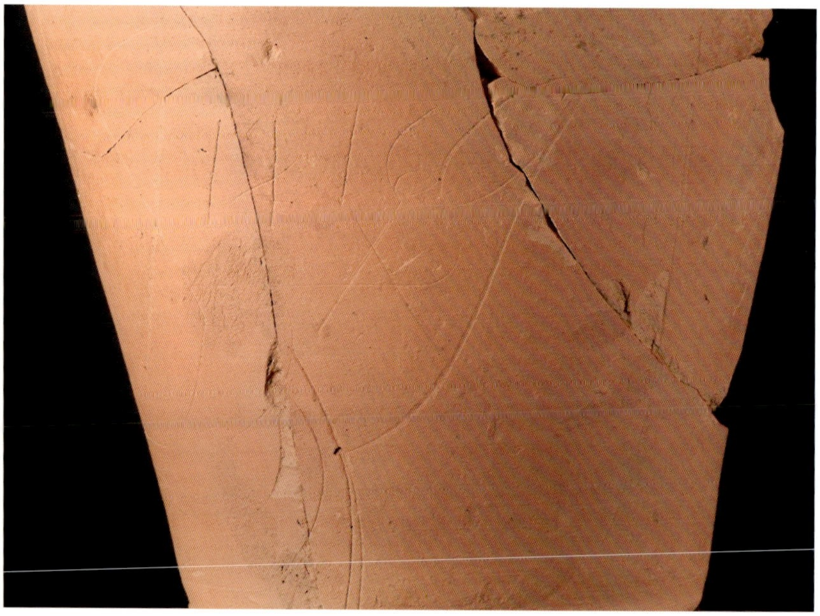

Figure 4 Detail of the scratches on the side of the flowerpot illustrated at the far right in fig. 1. Another pot (not shown) is marked with a scratched "X." The meaning of these scratches is not clear, but they may have recorded information about the plant, as gardening manuals advised.

assumed that, like other utilitarian earthenware, flowerpots were acquired from one of the region's local potters before the industrialization of the flowerpot industry after 1861; they were occasionally imported from overseas, however, as indicated in an advertisement in a Quebec newspaper for flowerpots from England.[8]

Identifying the source for Gore's pots would be very difficult without some mention in his personal papers. Early-nineteenth-century potters did not necessarily specialize in flowerpots. Hervey Brooks of Goshen, Connecticut, for example, was a farmer-potter who produced assorted utilitarian earthenwares between 1795 and 1867; flowerpots are mentioned only in some of his later lists of wares. John Worrell's study of Brooks's immediate neighborhood identified seven individuals who occasionally made and sold pottery between 1790 and 1810, only two of whom had been previously identified by scholars and none of whom was identified in period sources as primarily a potter.[9]

Other eastern Massachusetts potters known to have produced planting pots during this period include Benjamin Ingalls of Taunton, Stephen Bradford of Kingston, several generations of the Hews family of Weston and North Cambridge, and John Runey of Charlestown.[10] Of these, the Hews company is the best documented, since it continued into the mid-twentieth century. Flowerpots became the company's primary product but were mentioned in the company records as early as 1807. Examination of pots attributed to Hews that are owned by the Weston Historical Society, however, did not detect any diagnostic similarities between those and the pots from Gore Place.[11] Receipts indicate that John Runey sold pots to the MSPA's Botanic Garden, and Gore, as a founding member of the MSPA and visitor to the Botanic Garden, may have been aware of Runey as a source; however, no attributed Runey pots are known to the authors.[12]

Intensive Horticultural Practices

Pot sizes were variously described by contemporary sources by the cost of the pot, the number of pots of the size made from a set amount of clay, or by rim diameter. Several nineteenth-century gardening treatises attempted to create a standard terminology to describe pot sizes, but none seemed to gain widespread acceptance.[13]

Most of the pots in the Gore Place collection were between roughly 3⅞" to 8¹³⁄₁₆" in rim diameter, with the largest concentration between about 4⁵⁄₁₆" to 6⁵⁄₁₆". Nineteenth-century gardening manuals recommended constant repotting into gradually, incrementally larger pots as a process integral to plant health, so having a range of mid-size pots was important. The range of pot sizes on hand in the greenhouse, therefore, speaks not only to the owner's initial investment in purchasing the plants but to the continued investment in their care and display. Caring for greenhouse plants involved specialized labor, professional English gardeners, and, in the Gores' case, construction and maintenance of the greenhouse as well as upkeep of the plants by watering, applying pesticides, removing dead leaves, and repotting.

The smallest and largest of the pots speak to the level of intensity and specialization of the horticultural enterprise at Gore Place. The smallest ones, called "thumb" or "thimble" pots, include two pots with rim diameters of about ¹³⁄₁₆" or 1⅝" and seven bases 2" in diameter. These small pots were used for propagating plants from seeds or clippings and indicate that their owners were not only purchasing mature plants but raising new ones, a more intensive activity. C. M. Hovey, who tried to institute a standard numeric system for pot sizes, did not include thumb pots ("little use is made of [them], except by propagators") nor pots greater than 9" in diameter, which he called "Extras," as "[t]hey are scarcely ever made by the manufacturers, unless expressly ordered by persons who desire such for particular purposes."[14] The Gore Place collection includes six pots with rim diameters of 10" or larger, again suggesting large pots that could be used for shrubs or potted fruit trees.

ACKNOWLEDGMENTS The authors wish to thank the Gore Place Society for its support. Gore Place is undertaking a major research effort to understand the agricultural and horticultural activities that have taken place on its property, of which archaeology is just one part.

Christa M. Beranek and Rita A. DeForest, Fiske Center for Archaeological Research, University of Massachusetts, Boston, christa.beranek@umb.edu, radefor@verizon.net

1. Other archaeological studies of planting pots include Geneviève Duguay, "Utensils for the Growing of Plants," in *Under the Boardwalk in Québec City: Archaeology in the Courtyard and Gardens of the Château Saint-Louis*, edited by Pierre R. Beaudet (Montreal, Quebec: Guernica; Sillery, Quebec: Septentrion, 1990), pp. 109–21; James Goodwin and Eleanor Breen, "Flowerpots of Mount Vernon's Upper Garden: 44FX762/43," unpublished report, Mount Vernon Archaeology Department, 2005 (available online at www.mountvernon.org/files/flowerpotreport2.pdf; accessed January 31, 2011); and William Pittman and Robert Hunter, "A

Cache of Eighteenth-Century Flowerpots in Williamsburg," in *Ceramics in America*, edited by Hunter (Hanover, N.H.: University Press of New England for the Chipstone Foundation, 2002), pp. 209–13. Duguay characterizes the 548 pots and 176 saucers from five periods between 1740 and 1854 from the Chateau Saint-Louis in Quebec City; Goodwin and Breen cover 87 pots and 2 saucers from Mount Vernon dating to between 1775 and the modern period. Pittman and Hunter study 18 eighteenth-century pots from the John Page House in Virginia. Primarily documentary studies of flowerpots include Hazel H. Lathrop, "The Culture of Flowerpots" (master's thesis, Department of Anthropology, University of Massachusetts, Boston, 2000); and C. K. Currie, "The Archaeology of the Flowerpot in England and Wales, circa 1650–1950," *Garden History* 21, no. 2 (1993): 227–46. The current article is based on analysis conducted by Rita A. DeForest, "'A Good Sized Pot': Early 19th-Century Planting Pots from Gore Place, Waltham, Massachusetts" (master's thesis, Department of Anthropology, University of Massachusetts, Boston, 2010).

2. Tamara Plakins Thornton, *Cultivating Gentlemen: The Meaning of Country Life among the Boston Elite, 1785–1860* (New Haven: Yale University Press, 1989).

3. J[ohn] C[laudius] Loudon, *The Green-House Companion: Comprising a General Course of Green-House and Conservatory Practice throughout the Year* . . . , 2nd ed. (London: Printed by Richard Taylor, 1825), pp. 1–2.

4. Although there is detailed information on Gore's field crops, we do not know what was grown in the greenhouse except for a reference at Mrs. Gore's death in 1834 to the sale of roses and geraniums as well as orange, variegated orange, and lime trees; quoted in Lucinda A. Brockway, "Landscape History, Gore Place, Waltham, MA," unpublished report, Past Designs for The Halvorson Group (2001), pp. 26, 28.

5. Goodwin and Breen, "Flowerpots of Mount Vernon's Upper Garden," p. 5.

6. Mrs. [Jane] Loudon, *Gardening for Ladies; and Companion to the Flower-Garden*, 2nd American ed., from the 3rd London ed. (New York: Wiley & Halsted, 1857), p. 81. Saucers, or plant trays, were numerous in the collection from the eighteenth- and nineteenth-century greenhouses around the Château Saint-Louis in Quebec City, however. See Duguay, "Utensils for the Growing of Plants," pp. 112–15.

7. Currie, "Archaeology of the Flowerpot"; and Duguay, "Utensils for the Growing of Plants."

8. Duguay, "Utensils for the Growing of Plants," p. 113.

9. John Worrell, "Ceramic Production in the Exchange Network of an Agricultural Neighborhood," in *Domestic Pottery of the Northeastern United States, 1625–1850*, Studies in Historical Archaeology, edited by Sarah Peabody Turnbaugh (Orlando, Fla.: Academic Press, 1985), pp. 153–69.

10. For Ingalls, see Lura Woodside Watkins, *Early New England Potters and Their Wares* (Cambridge, Mass.: Harvard University Press, 1950), pp. 43–47. For Bradford, see ibid. For Hews, see A. H. Hews, "Pots," in *Cyclopedia of American Horticulture*, edited by L[iberty] H[yde] Bailey and Wilhelm Miller, 7th ed., 4 vols. (New York: Macmillan Company, 1901), 3:1423–24; and Barbara Gorely Teller, "The Founding of a Dynastic Family Industry: The Hews, Terracotta Potters of Massachusetts," in Turnbaugh, ed., *Domestic Pottery of the Northeastern United States*, pp. 249–63. For Runey, see Watkins, *Early New England Potters*, p. 29, and Massachusetts Society for Promoting Agriculture Records, Massachusetts Historical Society, Boston, box 13.

11. DeForest, "'A Good Sized Pot,'" pp. 29–31.

12. Massachusetts Society for Promoting Agriculture Records, Massachusetts Historical Society, Boston: Cambridge Botanic Garden, list of expenses, 1822, drawer C, folder XXXI, doc. 54; John Runey to Cambridge Botanic Garden, receipt, March 29, 1822, drawer C, folder XXXI, doc. 55; and John Runey to William Carter, receipt, March 16, 1830, drawer C, folder XXXII, doc. 77.

13. For example, C. M. Hovey, "Some Remarks on the Sizes of Flower Pots Usually Employed for Plants, with Hints upon the Importance of Having Some Standard for Classifying the Various Sizes," in *The Magazine of Horticulture, Botany, and All Useful Discoveries and Improvements in Rural Affairs*, edited by Hovey (Boston: Hovey and Co., 1837–), 5:46–50; J. C. Loudon, *The Horticulturist; or, an Attempt to Teach the Science and Practice of the Culture and Management of the Kitchen, Fruit, & Forcing Garden*. . . . (London: Henry G. Bohn, 1860), p. 420.

14. Hovey, "Some Remarks on the Sizes of Flower Pots," 5:48–49.

Al Luckenbach and Diana Edwards

An Unusual Drabware Strainer from the Chew Site, Anne Arundel County, Maryland

▼ BETWEEN 2000 and 2002 the Anne Arundel County Lost Towns Project conducted field investigations searching for traces of the mid-seventeenth-century town of Herrington, which led to the rediscovery of this "lost" town's location.[1] A 2006 study searched for the home of Samuel Chew, one of the town's founders and arguably its most prominent resident.[2]

An inset to the 1732 Hoxton map of the Chesapeake demonstrates how to avoid shoals and shallow water while sailing into Herring Bay. Navigating these tricky waters involved lining up the "Great Tree" with three structures, two of which were the homes of Samuel Chew Sr. and Samuel Chew Jr. The most prominent structure, the mansion of Chew Sr., is shown as a two-story building with hipped roof, paired chimneys, and a cupola (fig. 1), and the primary initial goal of the Lost Towns Project excavations was to discover the building's footprint.

With the excavation of fifty-eight 5-foot-square test areas, that footprint is beginning to come into focus. The most significant discovery is that the

Figure 1 Detail from 1732 Hoxton map showing the location of three homes, including those of the Chews. (Courtesy, Maryland State Archives.)

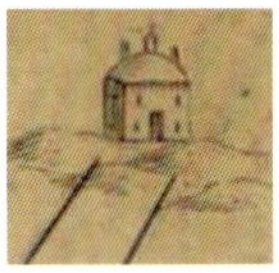

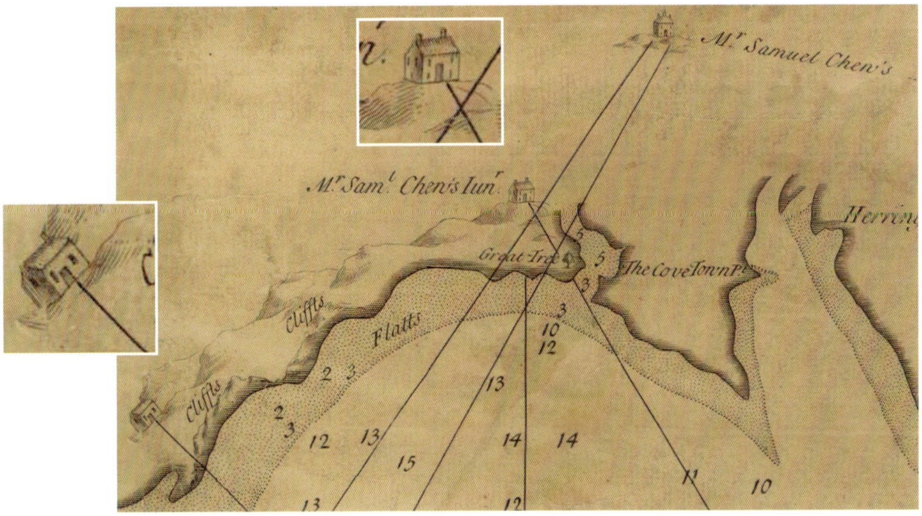

Chew mansion was far larger than anticipated. Its basic size—66 feet by 56 feet—places it among the largest known structures in the Chesapeake and, given its presumed construction date of circa 1690–1718, was apparently the largest building in both Maryland and Virginia when erected. Although there is still some question as to the exact construction date (a 1694 dated window lead may be the most definitive clue), there is no question about the date of its demise. The April 16, 1772, *Maryland Gazette* reported its "total destruction" by fire four days earlier.

Large quantities of artifacts were taken from the destruction and demolition levels, and the ceramics recovered range from high-end Chinese porcelains

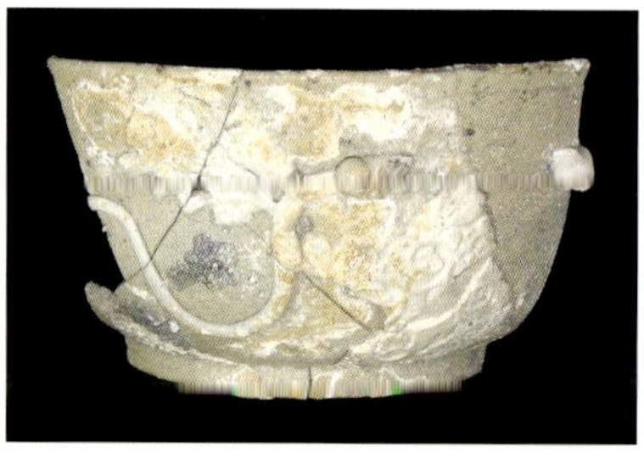

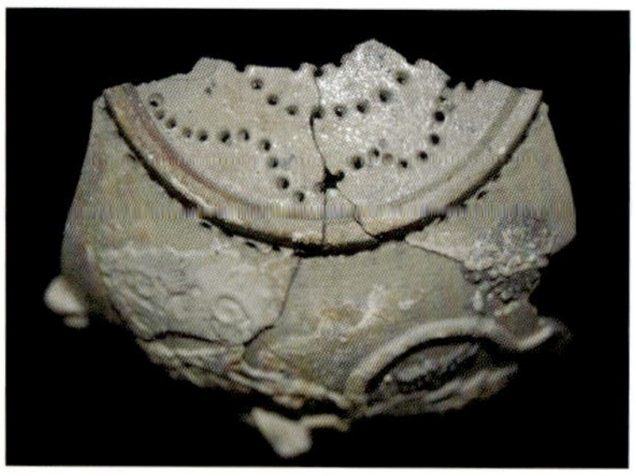

Figure 2 Strainer, probably Staffordshire, England, ca. 1745–1750. Drab salt-glazed stoneware. H. 2⅛". (Courtesy, Lost Towns Project; photo, Shawn Sharpe.)

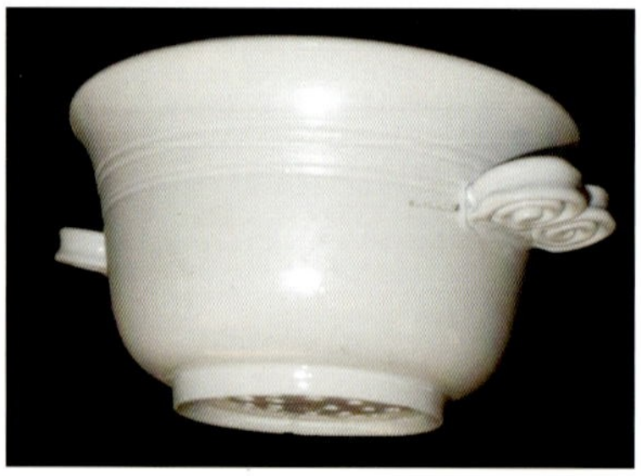

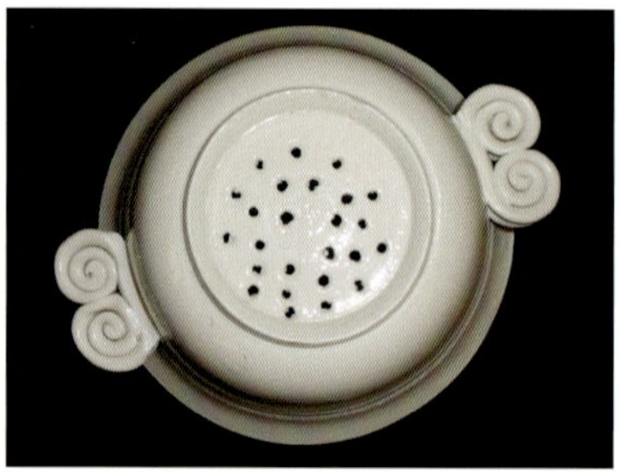

Figure 3 Strainer, Staffordshire, England, ca. 1765–1775. White salt-glazed stoneware. H. 1⅞". (Courtesy, Strawbery Banke Museum; photo, Tara Vose.)

to large utilitarian earthenware and stoneware vessels. A particularly intriguing artifact is a perforated, drab salt-glazed stoneware bowl-like object with white floral sprigging and touches of cobalt (fig. 2). Holes form both a concentric ring above the foot rim and a floral pattern on the base. The holes above the foot rim suggest that cloth or muslin was sewn in to further filter impurities from a liquid. The question is, what liquid was the vessel intended to strain?

In the collection of the Strawbery Banke Museum in Portsmouth, New Hampshire, is a white salt-glazed punch strainer (fig. 3) that is a close match to the stoneware "bowl" recovered at the Chew site. Popular in the seventeenth and eighteenth centuries, alcoholic punches were made from various combinations of rum or brandy, water, sugar, and citrus fruits and served from bowls made of porcelain, delftware, or silver. Often a pierced ladle was kept handy, to strain seeds and fruit. Silver punch strainers were small, one- or two-handled porringer-like vessels with rounded bases and, like most ceramic strainers, did not have foot rims.[3]

By the mid-eighteenth century white salt-glazed stoneware and delftware punch pots were used to serve the enlivening drink. The Chew strainer might have been used in conjunction with one of these punch pots, although its highly decorative nature suggests it was used in the drawing room, not the kitchen. Possibly it was used to strain sediment from popular fortified wines, such as port and Madeira; a strainer with sewn-in muslin would have been an excellent filter for wine dregs. The unfortunate lack of pictorial evidence of similar ceramic strainers in prints or paintings makes it impossible at this time to identify with certainty the use of the Chew example.

Al Luckenbach, Ph.D., Director, Anne Arundel County's Lost Towns Project, Alluck@aol.com

Diana Edwards, English Ceramics Scholar, dmurnag@verizon.net

1. Supported by Anne Arundel County and a grant from the Maryland Historical Trust. John Kille and Jason Moser, "An Elusive Discovery: The 17th Century Town of Herrington," *Maryland Archeology* 38 (2002): 1.

2. Funded in part by grants from the Annapolis, London Town and South County Heritage Area, as well as by the Crandall and Chew families.

3. Punch strainers are sometimes referred to in early records as lemon or orange strainers, and occasionally there are specific references to ceramic imports. For instance, the *Virginia Gazette*, July 25, 1766, carried an advertisement for Norfolk merchants Balfour and Barraud, who offered imported ceramic items such as " colliflower, . . . enamel, tortoise . . . pepper castors and punch strainers. . . ."; cited in Ivor Noël Hume, *A Guide to Artifacts of Colonial America* (New York: Alfred A. Knopf, 1978), p. 33.

William Pittman, David Brown, and Thane Harpole

Langerwehe Stoneware Vessel from Gloucester Courthouse Jail, Gloucester, Virginia

The German town of Langerwehe is located between the cities of Aachen and Cologne in Westphalia. Potters, working with local clays, produced a wide variety of lead-glazed earthenware and salt-glazed stoneware vessels beginning in the late twelfth century and continuing through the nineteenth century.[1] Inexpensive, mass-produced glass bottles caused a decline in the demand for stoneware storage vessels and eventually led to the end of commercial production. Langerwehe salt-glazed stoneware vessels were similar in form to those made at other production centers in the Rhineland but usually were undecorated (fig. 1). Most of these plain vessels were sturdy, utilitarian forms.[2]

The Langerwehe vessel found near the Gloucester County courthouse in Virginia is a brown salt-glazed stoneware storage jar or butter churn with a distinctive base pinched to make the pot sit securely on dirt floors (fig. 2). It was wheel thrown with a rounded footring and a base, which is slightly domed on the exterior. A foot rim was added by attaching a projecting

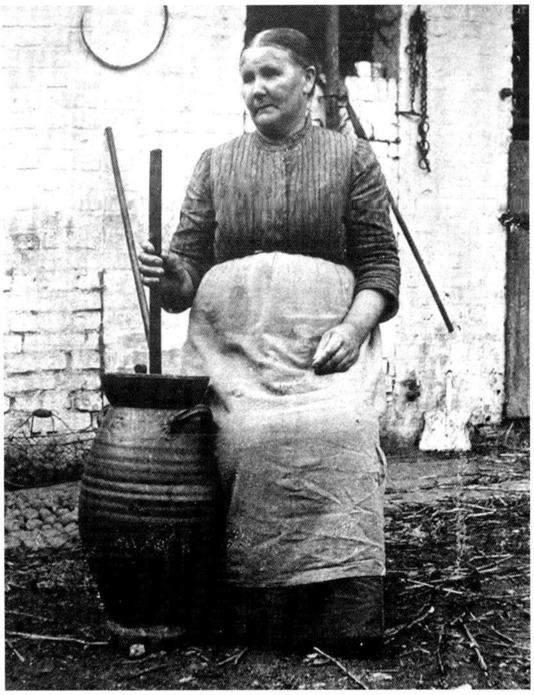

Figure 1 The woman in this early-twentieth-century photograph is making butter in a Langerwehe brown salt-glazed stoneware churn with a scalloped foot. (Courtesy, Töpferei-museum, Langerwehe, Germany.)

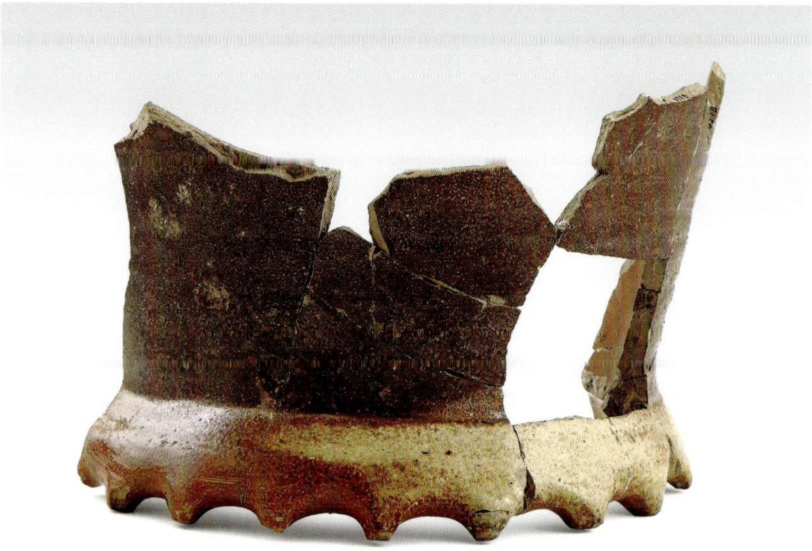

Figure 2 Storage jar or churn fragment, Langerwehe, Germany, probably mid-eighteenth century. Salt-glazed stoneware. H. 5". (Courtesy, Gloucester County; photo, Robert Hunter.)

Figure 3 Base of the vessel fragment illustrated in fig. 2.

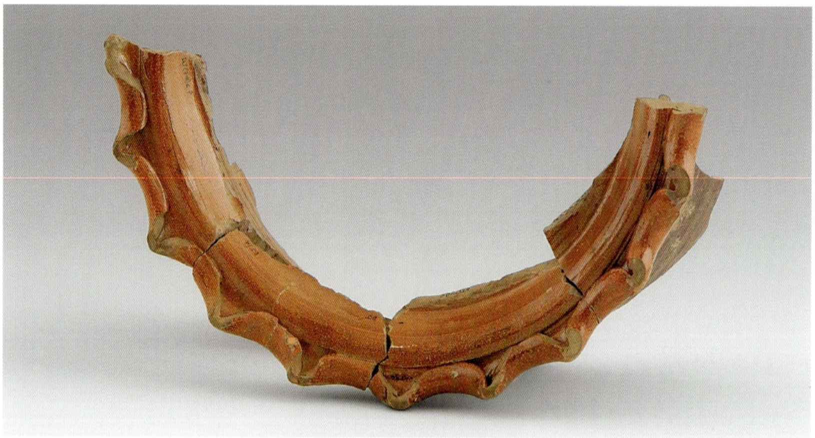

band of clay to the bottom of the pot, which was imperfectly smoothed at the joints. The footrim was folded downward and pinched to form a series of somewhat evenly spaced scalloped projections that terminate in points (fig. 3). The entire jug is thinly salt-glazed, the characteristic orange-peel texture of which is most apparent on the interior surfaces. The uniformity of the interior glazed surfaces suggests that the pot had a wide rim that allowed the salt glaze to adhere evenly. The interior and exterior surfaces were smoothed before firing but still show some potting rings and fingermarks.

Archaeologist Douglas Sanford, director of the former Gloucester County Archaeology Project, recovered the fragmented vessel in 1976. He found

primarily base fragments within a mid- to late-eighteenth-century archaeological context associated with the destruction of a building located in close proximity to the courthouse.[3] In the early nineteenth century the context was impacted by the construction of a debtors' jail. Initial interpretation of the complex of features found beneath and to the west of the jail leaned toward a tavern or ordinary—perhaps the building marked "Ordinary" or "Clubhouse" on, respectively, the 1754 and 1774 plats of the courthouse area. This establishment served people on court days as well as travelers passing through along "the great roade," the primary thoroughfare in the county.

German storage jars of this type are virtually unrecorded in Chesapeake archaeological sites, and this vessel represents a unique specimen of this medieval form in a mid-eighteenth-century Virginia context. However, Martha Zierden, Curator of Historical Archaeology at the Charleston Museum in South Carolina, reports the recovery of a similar jar from the James Stobo plantation (1741–1767) at Willtown.[4] It is hoped that, with the publication of this specimen from Gloucester, Virginia, other examples from American sites will be recognized.

William Pittman, David Brown, and Thane Harpole, Fairfield Foundation, Gloucester, Virginia, Fairfield@inna.net

1. David R. M. Gaimster with Robin Hildyard, *German Stoneware, 1200–1900: Archaeology and Cultural History, Containing a Guide to the Collections of the British Museum, Victoria & Albert Museum, and Museum of London* (London: British Museum Press, 1997).
2. The Töpfereimuseum (Pottery Museum) in Langerwehe, which displays several jugs with the identical scalloped foot, states that these storage jars were used extensively to store such foods as honey, eggs, and sauerkraut.
3. David Brown, "The Curation and Archaeological Assessment of the Gloucester County Archaeological Project (GCAP)," Gloucester Historical Committee, 2001, p. 13 and Appendix A.
4. Martha Zierden, Suzanne Linder, and Ronald Anthony, *Willtown: An Archaeological and Historical Perspective*, Archaeological Contributions 27 (Columbia: South Carolina Department of Archives and History, 1999), p. 205.

Karen K. Shriver

A Tin-glazed Earthenware Salt from Minges Ferry

▼ MINGES FERRY is one of many culturally significant archaeological sites located at Flowerdew Hundred in Prince George County, Virginia. On the south bank of the James River, this plantation was granted to Sir George Yeardley in 1619 and named for his wife, Temperance (née Flowerdew). In the 1980s the roughly crescent-shaped site, which was first recognized in the mid-1970s during spring plowing, yielded more than two thousand pipe-stem fragments in a controlled surface survey. Dating from circa 1650 to 1770, Minges Ferry has had the longest occupation of any site at Flowerdew Hundred.[1]

In 1673 the one-thousand-acre Flowerdew Hundred tract was divided into two equal parts by James Minge, surveyor and clerk of the House of Burgesses, for sisters who had inherited it from their brother John Barker. The lower dividend was attained by Sara and her husband, Robert Lucy. The upper dividend, which included Minges Ferry, was owned by Elizabeth and her husband, Philip Limbrey.[2]

More than three decades ago, a small blue-on-white tin-glazed earthenware vessel was recovered during a random surface survey at Minges Ferry. Recently returned to the Flowerdew Hundred collection, the form is unique among the thousands of ceramic objects found at Flowerdew. A small footed dish for holding salt, known as a salt or saltcellar, this vessel type was used on the tables of the economically and socially advantaged when salt was a precious commodity (fig. 1).[3]

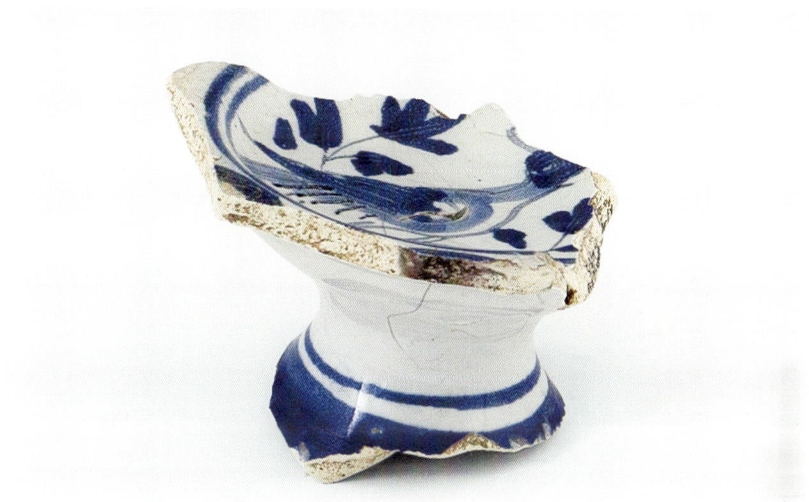

Figure 1 Salt, probably London, England, ca. 1650–1700. Tin-glazed earthenware. H. 2½". (Courtesy, Flowerdew Hundred Collection, University of Virginia Library; photo, Robert Hunter.)

The Minges Ferry salt is hand painted under the glaze in two tones of cobalt blue, with a bird-and-foliage design on the interior of the shallow bowl (fig. 2). Two concentric bands surround the central design just below the rim. The fragmented hollow pedestal foot is encircled with two horizontal stripes, and traces of a swaglike motif decorate the flat rim. It is impossible to determine whether rim knobs or supports are missing, but if they are, only three would have surmounted a circular salt such as this.[4]

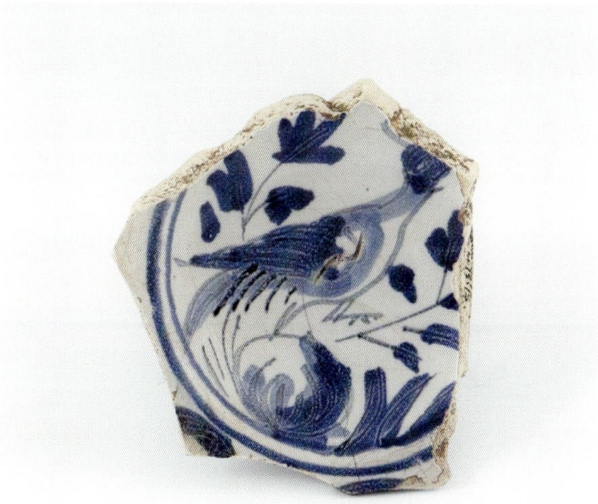

Figure 2 View of the shallow bowl of the salt illustrated in fig. 1, with hand-painted, cobalt-blue bird-and-foliage decoration.

Despite a wealth of published knowledge and consultation with delft experts, the date and origin of the salt have yet to be firmly identified. The object resembles a London delftware saltcellar, formerly in the Longridge collection, that is decorated in rich cobalt blue and is dated "1675."[5] Likewise, it is similar to a Chinese porcelain example made for the European market in Jingdezhen during the Kangxi period, circa 1662–1722.[6]

Although antique salts in diverse forms and materials are found in many English, European, and American museums, they are unusual in colonial American archaeological assemblages. Nevertheless, several English delftware examples were found on other seventeenth-century sites in nearby James City County, where extensive excavations were carried out in the second half of the twentieth century. These include one each at Kingsmill Tenement, the Joseph Petitt site, and Martin's Hundred.[7] It is tantalizing to think that the Minges Ferry salt, probably made in London sometime in the second half of the seventeenth century to circa 1700, might once have graced the table of Elizabeth Limbrey.[8]

Karen K. Shriver, Curator, Flowerdew Hundred Collection, University of Virginia Library, ks2kc@virginia.edu

1. Brenda Butler, *Field Notes, 1983–1986* (Charlottesville: Flowerdew Hundred Collection, University of Virginia Library). See also "Flowerdew Hundred Historic Site Survey, 44PG103," unpublished report, Flowerdew Hundred Collection, University of Virginia Library, Charlottesville, 1983.

2. Eve S. Gregory, "Flowerdew Hundred—an Historical Report," Appendix D, unpublished report, Flowerdew Hundred Dividend, 1976, Flowerdew Hundred Collection, Morven Guest House.

3. For a link to saltcellars dating from the first century to circa 1600, see www.larsdatter.com/saltcellars.htm (accessed March 7, 2011).

4. Leslie B. Grigsby, *The Longridge Collection of English Slipware and Delftware*, vol. 2, *Delftware* (London: Jonathan Horne Publications, 2000), p. 232. This reference includes a comprehensive discussion of delftware salt forms.

5. Ibid., pp. 232–33.

6. A photograph and catalog record of this item in the British Museum can be viewed online at http://collections.vam.ac.uk/item/O494256/salt-cellar/ (accessed March 7, 2011).

7. Ivor Noël Hume, *Early English Delftware from London and Virginia* (Williamsburg, Va.: Colonial Williamsburg Foundation, 1977), p. 29, pl. 16, p. 67, fig. VI, nos. 14, 15. See also John C. Austin, *British Delft at Williamsburg* (Williamsburg, Va.: Colonial Williamsburg Foundation, in association with Jonathan Horne Publications, 1994), p. 189.

8. John C. Austin, former curator of ceramics for the Colonial Williamsburg Foundation, suggests the salt was probably made in London circa 1700 (personal communication, December 2010). Michael Archer, author of many books on English and Dutch delftware, believes it is most likely Dutch and probably was made in the second half of the seventeenth century (email communication, January 2011).

Richard Veit

Fragments of Forgotten Empires: A French Creamware Bowl from Joseph Bonaparte's Estate at Point Breeze in Bordentown, New Jersey

▼ ON JANUARY 4, 1820, an enormous conflagration consumed the house of lawyer and diplomat Joseph Bonaparte (1768–1844), former king of Naples and Spain and elder brother of Napoléon Bonaparte. Joseph arrived in time to see the roof of the house collapse as servants and neighbors rushed to retrieve his possessions. Later he wrote, "All the furniture, statues, pictures, money, plate, gold, jewels, linen, books, in short everything that was not consumed, has been most scrupulously delivered into the hands of the people of my house."[1]

Following his brother's defeat at the battle of Waterloo in 1815, Joseph had fled to the United States, where initially he settled in Philadelphia and then, in 1816, acquired a small estate called Point Breeze in Bordentown, New Jersey, on the Delaware River.[2] Over a period of three years he considerably expanded the house and transformed the property into one of America's first purposefully designed picturesque landscapes. After his first house was destroyed by fire, Joseph converted the former stables into a second grand house, where he lived with short interregna until he returned to Europe permanently, in 1838.

The archaeological traces of his first house lay hidden beneath manicured lawns until 2007, when Divine Word Missionaries, the current owners of the property, invited Monmouth University to conduct an archaeological survey of the property. Fieldwork identified the rubble-filled foundation of Joseph's first mansion, several outbuildings, and numerous landscape features, ultimately recovering approximately 23,000 artifacts.

Among the items found in the burned ruins, under a stratum of carbonized floor joists, were several fragments of a shallow ceramic bowl (fig. 1). The sherds were stained black from the charcoal, but careful cleaning revealed a cream-colored refined earthenware decorated with a detailed black transfer print. On the lip of the vessel is a border of acorns; at center is a neoclassical image of a woman seated next to a furled flag and sail. Although the heat of the fire erased some of the image, the printed phrase "Histoire Romaine" is visible on the interior of the bowl. A maker's mark is present but unreadable. Similar vessels were produced at the Creil et Montereau faience factory and other manufactories in France in the early nineteenth century.[3] French creamware, or fine faience, rare in American contexts, "followed the classical forms of Wedgwood, but was decorated with elegant transfer prints, usually in black of mythological scenes or landscapes. . . . The shapes are extremely refined and the overglaze printing restrained, the whole providing a suitable utilitarian rival to French porcelain of the period."[4] This vessel appears to be part of a series depicting allegorical historical scenes that was produced by the Creil faience factory and decorated by the Paris firm of Stone, Coquerel, et Le Gros.[5]

This shattered and stained vessel provides a glimpse of how Joseph Bonaparte, an exiled king living in America, once set his table. It is particularly appropriate that it is decorated with a classical scene celebrating Roman history. Joseph's younger, more famous brother Napoléon is reputed to have said of him, "If I was in his place, I would found a great empire of all Spanish America, but you will see that he will be a bourgeois

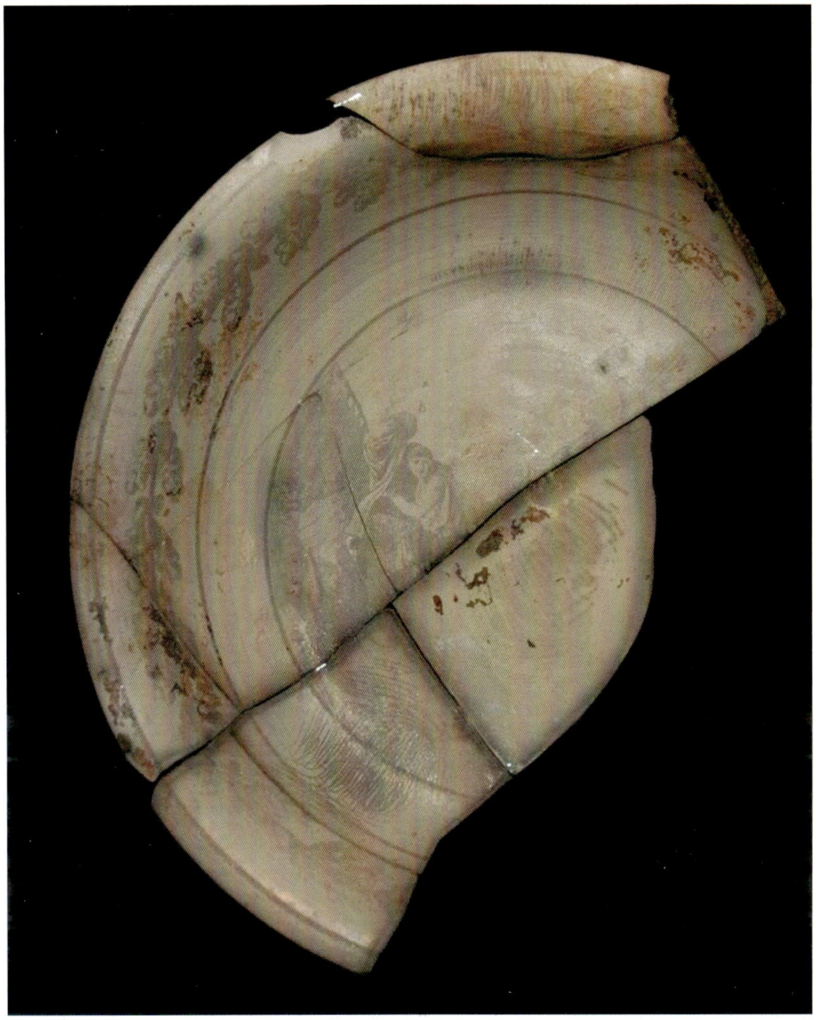

Figure 1 Bowl, France (possibly Creil faience factory), 1808–1818. Creamware. D. 8¼". (Photo, Richard Veit.) This vessel is decorated with a classical design named *Histoire Romaine* (Roman History).

American and spend his fortune in making gardens."[6] Napoléon's prediction came true—Joseph did spend a fortune making gardens—but it is clear he also set his table with fine French earthenwares, material reminders of the Bonapartes and their dashed imperial aspirations.

ACKNOWLEDGMENTS A special thanks to Divine Word Missionaries, Father Ray Lennon, and Dr. Andrew Cosentino for the opportunity to study Joseph Bonaparte's Point Breeze. Thanks to Peter Tucci Esq. for his continuing support of this project. The assistance of Michael Gall, Sean McHugh, Gerry Scharfenberger, and Bill Schindler with the excavations is gratefully acknowledged. Andrew Cosentino, Donna Corbin, and Alexandra Kirtley all shared their insights into the piece's history and suggested valuable sources.

Richard Veit, Monmouth University, Department of History and Anthropology, West Long Branch, New Jersey 07764-1898, rveit@monmouth.edu

1. E[van] M[orrison] Woodward, *Bonaparte's Park and the Murats* (Trenton, N.J.: MacCrellish and Quigley, 1879), p. 41.

2. Patricia Tyson Stroud, *The Man Who Had Been King: The American Exile of Napoleon's Brother Joseph* (Philadelphia: University of Pennsylvania Press, 2005), pp. 22–24.

3. Dorothée Guillemé-Brulon, *La faïence fine française, 1750–1867* (Paris: Éditions Massin, 1995).

4. Howard Coutts, *The Art of Ceramics: European Ceramic Design, 1500–1830* (New Haven: Yale University Press, 2001), p. 220.

5. For other vessels in the Roman History series, see the collections of the Victoria and Albert Museum, London (http://collections.vam.ac.uk/name/stone-coquerel-and-le-gros/2664/), and the French Ministry of Culture (www.culture.gouv.fr/public/mistral/joconde_fr?ACTION=CHERCHER&FIELD_98=AUTR&VALUE_98=%20LEGROS%20D%27ANIZY%20&DOM=All&REL_SPECIFIC=1 (both accessed February 26, 2011).

6. Stroud, *The Man Who Had Been King*, p. 77.

Book Reviews

David Barker and Steve Crompton. *Slipware in the Collection of the Potteries Museum and Art Gallery*. London: A. and C. Black, 2007. 192 pp.; 300 color illus. £30.00 (hardcover).

The modern city of Stoke-on-Trent resulted from the amalgam of six contiguous towns in the early twentieth century, not five as suggested by the title of Arnold Bennett's famous 1902 social-realist novel, *Anna of the Five Towns*. The city's museum collection was thus formed from the amalgamation of the collections of five separate museum services. In 1956 these were brought together in a single building in Hanley to create the Potteries Museum and Art Gallery, a place of pilgrimage for the world's ceramists. At the core of the current collection are several large private collections, many of which were formed by local pottery manufacturers. These bequests continue to be supplemented by occasional purchases. The vessels in the collection represent the relatively small and finite number of vessels that have survived as antiques. The odd vessel might have been recovered from house clearance deposits in cesspits recovered by building laborers in the days before powered machines. Unfortunately, such finds are rare in Britain.

Stoke-on-Trent and its surrounding area in North Staffordshire produced a wide range of pottery between the medieval period and the present. Among its most distinctive and famous products are its slip-decorated wares, which also influenced wares produced in other pottery centers, such as Bristol. *Slipware* concentrates primarily on the classic wares of the late seventeenth and eighteenth centuries but also covers the later artisan, industrial, and studio wares. While the emphasis is on local products from Stoke and North Staffordshire, also included are comparative pieces from elsewhere in Britain. The gallery's collection is clearly biased toward the more decorative pieces, those intended to be treasured rather than used. To obtain a realistic picture of what was produced for everyday use, one would have to view the vast number of sherds, collected from the waster heaps upon which the city is literally built, in the museum's archaeological storage area.

Slipware illustrates 217 vessels, each of which is reproduced in color photographs taken by professional photographer Steve Crompton with multiple views and detailed shots of decoration where appropriate. A few additional items—for example, a dish mold and slip-trailing tools—are also included. Each vessel has a brief description of form, glaze, fabric, and decoration. Likely provenance and date of manufacture, a metric measurement (height or width), and the museum accession number also are noted. The volume is prefaced by a brief but elegant and informative essay written by freelance ceramics specialist David Barker, former keeper of archaeology and city archaeologist. Barker discusses the technology of making classic slipwares of the seventeenth and eighteenth centuries. He draws attention to the finds of both thrown and press-molded wares from English Civil War deposits of the 1640s, including Dudley and Eccles Hall castles, both in the West Midlands. As new refined earthenwares and stonewares came to the fore, North Staffordshire production of slipwares declined beginning in the 1720s, although slipwares persisted in other regions, sometimes into the nineteenth century. Briefly discussed are the slip-decorated wares (mocha, marbled, and so forth) of the industrial period and the studio slipwares of the 1920s onward.

The bulk of the photographic catalog is given over to traditional slipwares organized into sections by form: cups and mugs, posset pots, fuddling cups, jugs, jars, salt kits, roasters, cradles, flasks and kegs, nested cups, egg holders, chamber pots, money boxes, miscellaneous forms, and dishes. Many of the pieces are of such quality or such an unusual form that one would be lucky to see them at all in a professional lifetime as an excavator, even as sherds. Such examples, clearly treasured across the generations, include the Toft plates and a number of slip-decorated model cradles (some with dates and names), made either as toys or to commemorate births or weddings. Other photographs show complete and undamaged examples of the everyday wares found in vast quantities on either side of the Atlantic.

Most of the pieces are North Staffordshire in origin with the exception of the jugs. The only illustrated jug known to be from North Staffordshire is the famous owl jug discovered in 1990 on *The Antiques Roadshow* on BBC and now perhaps the Pottery Museum's most iconic piece. Other jugs include North Devon sgraffito harvest jugs and puzzle jugs from Donyatt in Somerset.

Among the most spectacular products of the seventeenth and eighteenth centuries were thrown and press-molded dishes. The popularity of decorated flatwares in many parts of Western Europe in this period was probably linked to the popularity of the "dresser," a cupboard with plate racks used for displaying such items. Four Thomas Toft dishes, two by Ralph Toft, and several other highly decorated pieces of the late seventeenth century are illustrated, as well as a modern fake of a Thomas Toft dish and a considerable number of eighteenth- and nineteenth-century (largely press-molded) dishes. The dates and provenance of later and less spectacular dishes are often difficult to establish as such pieces were produced by many potting centers.

Twenty-five pieces of industrial slipwares from North Staffordshire are selected to show the range of decorative techniques. Sixteen studio pots from the Winchecombe pottery in Gloucestershire and two vessels by Bernard Leach and three by Shoji Hamada, both then working at the St. Ives pottery in Cornwall, are presented. All of the studio wares come from a donation made in 1948 of 556 pots by Henry Vanderveer Bergen, an American who collected and decorated pots.

Local government museums in Britain increasingly have had to modernize and adapt themselves to centrally designed social and educational agendas. This has undoubtedly helped their survival in an increasingly difficult financial climate and made museums more relevant to a wider part of the population. The cost has been that many small museums have become identical clones of each other, and the traditional scholar-curator has become an endangered species. Museum catalogs seem to be increasingly regarded as an elitist luxury, if not a sin. Stoke's museum is undoubtedly of international importance, and its insistence on visual quality has enabled it to produce, with varied funding sources, this beautifully illustrated volume. It is clearly aimed at a popular audience and deserves to sell many copies. Nevertheless, one cannot help but feel that the lack of even a short bibliography owes more to political correctness than a realistic profile of those actually prepared to spend £30 on a ceramics volume.

In conclusion, if you have the slightest interest in early modern English ceramics, buy this handsome and useful volume before it goes out of print and becomes an expensive rarity. It makes available a very important reference collection at a modest price.

Paul Courtney
Leicester, U.K.
paul.courtney2@ntlworld.com

Peter Roden. *Copyhold Potworks and Housing in the Staffordshire Potteries, 1700–1832*. Cumbria, Eng.: Wood Broughton Publications, 2008. 433 pp.; b/w illus., appendixes, glossaries, bibliographies, indexes. £25.00 (softcover).

It has taken Peter Roden fifteen years to compile *Copyhold Potworks*, a lengthy publication containing many reproductions of maps and document excerpts.[1] It examines the copyhold tenure of potworks and housing within the manor of Newcastle-under-Lyme and is therefore divided into three manageable parts: Part 1, Copyhold Tenure; Part 2, Potworks; and Part 3, Housing Developments.

Part 1 explains the process of copyhold tenure. In short, copyhold dates back to medieval times and was how land was tenanted within a manor. Records of the manor court dealt with changes of ownership and the mortgages of land and buildings by tenants (or copyholders). Roden's methodology has been to transcribe fully the court minutes (preserved in the National Archives, Kew) from 1700 to 1799. Owing to the daunting quantity of records from 1800 to 1833, Roden has transcribed two-thirds of these, selected on the basis of continuity with and relevance to the eighteenth-century records.

To familiarize readers with Newcastle-under-Lyme's medieval and copyhold tenure boundaries, a clear map and explanation are provided. The potworks and housing developments included are within the districts of Hanley, Shelton, Stoke-upon-Trent, and Penkhull. The copyhold manor did not include Burslem, Tunstall, Fenton, or Longton; still, the book refers to more than fifty different potworks, and nearly two thousand people are indexed.

It has often been pointed out that there are few comprehensive or accurately dated lists of Staffordshire potteries before Allbut's directory and map of 1802. Likewise, the *Staffordshire Advertiser* (usefully indexed by Rodney Hampson) began only in 1795, leaving Simeon Shaw's *Rise of the Staffordshire Potteries* of 1829 to provide more insight into earlier times.[2] Therefore, one value of Roden's *Copyhold Potworks* is the significant contribution it makes to understanding the eighteenth-century activities of pottery families such as Astbury, Twyford, Palmer, Minton, and Spode. While the book does not reveal information about the actual types of wares produced (beyond references to earthenware, china, or both), the book enhances what is known about the ownership of potteries, dates of production, growth of factories, various property transactions, and other business activities. Roden usefully applies colored lines to highlight key points or boundaries when reproducing original maps and documents.

A real triumph of Part 2 of this book is the documentation of many lesser-known individuals. For example, in the case of Yates, potters of Shelton from at least the 1780s, Roden notes:

> It is surprising that no research yet seems to have produced any paper devoted to them in the ceramic literature. The manor court records contain a significant number of transactions involving them and other members of the Yates family and they involve a relatively small number of first names. (p. 148)

Copyhold Potworks attempts to index individuals separately, but, as Roden candidly admits, ambiguity in the records necessarily limits the results. Intermarriages and repetitive Christian names make it hard to distinguish between certain individuals. Nevertheless, ceramic historians can appreciate the complexity and interconnections of these family dynasties, which helped shape the pottery industry of the nineteenth century: included in the Taylor family, potters in Shelton, for instance, are George and John Taylor, described as "potsellers" in Edinburgh in the 1780s (p. 155).

Roden's inclusion of a "Bibliography of Potworks" shows which manufacturers have been extensively researched and which have not. Apparently nothing has been written about Joseph Turner, a potter in Hanley working in the early eighteenth century; Moses Glass, potter on Marsh Street, Shelton, active until the end of the 1760s; or the Ratcliffe family (listed in Allbut's 1802 directory as Mrs. Ratcliffe of Stoke-lane). One of Humphrey Ratcliffe's sons was described as Thomas Ratcliffe, potter of Warrington (near Liverpool). When Humphrey Ratcliffe died in 1798, his wife was given life interest in the pottery, which accounts for Allbut's reference to Mrs. Ratcliffe (pp. 60, 98, 176–77). This cross-referencing of factory sites with Allbut's directory of 1802 is invaluable.

Part 3 of the book deals with "Housing Developments." Roden argues that, in the main, it was speculators rather than pottery entrepreneurs who built housing, attempting to attract workers (p. 227). Nevertheless, these piecemeal transactions are equally intriguing. John Ridgway, manufacturer of earthenware, sold a plot of land on George Street, Shelton (now Hanley), to "Anson Floyd of the City of Kikenny in Ireland, Dealer in Delph and China" (p. 306). Was this transaction a reflection of trading networks, or perhaps Floyd returning to Staffordshire after a period in Ireland, selling ceramics?

Due to the existence in museums of black-printed creamware (often with American-related themes), we know of Staffordshire engravers such as F. Morris, T. Baddeley, and Bentley, Wear and Bourne.[3] Roden's book lists William Wear and William Bentley, engravers in Shelton, purchasing plots of land on Vine Street in Hanley in 1819. There are a number of other references to engravers and printers working in the 1810s. These include James Wooldridge of Stoke-upon-Trent and Richard Meigh and Samuel Wooliscroft of Shelton, all listed as printers on earthenware (pp. 296, 308, 339). Obviously, there are opportunities to use this impressive publication to research other aspects of the ceramic industry.

It is worth emphasizing that Roden has compiled useful appendixes, glossaries, bibliographies, and indexes of people, places, streets, fields, and organizations, all of which help readers navigate their way around the material. And if a researcher wants to pursue a line of enquiry, the "Surviving Records" appendix indicates by date which volume to consult at the National Archives, Kew.

Copyhold Potworks will serve as an invaluable reference, but it can just as easily be read systematically. By drawing attention to this wealth of fascinating material, Roden has created many opportunities not only to enhance existing knowledge but to pursue other avenues of research. It is meticulously researched, and the author must be congratulated for making such an extraordinary range of material accessible to ceramics scholars who are far removed from the Kew archives.

Neil Ewins
Senior Lecturer, Design History, University of Sunderland
neil.ewins@sunderland.ac.uk

1. Previous publications by Peter Roden have included research on the Spode business, published in Peter Roden, "Josiah Spode (1733–1797), His Formative Influences and the Various Potworks Associated with Him," *Journal of the Northern Ceramic Society* 14 (1997): 1–43. Roden is currently the treasurer of the Northern Ceramic Society.

2. Rodney Hampson, *Pottery References in the Staffordshire Advertiser, 1795–1865* (Hanley, Eng.: Potteries Museum and Art Gallery, Northern Ceramic Society, 2000).

3. Potteries Museum and Art Gallery, Stoke-on-Trent, or the S. Robert Teitelman Collection at Winterthur Museum, Delaware.

Janine E. Skerry and Suzanne Findlen Hood. *Salt-glazed Stoneware in Early America*. Williamsburg, Va.: Colonial Williamsburg Foundation, in association with University Press of New England, 2009. 271 pp.; 300 color illus., appendixes, bibliography, index. $75.00 (hardcover).

Mounting an exhibition of stoneware at Colonial Williamsburg was an excellent idea, and publishing a book about it was even better. The foundation has a large collection of English and European wares associated with those that were in use in Virginia in the eighteenth century, and illustrating all the pots in the historic buildings and in the reserve collections would make many rarely seen specimens available to other scholars and collectors. Janine E. Skerry and Suzanne Findlen Hood admirably handled this visual challenge in a book that is beautifully illustrated, a pleasure to hold, and a joy to display on one's coffee table. The authors were then faced with the problem of how to make their book needed by fellow curators and collectors.

For specialists a catalog could have been enough, a task easily accomplished by quoting from the accession cards of previous registrars and curators. This book was intended to reach a wider audience, however, but how wide was wide? Ceramics collectors are usually well versed in the literature of their specialty, be it seventeenth-century Rhenish jugs or the nineteenth-century products of Lambeth's Doulton & Watts. To satisfy such people, a book has to break new ground and thus become an indispensable addition to their libraries.

One part of this book does fulfill that need: Angelika Ruth Kuettner's appended study of white salt-glazed stoneware rim designs—a useful expansion of Arnold Mountford's classic *The Illustrated Guide to Staffordshire Salt-glazed Stoneware*. At his time of writing Mountford said most of what needed to be said about his subject, as did Adrian Oswald, R. J. C. Hildyard, and R. G. Hughes in their seminal *English Brown Stoneware, 1670–1900* and David Gaimster in his *German Stoneware, 1200–1900*.[1] To earn its spurs alongside those experts, *Salt-glazed Stoneware in Early America* had to provide new insights. To that end the authors elected to build their chapters around the words "in early America" and drew on archaeological evidence as their keystone. John C. Austin's previous Colonial Williamsburg book, *British Delft at Williamsburg*, was a catalog that consistently used fragments excavated in the city to justify the presence of purchased delftware.[2] Within those parameters the approach worked extremely well. But John Austin was writing about one type of ceramic ware and one town. The title of this new book leads purchasers to believe that they are buying a book that embraces all of early America and perhaps to assume that the focus is on stoneware found in its earth. Why else would the dust jacket feature a rare English mug not in the protectively gloved hand of a curator but in the dirty hands of an archaeologist? This would make sense if the mug were newly dug from an archaeological site, but it was not. It came from an antique dealer's shelf in 1958.

No less open to misunderstanding are the title words "Early America." How early and how late? Does that include French and Spanish America or is this only Colonial Williamsburg's thirteen-part version of it? The second map suggests limitations from Florida to Maine and is titled "Key Seventeenth- and Eighteenth-Century American Stoneware Archaeological Deposits." One might assume that the stoneware is of American manufacture, but that would be presumptuous. It means stoneware of any sort from anywhere as long as it was found on American sites—though not, for example, in French Louisiana, where the famed "Tunica Treasure" yielded diagnostically important German stoneware—or in Bermuda.

One might counter that Bermuda is only Americanized and not American. Nevertheless, the wreck of the Jamestown-bound *Sea Venture* in 1609 yielded the most closely datable examples of German stoneware discovered in the New World and might justify their inclusion in a book on stoneware found in early America. Critics inevitably will pounce on archaeological sherds that have been overlooked or choices that should or should not have been made, but there can be no denying that the authors have done a thorough job of noticing evidence from up and down the East Coast. However, the reader who does not care that "Similar examples have also been found in Virginia on the Governor's Land site (44JC637) and Martin's Hundred" may ask where are (or were) the Governor's Land or Martin's Hundred sites, and what do these sherds tell us about themselves and the places in which they were found?

The same can be said of every archaeological site in the erstwhile English colonies, be they in Michilimackinac or Williamsburg. They all absorbed goods imported in English and Dutch ships and so, except in rare cases, when you have seen one, you have seen them all. Thus a white salt-glazed plate fragment from Vermont can be expected to parallel another from Charleston, but with both imported from the same source neither has anything new to say. Merely listing stoneware fragments from hither and yon quickly loses the reader's attention, particularly if the description is insufficiently evocative and the fragment is not illustrated. The study of ceramics—any ceramics—is an essentially visual experience. Therefore, seeing precedes believing. Without a photograph or a drawing to illuminate what one is being told, the reader soon starts skipping.

Nevertheless, the photographs by Colonial Williamsburg's Craig McDougal and Hans Lorenz are spectacular and make the book a noble addition to any coffee table. Helen M. Olds's excellent layout is carefully considered, with images conveniently placed adjacent to the related text. One might wish for more expansive captions, particularly when one sees an unmentioned pinhole in the center of a medallion of a Westerwald jug that may have something to say about the method of attachment. On reaching the more decorative English salt-glazed wares in the Colonial Williamsburg collection, one wishes for detail images of the justifying fragments or a caption reference to their existence. This frustration is exacerbated by the inclusion of unnecessary space-stealing details alongside page-filling master images that clearly need no such reinforcement. More cross-referencing between images and chapters also might have been helpful. For example, the reference to Yorktown saggers on page 187 could have usefully drawn attention to the illustrated New Jersey example on page 206—particularly since no Yorktown sagger is pictured. It also might have been helpful to have included an appendix further identifying the discussed archaeological sites, along with their dates and significance in the story of imported stoneware.

These picayune quibbles aside, *Salt-glazed Stoneware in Early America* provides a valuable summary of earlier published research and has an excellent bibliography of the works cited in the text, although a surprising omission is the first major work on German stoneware in the English language: M. Solon's two-volume *The Ancient Art Stoneware of the Low Countries and Germany* published in 1892.[3]

In modern scholarship it has become customary to give credit wherever it might appear to be due. This handsome book is no exception, containing as it does five pages of helpers' names and even graciously including this reviewer and his wife, though we actually made no contribution.

Ivor Noël Hume

1. Arnold R. Mountford, *The Illustrated Guide to Staffordshire Salt-glazed Stoneware* (New York: Praeger, [1971]); Adrian Oswald, R. J. C. Hildyard, and R. G. Hughes, *English Brown Stoneware, 1670–1900* (London: Faber and Faber, 1982); David R. M. Gaimster with Robin Hildyard, *German Stoneware, 1200–1900: Archaeology and Cultural History, Containing a Guide to the Collections of the British Museum, Victoria & Albert Museum, and Museum of London* (London: British Museum Press, 1997).

2. John C. Austin, *British Delft at Williamsburg* (Williamsburg, Va.: Colonial Williamsburg Foundation, in association with Jonathan Horne Publications, 1994).

3. Louis Marc Solon, *The Ancient Art Stoneware of the Low Countries and Germany* (London: Chiswick Press, 1892).

Kieron Tyler, Ian Betts, and Roy Stephenson. *London's Delftware Industry: The Tin-Glazed Pottery Industries of Southwark and Lambeth*, MOLAS Monograph 40. London: Museum of London Archaeology Service, 2008. 138 pp.; 170 color illus. £15.95 (hardcover).

Students of English delftware fall into two groups: collectors and archaeologists. Collectors don't want it if it's chipped, archaeologists don't notice until it is. In recent years, however, both have been brought closer together by museum curators who recognize that carefully excavated fragments can help provide dates and provenances, and archaeologists understand that sherds make sense to collectors only when they can be shown in tandem with intact examples.

In this important new contribution from the Museum of London's Archaeology Service, the authors' focus is primarily an archaeological one on five pothouses on the south bank of the River Thames at London. The locations were chosen for having been subjected to what the volume's introduction calls "archaeological interventions," which the *Oxford English Dictionary* describes as "the act of intervening, stepping in, or interfering in any affair, so as to affect its course or issue." Given the absence of detailed site plans and stratigraphic sections, and the presence of only one wide-angle photograph, it seems that these archaeological interventions were conducted, on a limited scale, on sites that were being mechanically cleared for new construction, and building developers invariably view archaeologists as interference to progress. It follows, then, that artifacts and information retrieved, though valuable, tell only part of the complex story of these five sites.

Readers who are well versed in archaeology or archaeologists themselves will no doubt ask "Who needs to be told that?" A valid question if the readership is limited to their peers, but is this attractive volume so intended? The introductory chapters contain much information of general interest which, even though available in previous delftware studies, suggests that the volume offers more than an archaeological report and perhaps, therefore, is looking to the collectors' market.

In their summary the authors state, "The tin glaze industry of London was a remarkably homogeneous one: the pothouses were manufacturing much the same range of products as each other" (p. xiv). In their conclusion to chapter 8 they note that the "production of forms common to more than one pothouse is to be expected: this was a homogeneous industry, with staff moving between a series of pothouses concentrated along a strip of the south bank of the Thames. The ceramics of the London pothouses are virtually indistinguishable from each other in terms of clay fabric alone, not surprising given their common access to clay sources" (p. 115). Having read *London's Delftware Industry* in its entirety, this reviewer came to the same conclusion—bolstering one that he had reached more than forty years ago when working with similar material from just one of the pothouse sites.

Three of the five sites—Montague Close (ca. 1613–1755), Pickleherring (ca. 1618–1723), and Rotherhithe (1638–1684)—are in the Southwark/Bermondsey districts. The remaining two—Norfolk House (ca. 1680–1779) and Glasshouse Street (ca. 1743–1784 and 1823–1846)—are in Lambeth. Fortunately, the sites are sufficiently far apart to prevent a cross-fertilizing of their waster dumps. The biscuit (unglazed first fired) wares and the glazed wasters can safely be attributed to the specific pothouse locations. But, depending on their stratigraphic contexts, finished wares need to be treated with more caution, as they could be household trash rather than the products of factory accidents. Tiles with mortar on their backs clearly had been used in buildings and therefore could have been brought from somewhere else. In short, archaeological red herrings are more likely to be netted in building-site archaeology than when sites can be leisurely and fully excavated.

Of the five sites discussed in *London's Delftware Industry* the previously unexplored Rotherhithe pothouse is given the most space, and consequently the length of discussion and the quantity and quality of the illustrations are important additions to the corpus of English delftware knowledge. In contrast, fragments from the better-known and often investigated Montague Close site receive sparse attention, leaving this reviewer yearning for a drawing or photograph of the "failed biscuit ware puzzle jug"—a rarity in any ware in the seventeenth century. As the Montague Close delftware waste was recovered from a feature "interpreted on site as the remains of a cesspit" (p. 24), one might have expected the context to have included datable wine bottle and tobacco-pipe fragments, but presumably it did not, as no such artifacts are mentioned.

The most constructive aspect of the published study are the chemical analyses of clays from biscuit wares from specific sites. The concluding chapter by Michael J. Hughes suggests that the detected differences can "significantly open up future work on the post-medieval [ceramic] industry products of London" (p. 131). Although such testing can be conducted without serious damage to the subjects, one wonders whether the average collector, curator, or fellow archaeologist can afford the cost of clay analysis, and, if they can, what importance they will place on knowing that their dish was made at Pickleherring and not a third of a mile up the road at Montague Close.

The present study relates only to the products of London factories and does not attempt to distinguish them from Dutch tin-glazed wares of the same periods. Simplistic though it may seem, whether a dish or a tile is English or Dutch is perhaps one of the most frequently asked questions. Therefore it would have been helpful to identify those London designs that are *not* paralleled in the voluminous Dutch literature. Take but one example, that of Figure 30, pattern P7. The description of this bowl fragment includes the statement "A distinctive decoration type is a blue and white chequer pattern which also appears on a dish in the Museum of London's collection" (p. 41). The latter is illustrated to provide reinforcing evidence, but does it? Are we sure that the London Museum's plate is English and not Dutch? A seemingly identical dish in my own collection was found in Dutch Limburg and points to the need for the kind of clay analysis described by Michael Hughes.[1]

An average reader could use the volume's excellent color illustrations by drawing comparisons between the products of one pothouse and another. For one example: biscuit salts from Rotherhithe are shown in two forms, waisted *and* cylindrical, whereas those from Pickleherring are only of the former shape. From this one might deduce that the cylindrical type was peculiar to Rotherhithe, but it was not.

Between 1955 and 1962 preconstruction clearance on riverfront properties owned by the Hays Wharf company uncovered quantities of Pickleherring kiln waste similar to that illustrated in the present study. Included in it were examples of both waisted and cylindrical salt types. Fortunately, Sir David Burnett, the Hays Wharf CEO and a lifelong antiquary, had made it his business to salvage whatever archaeological material might be revealed on his properties. A delft collector himself, Sir David recognized the importance of Pickleherring wasters and offered them to the Victoria and Albert Museum. Its curator of ceramics, Robert Charleston, was prepared to accept specific items but not the collection as a whole. At that time the Museum of London did not exist, and nobody in the Guildhall or the London Museum offered to house and publish so large and varied an assemblage. It was for that reason that I persuaded the Colonial Williamsburg Foundation, which already had a significant seventeenth-century delftware collection, to accept the collection and to underwrite the publication of my report. Although the authors of the new volume make a passing reference to my *Early English Delftware from London and Virginia* (1977), they erroneously state that the Burnett Collection "proved invaluable in identifying London-made material excavated in Williamsburg" (p. 19).[2] In fact, no seventeenth-century Pickleherring-era delftware has been excavated in Williamsburg. The collection's relevance instead lies in the products of subsequent excavations at Martin's Hundred, Jamestown, and other seventeenth-century Virginia archaeological sites.

As the authors of *London's Delftware Industry* have felt it useful to draw on complete vessels from the Museum of London's collection to parallel and better explain their sherds, there might, perhaps, have been merit in including relevant examples from the previously published Burnett Collection. Indeed, future studies of Pickleherring products will do well to combine the contents of the 1977 and 2008 publications—though even seen together the assembled archaeological fragments only hint at the number and varieties of tin-glazed wares that were made in the Pickleherring factories. Much more specific is the 1699 inventory of the deceased John Robins, who owned two Pickleherring pothouses (one in Vine Yard, the other in Stoney Lane), and at the time of his death had 121,000 objects either completed or in production. In publishing the inventory, ceramic historian Frank Britton identified seven hundred different shapes and sizes.[3] It is significant that these were all in production at the close of the seventeenth century, whereas most of the recovered fragments date at least twenty years earlier.

London's Delftware Industry is a beguilingly packaged archaeological report and as such is a valuable addition to any ceramics library. Like virtually every other highly professional end-of-project report, however, the necessity to interlace it with innumerable field number references and bibliographic citations keeps it useful without aspiring to "can't put it down" reading.

Ivor Noël Hume

1. Other examples of the *schaarkbordmotief* are to be found in Pieter Biesboer, *Nederlandse Majolica, 1550–1650* (Haarlem: Frans Hals Museum, 1997), nos. 58–60, and Frits T. Scholten, *Dutch Majolica and Delftware, 1550–1700, from the Edwin van Drecht Collection* (Amsterdam: E. van Drecht, 1993), nos. 44, 45, wherein parallels are cited from kiln waste in Haarlem and Deventer, their design derived from Italian *albarelli*.

2. Ivor Noël Hume, *Early English Delftware from London and Virginia*, Colonial Williamsburg Occasional Papers in Archaeology 2 (Williamsburg, Va.: Colonial Williamsburg Foundation, 1977).

3. Frank Britton, "The Pickleherring Potteries: An Inventory," *Post-Medieval Archaeology* 24 (1990): 61–92.

Robert Copeland. *Manufacturing Processes of Tableware during the Eighteenth and Nineteenth Centuries*. Cumbria, U.K.: Northern Ceramic Society, 2009. 186 pp.; 185 color, 69 b/w illus., appendixes, index. £25 (softcover).

After listing several books written for potters that describe production technology, author Robert Copeland states, "My own contribution is aimed especially at museum curators, dealers, collectors and all those who are intrigued to know *how* pots were made" (p. 7). He has done a superb job of describing and illustrating ceramic production processes for nonpotters in a work that clearly will stand the test of time.

Robert was uniquely qualified for this task. His connection to the Spode company began when his great-grandfather William Copeland, father of

William Taylor Copeland, entered into a partnership with Josiah Spode II in 1805 and then went on to become owner of the company in 1833. When Robert joined the family business in 1947, he spent three years "at the bench," learning the craft of being a presser. He also worked at various other production processes, including the firing of a bottle kiln, and thereby gained hands-on knowledge of production before advancing into management and sales. This, in addition to his long-held interest in the history of the Staffordshire potteries, made him the ideal person to provide collectors, curators, and archaeologists with detailed descriptions of the many processes involved in the production of pottery and porcelain, from preparing the clay to firing the wares.

When I had the privilege of walking through part of the Spode works with Robert in 1986, I was amazed that he addressed almost all of the working potters by name; clearly he knew them very well. Their response to his "Hello, John," "Hello, Bill" was "Hello, Mr. Robert." In his acknowledgments Robert thanks no fewer than forty-five of his fellow workers for helping him to understand the production processes. Many of the workers are identified in the photographs of the various stages of pottery production, providing insights into Robert's intimate knowledge of his subject. The excellent section on pressing, for example, features pictures of Tom McCue making meat dishes and applying footrings. This is the man who taught Robert how to be a presser.

Most of the numerous color photographs appear to date to the late 1950s, just prior to the last coal firing of a bottle kiln at the Spode factory in 1960. We are fortunate to have these color reproductions, for had these photos been published at that time they no doubt would have been printed in black and white. The images enhance Robert's excellent description of the stages of production, including the preparation of clay, slip, flint, and burnt bones for making the clay bodies. He goes on to describe pugging clay, making molds, and throwing, pressing, casting, and turning wares.

The section on the placement of green ware in saggers and the various types of kiln hardware along with the loading and firing of the glost kiln is very detailed and helpful in identifying marks left by kiln hardware. Of particular interest are descriptions of all the steps necessary to keep bone china from warping or sagging in the firing process. Robert's text and the photographs reveal in detail the complex process of engraving copperplates and printing under the glaze. In an illuminating series of color photographs, engraver Paul Holdway demonstrates bat printing over the glaze using hand-held glue bat and copperplate.

There is very little to quibble about in this masterful work. An exception involves problematic information on the "Potter's Dozen" in Appendix I ("Potter Trade Sizes," pp. 179–81). Robert states that "The 'count to the dozen' was determined by the number of a given size that would fit on a standard size of work-board. . . . *It was the number and not the capacity that decided the count*" (emphasis added). While in the twentieth century the count related to the number of pieces on a standard work board, others have documented that the potter's dozen for hollow wares historically was based on vessel capacity. Leonard Whiter's book on Spode gives Dr. Plot's 1686 description of the potter's dozen (in which vessels holding a pint were counted twelve to the dozen, those holding a half pint were counted twenty-four to the dozen, and those holding two pints were counted six to the dozen) and also references an 1804 publication that describes the same system.[1] This form of the potter's dozen has been described as well by David Drakard and Paul Holdway.[2] The evolution of a capacity-based count of one dozen vessels to one based on the number that would fit on a work board apparently has erased the memory of the origins of the system. This minor error does not detract from the significant contribution that Robert Copeland's book makes to our understanding of pottery production.

Robert's documentation of these processes and his insights into them at the Spode factory appear at a sorrowful time for the pottery industry, since the Spode company has gone out of business, and the works, which had been in the same location in Stoke-on-Trent for more than two centuries, have been torn down. *Manufacturing Processes of Tableware during the Eighteenth and Nineteenth Centuries* joins some of his other major contributions to the ceramic literature, including *Spode's Willow Pattern and Other Designs after the Chinese* and *Spode and Copeland Marks and Other Relevant Intelligence*.[3] Robert, who was very generous in sharing information with other researchers or those just wanting to know about a piece of Spode, passed away in September 2010 and will be greatly missed by those of us who had the good fortune to have known him.

George L. Miller, Retired historical archaeologist
Miller.CCindex@gmail.com

1. Leonard Whiter, *Spode: A History of the Family, Factory and Wares from 1733–1833* (New York: Praeger Publishers, 1970), p. 63.
2. David Drakard and Paul Holdway, *Spode Transfer Printed Ware, 1784–1833* (Woodbridge, Suffolk, Eng.: Antique Collectors' Club, 2002), p. 70.
3. Robert Copeland, *Spode's Willow Pattern and Other Designs after the Chinese* (New York: Rizzoli/Christie's, 1980); Robert Copeland, *Spode and Copeland Marks and Other Relevant Intelligence* (London: Studio Vista, 1993).

Jonathan Goodwin and David Barker. *Small Pieces of History: Archaeological Ceramics from Tunstall, Stoke-on-Trent*. Stoke-on-Trent Archaeology Service, Monograph No. 2. Stoke-on-Trent, Staffordshire, Eng.: Stoke-on-Trent Archaeological Service, 2009. 69 pp.; 214 color illus., tables, references. £13.95 (softcover).

This slender volume is informative and the stunningly beautiful photos of ceramic sherds provide vivid accompaniment to the text. As the first of a projected series of reports on salvage excavations around the potteries in the town of Tunstall, Staffordshire, it sets a high standard for the future.

The study begins with a general history of the development of Tunstall and a detailed description of the eight pottery factories that operated during the nineteenth century in or adjacent to the areas of the salvage work. Many of the potteries changed hands multiple times, and the manufacturer names, including such well-known ones as Johnson Brothers, Podmore, Walker and Co., Wedgwood and Co., and Wood and Challinoris—are familiar to anyone who works with nineteenth-century ceramics. While much of this is very localized history, it provides an interesting view of the impact that family relationships and the local economy had on pottery production.

The heart of the report is the presentation of material from three "watching briefs" and a private collection. Each section begins with a brief description of the site and the method of discovery. While there are only four collections, two of them are divided into several different find locations within the site. The Scotia Road vicinity, for example, had seven distinct findspots. As might be expected with archaeological monitoring projects, there was little hand excavation involved; most of the material was collected from mechanically excavated trenches.

Presentation of the ceramic finds varies somewhat by collection but generally begins with a discussion of the manufacturers' marks, followed by classification into broadly defined ware types (equivalent to decoration types). The majority of the sherds collected are transfer printed, and these are described by manufacturer and pattern. Patterns that can be associated with a particular manufacturer and/or pottery are grouped in the text. Following this, other decorative types are described, among them hand painted, slipped, and sponge decorated. These are the types of wares that usually get little attention from archaeologists but formed an important part of any potter's business. The finds also included biscuit wares, wasters, and kiln furniture. Each site section concludes with an analysis of the date of the assemblage, most often based on manufacturers' marks, and a discussion of the significance of the finds and what they add to the history of the nearby pottery.

The bulk of the identified sherds are transfer-printed earthenwares, and many have makers' marks. Of these, there are a limited number of patterns, most of which are associated with a handful of companies known to have been working in the immediate vicinity. Figure 107 shows a blue-printed vessel with a romantic scene produced by John Wedgwood under the title Brussels; the presence of this well-known pattern strengthens the attribution of all the sherds in that deposit to Wedgwood and Co. The unidentified patterns are also well illustrated, which will help other researchers and lead to further identifications. Figure 129, for example, is a brown-printed vessel that exactly matches a pattern known as Acropolis. All of the collector books record this pattern, but it had never been associated with a particular manufacturer. Its discovery in a deposit entirely of wasters associated with the pottery of John Wedgwood and Co. suggests that this was one of their products. For patterns that were seldom if ever marked, this may be the only way to identify the manufacturer.

The one flaw in this study is its lack of organization, which tends to obscure its significant results. The narrative is hard to follow and would have benefited from a few tables summarizing the patterns from each site, number of vessels, and number of fragments. Placing the summary of all the ceramics in one section, separate from each find site, makes it difficult to know the extent and quantity of ceramic types in each collection. And while unidentified patterns are illustrated, it is almost impossible to determine what percentage of the material is unidentified. A few photographs or drawings of the archaeological fieldwork would have been useful in understanding the contexts.

Overall, however, this is an excellent source for students of ceramic history and archaeologists dealing with nineteenth-century sites. Future publications in this series will be highly anticipated.

Timothy B. Riordan
Historic St. Mary's City
tbriordan@smcm.edu

Index

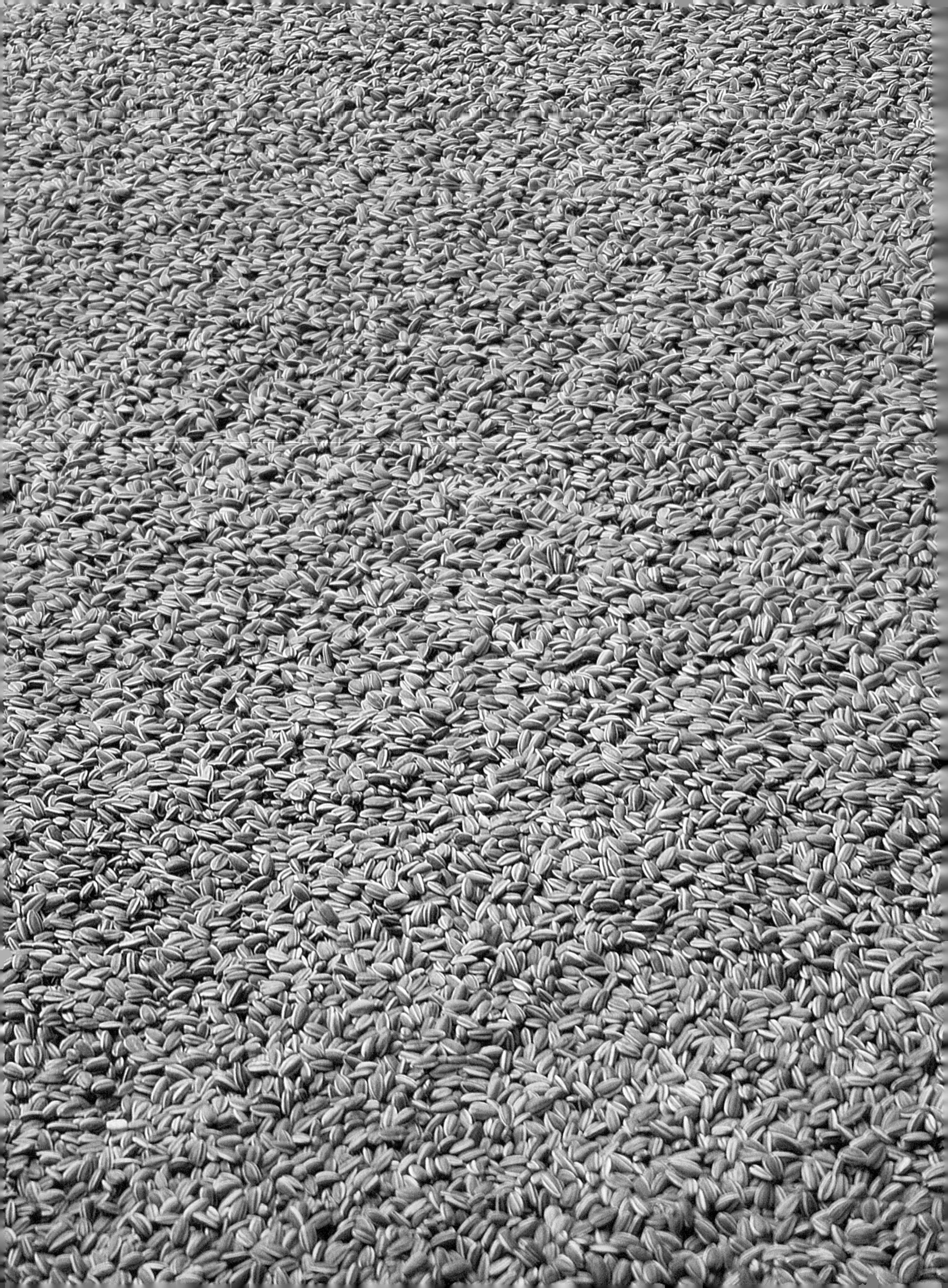